Space

45

Stars and Planets

Ian Ridpath

Illustrated by
Wil Tirion

Princeton University Press
Princeton, New Jersey

Acknowledgements and sources

This book is the result of an unusual collaboration between an author and an illustrator in different countries, united by their common fascination with the sky. International cooperation of another kind is apparent in the photographic illustrations in this book, which are the result of the efforts of astronomers from many nations working with some of the world's finest telescopes. Such awesome images rank with the most enduring works of art produced by humankind. It is fitting that the observatories concerned have made their results so readily available. The urge to study the sky transcends national boundaries, and so it should. The skies are open to us all.

The authors would like to thank those whose photographs are used in this book, in particular Nigel Sharp of the National Optical Astronomy Observatories, Tucson, Arizona. Pictures are credited individually where they appear. The U.S. Air Force Lunar Reference Mosaic was used for the photographic Moon maps, courtesy of the U.S. Defense Mapping Agency. Dimensions of formations are from the list by Ewen Whitaker in *Norton's Star Atlas*, 19th edition.

The following publications have been consulted extensively during the preparation of this edition: *The Hipparcos and Tycho Catalogues* by M. Perryman *et al.* (European Space Agency, Noordwijk); *The Millennium Star Atlas* by R.W. Sinnott and M. Perryman (European Space Agency; and Sky Publishing Corp., Cambridge, Mass.); *Sky Atlas 2000.0* (2nd edition) by Wil Tirion and R.W. Sinnott (Sky Publishing Corp.; and Cambridge University Press, Cambridge, England); *Sky Catalogue 2000.0*, Vols 1 and 2 (Sky Publishing Corp. and Cambridge University Press); *The Bright Star Catalogue*, 4th edition, by Dorrit Hoffleit (Yale University Observatory); *Burnham's Celestial Handbook* by Robert Burnham, Jr (Dover, New York; Constable, London); *Sky & Telescope* magazine (Cambridge, Mass.). Further details of the origin and mythology of the constellations may be found in *Star Tales* by Ian Ridpath (Lutterworth, Cambridge). For more on the meanings of individual star names see *Short Guide to Modern Star Names and their Derivations* by Paul Kunitzsch and Tim Smart (Otto Harrassowitz, Wiesbaden).

For this third edition, as the second, we are grateful for the attentive editorial skills of John Woodruff. We also appreciate the continuing support of our editors at HarperCollins, Myles Archibald and Katie Piper.

First published 1984. Second edition 1993. Third edition 2001

Published in North America by Princeton University Press,
41 William Street, Princeton, New Jersey 08540

In the United Kingdom published by HarperCollins*Publishers*

Library of Congress Catalog Card Number 00-110474
ISBN 0-691-08912-4 (cloth); ISBN 0-691-08913-2 (paperback)

Typesetting and layout by Ian Ridpath
Printed and bound by Printing Express Ltd., Hong Kong

www.pup.princeton.edu

10 9 8 7 6 5 4 3 2 1

Contents

SECTION I

Introduction

The night sky is one of the most beautiful sights in nature. Yet many people remain lost among the jostling crowd of stars, and are baffled by the progressively changing appearance of the sky from hour to hour and season to season. The charts and descriptions in this book will guide you to the most splendid celestial sights, many of them within the range of simple optical equipment such as binoculars, and all accessible with an average-sized telescope as used by amateur astronomers.

It must be emphasized that you do not need a telescope to take up stargazing. Use the charts in this book to find your way among the stars first with your own eyes, and then with the aid of binoculars, which bring the stars more readily into view. Binoculars are a worthwhile investment, being relatively cheap, easy to carry and useful for many purposes other than stargazing.

In the night sky, stars appear to the naked eye as spiky, twinkling lights. Those stars near the horizon seem to flash and change colour. The twinkling and flashing effects are due not to the stars themselves but to the Earth's atmosphere: turbulent air currents cause the stars' light to dance around. The steadiness of the atmosphere is referred to as the *seeing*. Steady air means good seeing. The spikiness of star images is due to optical effects in the observer's eye. In reality, stars are spheres of gas similar to our own Sun, emitting their own heat and light.

Stars come in various sizes, from giants to dwarfs, and in a range of colours according to their temperature. At first glance all stars appear white, but more careful inspection reveals that certain ones are somewhat orange, notably Betelgeuse, Antares, Aldebaran and Arcturus, while others such as Rigel, Spica and Vega have a bluish tinge. Binoculars bring out the colours more readily than the naked eye does. Section II of this book, starting on page 263, explains more fully the different types of star that exist.

By contrast, planets are cold bodies that shine by reflecting the Sun's light. Their physical nature is described in Section II, from page 298 onwards. The planets are constantly on the move as they orbit the Sun. Four of them can be easily seen with the naked eye: Venus, Mars, Jupiter and Saturn. Venus, the brightest of all, appears as a dazzling object in the evening or morning sky. The positions of Mars, Jupiter and Saturn for a 5-year period are given in the charts on pages 356–365.

About 2000 stars are visible to the naked eye on a clear, dark night, but you will not need to learn them all. Start by identifying the brightest stars and major constellations, and use these as signposts to the fainter, less prominent stars and constellations. Once you know the main features of the night sky, you will never again be lost among the stars.

Constellations The sky is divided into 88 sections, known as constellations, which astronomers use as a convenient way of locating and naming celestial objects. Each constellation is given a separate chart and description in this book. The main constellations of the sky were devised at the dawn of history, by Middle Eastern peoples who fancied that they could see a likeness to certain fabled creatures and mythological heroes among the stars. In particular, the 12 constellations of the zodiac, whose names are familiar from the astrological columns in newspapers and magazines, were of importance in ancient times. The zodiacal constellations are those that the Sun passes in front of during its yearly path around the heavens. However, it should be realized that the astrological 'signs' of the zodiac are not the same as the modern astronomical constellations, even though they share the same names.

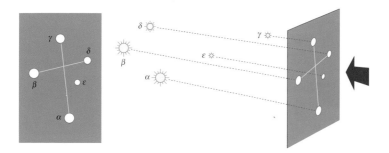

Stars in a constellation are usually unrelated to one another. Here the stars of Crux, the Southern Cross, are shown as they appear from Earth, left, and as they actually lie in space, right. (Wil Tirion)

Most of the stars in a constellation have no real connection with one another at all; they may lie at vastly differing distances from Earth, and simply form a pattern by chance. Some of the constellation patterns are easier to recognize than others, such as the magnificent Orion or the distinctive Cassiopeia and Crux. Others are faint and obscure, such as Lynx and Telescopium.

Our modern constellations derive from a catalogue of 48 compiled by the Greek astronomer Ptolemy in AD 150. This list was expanded by navigators and celestial mapmakers, notably the Dutchmen Pieter Dirkszoon Keyser (c. 1540–96) and Frederick de Houtman (1571–1627), the Pole Johannes Hevelius (1611–87) and the Frenchman Nicolas Louis de Lacaille (1713–62). Keyser and de Houtman introduced 12 new constellations, and Lacaille 14, in parts of the southern sky not visible from Mediterranean regions; Hevelius and others invented constellations to fill in the gaps between the figures recognized by the Greeks. The whole process sounds rather arbitrary, and indeed it was. A number of the newly

THE 88 CONSTELLATIONS

Name	Genitive	Abbrevn.	Area (square degs.)	Order of size	Origin*
Andromeda	Andromedae	And	722	19	1
Antlia	Antliae	Ant	239	62	6
Apus	Apodis	Aps	206	67	3
Aquarius	Aquarii	Aqr	980	10	1
Aquila	Aquilae	Aql	652	22	1
Ara	Arae	Ara	237	63	1
Aries	Arietis	Ari	441	39	1
Auriga	Aurigae	Aur	657	21	1
Boötes	Boötis	Boo	907	13	1
Caelum	Caeli	Cae	125	81	6
Camelopardalis	Camelopardalis	Cam	757	18	4
Cancer	Cancri	Cnc	506	31	1
Canes Venatici	Canum Venaticorum	CVn	465	38	5
Canis Major	Canis Majoris	CMa	380	43	1
Canis Minor	Canis Minoris	CMi	183	71	1
Capricornus	Capricorni	Cap	414	40	1
Carina	Carinae	Car	494	34	6
Cassiopeia	Cassiopeiae	Cas	598	25	1
Centaurus	Centauri	Cen	1060	9	1
Cepheus	Cephei	Cep	588	27	1
Cetus	Ceti	Cet	1231	4	1
Chamaeleon	Chamaeleontis	Cha	132	79	3
Circinus	Circini	Cir	93	85	6
Columba	Columbae	Col	270	54	4
Coma Berenices	Comae Berenices	Com	386	42	2
Corona Australis	Coronae Australis	CrA	128	80	1
Corona Borealis	Coronae Borealis	CrB	179	73	1
Corvus	Corvi	Crv	184	70	1
Crater	Crateris	Crt	282	53	1
Crux	Crucis	Cru	68	88	4
Cygnus	Cygni	Cyg	804	16	1
Delphinus	Delphini	Del	189	69	1
Dorado	Doradus	Dor	179	72	3
Draco	Draconis	Dra	1083	8	1
Equuleus	Equulei	Equ	72	87	1
Eridanus	Eridani	Eri	1138	6	1
Fornax	Fornacis	For	398	41	6
Gemini	Geminorum	Gem	514	30	1
Grus	Gruis	Gru	366	45	3
Hercules	Herculis	Her	1225	5	1
Horologium	Horologii	Hor	249	58	6
Hydra	Hydrae	Hya	1303	1	1
Hydrus	Hydri	Hyi	243	61	3
Indus	Indi	Ind	294	49	3
Lacerta	Lacertae	Lac	201	68	5
Leo	Leonis	Leo	947	12	1
Leo Minor	Leonis Minoris	LMi	232	64	5
Lepus	Leporis	Lep	290	51	1
Libra	Librae	Lib	538	29	1

Name	Genitive	Abbrevn.	Area (square degs.)	Order of size	Origin*
Lupus	Lupi	Lup	334	46	1
Lynx	Lyncis	Lyn	545	28	5
Lyra	Lyrae	Lyr	286	52	1
Mensa	Mensae	Men	153	75	6
Microscopium	Microscopii	Mic	210	66	6
Monoceros	Monocerotis	Mon	482	35	4
Musca	Muscae	Mus	138	77	3
Norma	Normae	Nor	165	74	6
Octans	Octantis	Oct	291	50	6
Ophiuchus	Ophiuchi	Oph	948	11	1
Orion	Orionis	Ori	594	26	1
Pavo	Pavonis	Pav	378	44	3
Pegasus	Pegasi	Peg	1121	7	1
Perseus	Persei	Per	615	24	1
Phoenix	Phoenicis	Phe	469	37	3
Pictor	Pictoris	Pic	247	59	6
Pisces	Piscium	Psc	889	14	1
Piscis Austrinus	Piscis Austrini	PsA	245	60	1
Puppis	Puppis	Pup	673	20	6
Pyxis	Pyxidis	Pyx	221	65	6
Reticulum	Reticuli	Ret	114	82	6
Sagitta	Sagittae	Sge	80	86	1
Sagittarius	Sagittarii	Sgr	867	15	1
Scorpius	Scorpii	Sco	497	33	1
Sculptor	Sculptoris	Scl	475	36	6
Scutum	Scuti	Sct	109	84	5
Serpens	Serpentis	Ser	637	23	1
Sextans	Sextantis	Sex	314	47	5
Taurus	Tauri	Tau	797	17	1
Telescopium	Telescopii	Tel	252	57	6
Triangulum	Trianguli	Tri	132	78	1
Triangulum Australe	Trianguli Australis	TrA	110	83	3
Tucana	Tucanae	Tuc	295	48	3
Ursa Major	Ursae Majoris	UMa	1280	3	1
Ursa Minor	Ursae Minoris	UMi	256	56	1
Vela	Velorum	Vel	500	32	6
Virgo	Virginis	Vir	1294	2	1
Volans	Volantis	Vol	141	76	3
Vulpecula	Vulpeculae	Vul	268	55	5

* Origin:

1 One of the original 48 Greek constellations listed by Ptolemy. The Greek figure of Argo Navis has since been divided into Carina, Puppis and Vela.
2 Considered by the Greeks as part of Leo; made separate by Gerardus Mercator in 1551.
3 The 12 southern constellations of Keyser and de Houtman, c. 1600.
4 Four constellations added by Petrus Plancius.
5 Seven constellations of Hevelius.
6 The 14 southern constellations of Lacaille, who also divided the Greeks' Argo Navis into Carina, Puppis and Vela.

devised patterns fell into disuse, leaving a total of 88 constellations that were officially adopted by the International Astronomical Union, astronomy's governing body, in 1922 (see the table on pages 6–7).

As well as the officially recognized constellations, you can find other patterns among the stars called *asterisms*. An asterism can be composed of stars belonging to one or more constellations. Well-known examples are the Plough or Big Dipper (part of Ursa Major), the Square of Pegasus, the Sickle of Leo and the Teapot of Sagittarius.

Star names   The main stars in each constellation are labelled with a letter of the Greek alphabet, the brightest star usually (but not always!) being termed α (alpha). Notable exceptions in which the stars marked β (beta) are in fact the brightest include the constellations Orion and Gemini. (For reference, the entire Greek alphabet is given in the table below.) Particularly confusing are the constellations Vela and Puppis, which were once joined with Carina to make the extensive figure of Argo Navis, the Ship of the Argonauts. As a result of Argo's subsequent trisection, neither Vela nor Puppis possesses stars labelled α or β, and there are gaps in the sequence of Greek letters in Carina as well.

THE GREEK ALPHABET

α	alpha	ι	iota	ρ	rho
β	beta	κ	kappa	σ	sigma
γ	gamma	λ	lambda	τ	tau
δ	delta	μ	mu	υ	upsilon
ε	epsilon	ν	nu	φ or ϕ	phi
ζ	zeta	ξ	xi	χ	chi
η	eta	ο	omicron	ψ	psi
θ or ϑ	theta	π	pi	ω	omega

The system of labelling stars with Greek letters was introduced by Johann Bayer, so these designations are often known as Bayer letters. The genitive (possessive) case of the constellation's name is always used when referring to a star within it; hence Canis Major, for instance, becomes Canis Majoris, and the name α Canis Majoris means 'the star α in Canis Major'. All constellation names have standard three-letter abbreviations; for instance, the abbreviated form of Canis Major is CMa.

In heavily populated constellations, where Greek letters ran out, fainter stars were assigned Roman letters, both lower-case and capital, such as l Carinae, P Cygni and L Puppis. An additional system of identifying stars is that of Flamsteed numbers, originating from their order in a star catalogue drawn up by England's first Astronomer Royal, John Flamsteed (1646–1719). Examples are 61 Cygni and 70 Ophiuchi.

Before 1930, there were no officially recognized constellation boundaries; some constellations overlapped, and some stars were shared between constellations. In that year, the International Astronomical Union published definitive boundaries for all constellations. In the process, certain stars allocated by the Bayer and Flamsteed systems to one constellation found themselves transferred to a neighbour, leading to gaps in the sequence of letters and numbers.

Prominent stars also have proper names by which they are commonly known. For example, α Canis Majoris, the brightest star in the sky, is better known as Sirius. Stars' proper names originate from several sources. Some, such as Sirius, Castor and Pollux, date back to ancient Greek times. Many others, such as Aldebaran, are of Arabic origin. Still others were added more recently by European astronomers who borrowed Arabic words in corrupted form; an example is Betelgeuse, which in its current form is meaningless in Arabic. To add to the confusion, spelling of names can vary from list to list, and some stars have more than one proper name. For the maps in this book we have adopted the names used by the *Millennium Star Atlas* (European Space Agency and Sky Publishing Corp., 1997).

Star clusters, nebulae and galaxies have a different system of identification. The most prominent of them are given numbers prefixed by the letter M from a catalogue compiled by the Frenchman Charles Messier. For example, M1 is the Crab Nebula and M31 the Andromeda Galaxy. Messier's catalogue contained 103 objects, and a few more were added later by other astronomers bringing the total to 110. A far more comprehensive listing, containing many thousands of objects, is the *New General Catalogue* (NGC) compiled by J.L.E. Dreyer, with two supplements called the *Index Catalogues* (IC).

Messier numbers and NGC numbers remain in use by astronomers, and both systems are used in this book. On the charts, such objects are labelled with their Messier number if they have one, or otherwise by their NGC number (without the 'NGC' prefix) or IC number (prefix 'I').

Star brightness　Stars appear of different brightnesses in the sky, for two reasons. Firstly, they give out different amounts of light. But also, and just as importantly, they lie at vastly differing distances. Hence, a modest star that is quite close to us can appear brighter than a tremendously powerful star that is a long way away.

Astronomers call a star's brightness its *magnitude*. The magnitude scale was introduced by the Greek astronomer Hipparchus in 129 BC. Hipparchus divided the naked-eye stars into six classes of brightness, from 1st magnitude (the brightest stars) to 6th magnitude (the faintest visible to the naked eye). In his day there was no means of measuring star brightness precisely, so this rough classification sufficed. But with the coming of technology it was possible to measure star brightnesses to an exact fraction of a magnitude.

In 1856 the English astronomer Norman Pogson put the magnitude scale on a precise mathematical footing by defining a star of magnitude 1 as being exactly 100 times brighter than a star of magnitude 6. Since, on this scale, a difference of five magnitudes corresponds to a brightness difference of 100 times, a step of one magnitude is equal to a brightness difference of just over 2.5 times (the fifth root of 100).

Objects more than 250 times brighter than 6th magnitude are given negative (minus) magnitudes. For example, Sirius, the brightest star in the sky, is of magnitude −1.44. At the other end of the scale, stars fainter than magnitude 6 are given progressively larger positive magnitudes. The faintest objects detected by telescopes on Earth are around magnitude 27.

Any object of magnitude 1.49 or brighter is said to be of first magnitude; objects from 1.50 to 2.49 are termed second magnitude; and so on. The magnitude system may sound confusing at first, but it works well in practice and has the advantage that it can be extended indefinitely in both directions, to the very bright and the very faint.

When used without any further qualification, the term 'magnitude' refers to how bright a star appears to us in the sky; strictly, this is the star's *apparent magnitude*. But because the distance of a star affects how bright it appears, the apparent magnitude bears little relation to its actual light output, or *absolute magnitude*. A star's absolute magnitude is defined as the brightness it would appear to have if it were at a standard distance of 10 parsecs from us (the parsec is explained in the following section). Astronomers calculate the absolute magnitude from knowledge of the star's nature and its distance.

Absolute magnitude is a good way of comparing the intrinsic brightness of stars. For instance, our daytime star the Sun has an apparent magnitude of −26.7, but an absolute magnitude of 4.8 (when no sign is given the magnitude is understood to be positive). Deneb (α Cygni) has an apparent magnitude of 1.3, but an absolute magnitude of −8.7. From comparison of these absolute magnitudes we deduce that Deneb gives out over 250,000 times as much light as the Sun and hence is one of the most luminous stars known, even though there is nothing at first sight to mark it out as extraordinary.

The star magnitudes given in this book come from *The Hipparcos Catalogue* (European Space Agency, 1997). A number of stars actually vary in their light output, for various reasons. The nature of such so-called variable stars is discussed in Section II.

Star distances In the Universe, distances are so huge that astronomers have abandoned the puny kilometre (km) and have invented their own units. Most familiar of these is the *light year* (l.y.), the distance that a beam of light travels in one year. Light moves at the fastest known speed in the Universe, 299,792.5 km per second. A light year is equivalent to 9.46 million million km. On average, stars are several light years apart. For instance, the closest star to the Sun, Proxima Centauri (actually a

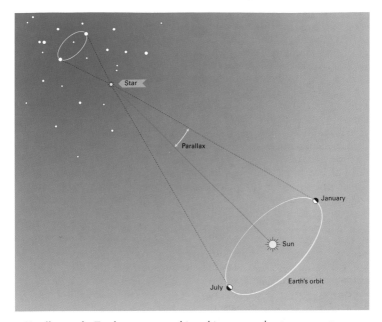

Parallax: as the Earth moves around its orbit, so a nearby star appears to change in position against the celestial background. The shift in position is known as the star's parallax. The nearer the star is to us, the greater its parallax. The amount of parallax is exaggerated for clarity. (Wil Tirion)

member of the α Centauri triple system), is 4.2 light years away. Sirius is 8.6 l.y. away and Deneb over 3000 l.y. away.

The distance of the nearest stars can be found directly in the following way. A star's position is measured accurately when the Earth is on one side of the Sun, and then remeasured six months later when the Earth has moved around its orbit to the other side of the Sun. When viewed from two widely differing points in space in this way, a nearby star will appear to have shifted slightly in position with respect to more distant stars (see diagram above).

This effect is known as *parallax*, and applies to any object viewed from two vantage points against a fixed background, such as a tree against the horizon. A star's parallax shift is so small as to be unnoticeable under normal circumstances – in the case of Proxima Centauri, which has the greatest parallax shift of any star, the amount is about the same as the width of a small coin seen at a distance of 2 km. Once the star's parallax shift has been measured, a simple calculation reveals how far away it is.

An object close enough to us to show a parallax shift of 1″ (one second of arc) would, in the jargon of astronomers, be said to lie at a distance of one *parsec*. In practice, no star is this close; the parallax of Proxima Centauri is 0″.77. One parsec is equivalent to 3.26 light years. Astronomers frequently use parsecs in preference to light years because of the ease of converting parallax into distance: a star's distance in parsecs is simply the inverse of its parallax in seconds of arc. For example, a star 2 parsecs away has a parallax of 0″.5, 4 parsecs away its parallax is 0″.25, and so on.

The farther away a star is, the smaller its parallax. Beyond about 50 light years a star's parallax becomes too small to be measured accurately by telescopes on Earth. Before the launch of the European Space Agency's astrometry satellite Hipparcos in 1989, astronomers had been able to establish reliable parallaxes for fewer than 1000 stars from Earth; from space, Hipparcos has increased the number of reliable parallaxes to over 100,000.

For stars that are too distant to have their parallax measured even by Hipparcos, astronomers first of all estimate the star's absolute magnitude by studying the spectrum of its light. They then compare this estimated absolute magnitude with the observed apparent magnitude to determine the star's distance. The distance obtained in this way is open to considerable error, and the values quoted in various books and catalogues often differ widely as a result.

Distances of stars given in this book are all expressed in light years. Like the star magnitudes, they are taken from *The Hipparcos Catalogue*. Most of these distances are accurate to better than 10 per cent, with the uncertainties tending to become greater for the more distant stars.

As the Earth moves along its orbit, the Sun is seen in different directions against the star background. The Sun's path against the stars is known as the ecliptic. The constellations that the Sun passes in front of during the year are known as the constellations of the zodiac. (Wil Tirion)

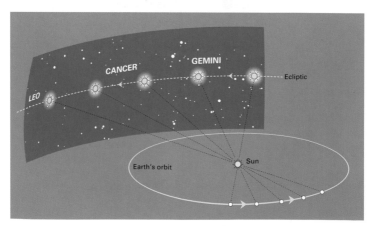

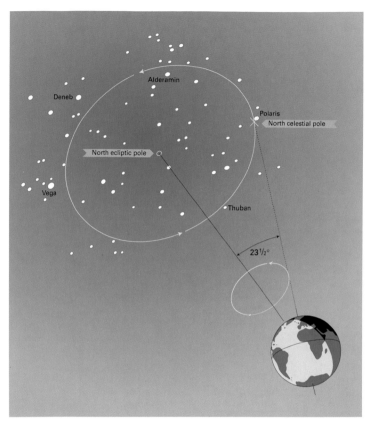

The Earth is very slowly wobbling in space like a tilted spinning top, an effect known as precession. As a result of precession, the celestial poles trace out a complete circle on the sky every 26,000 years. Only the north celestial pole is shown here, but the effect applies to the south pole as well. (Wil Tirion)

Star positions   To determine positions of objects in the sky, astronomers use a system of coordinates similar to latitude and longitude on Earth. The celestial equivalent of latitude is called *declination*, and the equivalent of longitude is called *right ascension*. Declination is measured in degrees, minutes and seconds (abbreviated °, ′ and ″) of arc from 0° on the celestial equator to 90° at the celestial poles. The celestial poles lie exactly above the Earth's poles, while the celestial equator is the projection onto the sky of the Earth's equator. Right ascension is measured in hours, minutes and seconds (abbreviated h, m and s), from 0h to 24h. The 0h line of right ascension, the celestial equivalent of the Greenwich meridian, is defined as the point where the Sun crosses the celestial equator on its way north each year. Technically, this point is known as the vernal (or spring) equinox.

The Sun's path around the sky each year is known as the *ecliptic*. This path is inclined at 23½° to the celestial equator, because the Earth's axis is inclined at 23½° to the vertical. The most northerly and southerly points that the Sun reaches each year are called the *solstices*, 23½° north and south of the equator. If the Earth's axis were directly upright with respect to its orbit around the Sun, then the equator and ecliptic would coincide. One result would be that we would have no seasons on Earth, for the Sun would always remain directly above the equator.

One additional effect that becomes important over long periods of time is that the Earth is slowly wobbling on its axis, like a spinning top. The axis remains inclined at an angle of 23½°, but the position in the sky to which the north and south poles of the Earth are pointing moves slowly. The Earth's poles describe a large circle on the sky, taking 26,000 years to return to their starting places. Hence the position of the celestial pole is always changing, albeit imperceptibly, as are the two points at which the Sun's path (the ecliptic) cuts the celestial equator. This wobbling of the Earth in space is termed *precession*. As an example of the effects caused by precession, whereas Polaris is the Pole Star today, in 11,000 years' time the north celestial pole will lie near Vega. And the vernal equinox which lay in Aries between 1865 BC and 67 BC now lies in Pisces.

The effect of precession means that the coordinates of all celestial objects – the catalogued positions of stars, galaxies, and even constellation boundaries – are continually drifting. Astronomers usually draw up catalogues and star charts for a standard reference date, or *epoch*. The epoch of the star positions in this book is the year 2000. For most general purposes, precession does not introduce a noticeable error until after about 50 years, so the charts in this book will be usable without amendment until halfway through the 21st century.

Proper motions   All the stars visible in the sky are members of a vast wheeling mass of stars called the Galaxy. Those stars visible to the naked eye are among the nearest to us in the Galaxy. More distant stars in the Galaxy crowd together in a hazy band called the Milky Way, which can be seen crossing the night sky.

The Sun and the other stars are all orbiting the centre of the Galaxy; the Sun takes about 250 million years to complete one orbit. Other stars move at different speeds, like cars in different lanes on a highway. As a result, stars are all very slowly changing their positions relative to one another. Such stellar movement, termed *proper motion*, is so slight that it is undetectable to the naked eye even over a human lifetime, but it can be measured through telescopes. As with many other aspects of stellar positions, our knowledge of proper motions has been radically improved by the Hipparcos satellite.

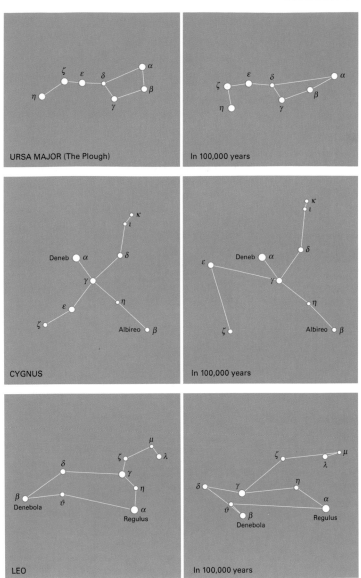

Three familiar star patterns as they appear today, at left of diagram, and as they will appear in 100,000 years' time, right. The changes in appearance are due to the proper motions of the stars. (Wil Tirion)

If ancient Greek astronomers could be transported 2000 years forward in time to the present day, they would notice little difference in the sky, with the exception of Arcturus, a fast-moving bright star, which has drifted more than two Moon diameters from its position then. Over very long periods of time the proper motions of stars considerably distort the shapes of all constellations. The diagrams on the previous page show some examples.

An additional long-term effect of stellar motions is to change the apparent magnitudes of stars as they move closer to us or away. For example, Sirius will brighten by 20 per cent over the next 60,000 years as its distance shrinks by 0.8 light years. Then, as its distance increases again, it will be overtaken as the brightest star in the sky by Vega, which will peak at magnitude −0.8 nearly 300,000 years from now.

Appearance of the sky   Three factors affect the appearance of the sky: the time of night, the time of year and your latitude on Earth. Firstly, let's consider the effect of latitude. An observer at one of the Earth's poles, latitude 90°, would see the celestial pole directly overhead, and as the Earth turned all stars would circle around the celestial pole without rising or setting (see the top diagram on the facing page). At the other extreme, an observer stationed exactly on the Earth's equator, latitude 0°, would see the celestial equator directly overhead, as shown in the middle diagram. The north and south celestial poles would lie on the north and south horizons respectively, and every part of the sky would be visible at one time or another. All stars would rise in the east and set in the west as the Earth rotated.

For most observers, the real sky appears somewhere between these two extremes: the celestial pole is at some intermediate altitude between horizon and zenith, and the stars closest to it circle around it without setting (they are said to be *circumpolar*) while the rest of the stars rise and set. The exact angle of the celestial pole above the horizon depends on the observer's latitude. For someone at latitude 50° north, for instance, the north celestial pole is 50° above the northern horizon (bottom diagram opposite). As another example, if you were at latitude 30° south, the south celestial pole would be 30° above the southern horizon. In other words, the altitude of the celestial pole above the horizon is exactly equal to your latitude, a fact long appreciated by navigators.

As the Earth turns, the stars march across the heavens at the rate of 15° per hour (the Earth rotates through 360° in 24 hours). Therefore the appearance of the sky changes with the time of night. An added complication is that the Earth is also orbiting the Sun each year, so the constellations visible will change with the seasons. For example, a constellation such as Orion, splendidly seen in December and January, will be in the daytime sky six months later and hence be invisible. The maps on pages 24–71 will help you find out which stars are on view whenever and wherever you want to observe.

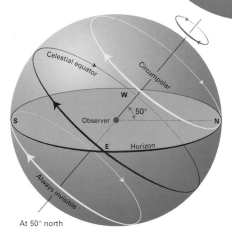

The changing appearance of the sky as seen from different latitudes on Earth.

Left: At the pole, only one half of the sky is ever visible, the other half being permanently below the horizon.

Right: At the equator, by contrast, all the sky is visible; as the Earth rotates, stars appear to rise in the east and set in the west.

At the equator

Left: At intermediate latitudes, the situation is between the two extremes. Part of the sky is always above the horizon (the part marked 'Circumpolar'), but an equal part is always below the horizon and hence is invisible. Stars between these two regions rise and set during the night. (Wil Tirion)

At 50° north

At the north pole

17

Star trails over the dome of the William Herschel Telescope on La Palma in the Canary Islands, created as the Earth rotated during a time exposure. The bright star near the centre is Polaris. (Nik Szymanek)

The star charts

The hemisphere charts   On the following four pages are charts showing the complete northern and southern hemispheres of the sky. As well as the main stars of each hemisphere, the charts depict the hazy band of the Milky Way; the dashed line is the ecliptic, the Sun's path in the heavens. When planets are visible, they will be found near the ecliptic.

Around the rim of each chart are listed the months of the year, to help you find which constellations are best placed at about 10 p.m. local time each month, or 11 p.m. when daylight-saving time (DST, or summer time) is in operation. Observers in mid-northern latitudes should take the northern hemisphere chart and turn it so that the month of observation is at the bottom. The chart will show the sky as visible when you face due south that evening. Rotate the chart 15° anticlockwise for each hour after 10 p.m., and turn it clockwise for each hour before 10 p.m.

Observers in mid-southern latitudes should take the southern hemisphere chart and turn it so that the month of observation is at the bottom. The chart will show the stars as they appear when you are facing due north. Turn the chart 15° clockwise for each hour after 10 p.m., and 15° anticlockwise for each hour before. (In all cases above, for 10 p.m. read 11 p.m. when daylight-saving time, DST, is in operation.)

The monthly charts   Next comes a series of maps showing the sky as seen when facing north or south at 10 p.m. (11 p.m. daylight-saving time) in mid-month from various latitudes. The first set of maps is usable from latitude 60° north to 10° north; the second set ranges from the equator to 50° south. (They will also be usable for about 10° either side of this range without significant error.) Curved lines on each map depict the horizon for each latitude. Taking these in conjunction with the maps of the complete celestial hemispheres, you should be able to identify the stars in the sky no matter where you are on Earth.

The constellation charts   The centrepiece of this book consists of individual charts of each constellation with descriptions of the brightest stars and main objects of interest. All stars within each constellation down to magnitude 6.0 are shown, with the sizes of the symbols graded in half-magnitude steps, plus the most prominent examples of what are termed *deep-sky objects* (star clusters, nebulae and galaxies). The total number of stars shown is about 5000. All constellation maps are to the same scale, with the exception of the rambling Hydra, the largest constellation of all, which is drawn to a significantly smaller scale to fit the page. Areas of particular interest in certain constellations, such as the Hyades and Pleiades clusters in Taurus, and the Orion Nebula, are shown to a larger scale in special detail charts.

We hope that the charts and descriptions in this book will serve as trusty companions for many nights of exploration under the stars. Good stargazing!

NORTHERN HEMISPHERE

Overview

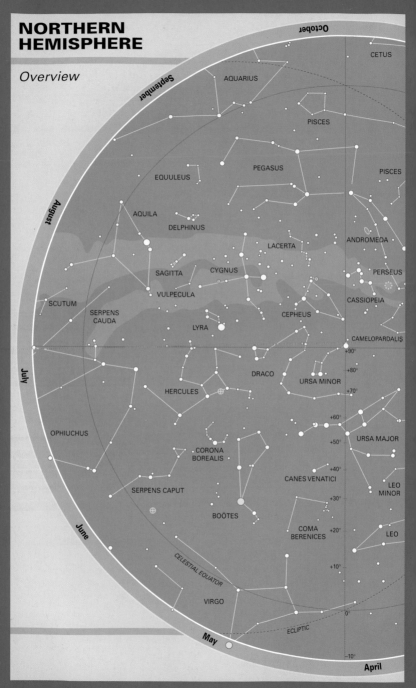

October

September

August

July

June

May

April

CETUS

AQUARIUS

PISCES

PEGASUS

EQUULEUS

PISCES

AQUILA

DELPHINUS

LACERTA

ANDROMEDA

SAGITTA

CYGNUS

PERSEUS

VULPECULA

CEPHEUS

CASSIOPEIA

SCUTUM

LYRA

CAMELOPARDALIS

SERPENS
CAUDA

+90°

DRACO

URSA MINOR

+80°

HERCULES

+70°

+60°

URSA MAJOR

OPHIUCHUS

+50°

CORONA
BOREALIS

+40°

CANES VENATICI

LEO
MINOR

SERPENS CAPUT

+30°

BOÖTES

COMA
BERENICES

+20°

LEO

+10°

CELESTIAL EQUATOR

0°

VIRGO

ECLIPTIC

−10°

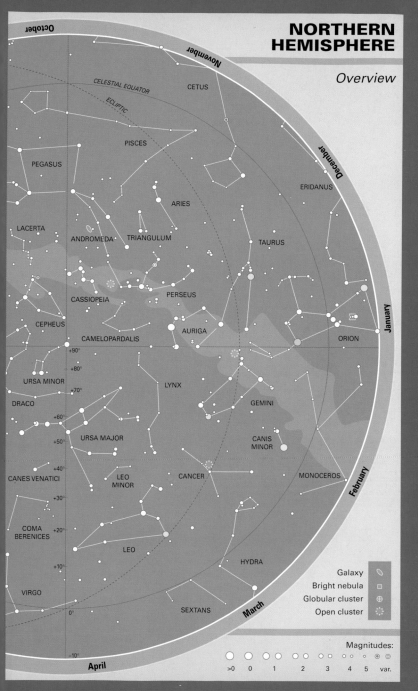

NORTHERN HEMISPHERE

Overview

October
November
December
January
February
March
April

CELESTIAL EQUATOR
ECLIPTIC

CETUS
PISCES
PEGASUS
ARIES
ERIDANUS
LACERTA
ANDROMEDA
TRIANGULUM
TAURUS
CASSIOPEIA
PERSEUS
CEPHEUS
AURIGA
CAMELOPARDALIS
ORION
+90°
+80°
URSA MINOR
+70°
LYNX
DRACO
GEMINI
+60°
+50°
URSA MAJOR
CANIS
MINOR
+40°
CANES VENATICI
LEO
MINOR
CANCER
MONOCEROS
+30°
COMA
BERENICES
+20°
LEO
HYDRA
+10°
VIRGO
0°
SEXTANS
−10°

Galaxy
Bright nebula
Globular cluster
Open cluster

Magnitudes:

| >0 | 0 | 1 | 2 | 3 | 4 | 5 | var. |

SOUTHERN HEMISPHERE

Overview

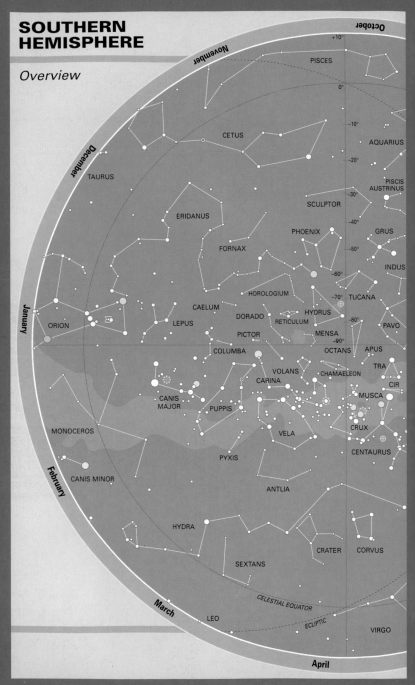

October
November
December
January
February
March
April

PISCES
CETUS
TAURUS
AQUARIUS
PISCIS AUSTRINUS
SCULPTOR
ERIDANUS
GRUS
PHOENIX
FORNAX
INDUS
HOROLOGIUM
TUCANA
CAELUM
HYDRUS
DORADO
PAVO
LEPUS
RETICULUM
PICTOR
MENSA
ORION
COLUMBA
OCTANS
APUS
VOLANS
TRA
CHAMAELEON
CIR
CARINA
MUSCA
CANIS MAJOR
PUPPIS
CRUX
MONOCEROS
VELA
CENTAURUS
PYXIS
CANIS MINOR
ANTLIA
HYDRA
CRATER
CORVUS
SEXTANS
CELESTIAL EQUATOR
MARCH
LEO
ECLIPTIC
VIRGO
April

+10°
0°
−10°
−20°
−30°
−40°
−50°
−60°
−70°
−80°
−90°

22

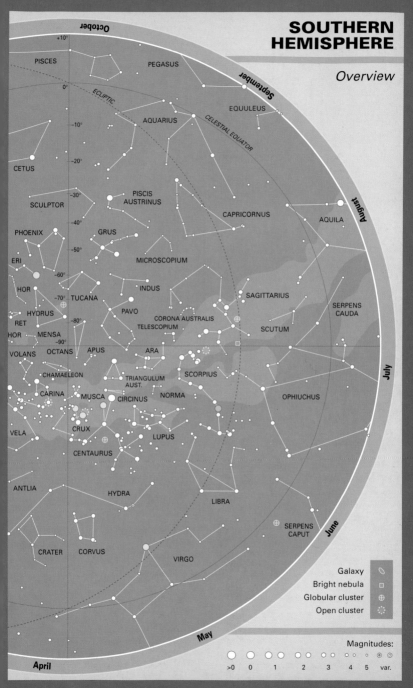

SOUTHERN HEMISPHERE

Overview

October

+10°

PISCES

PEGASUS

September

0°

ECLIPTIC

EQUULEUS

AQUARIUS

CELESTIAL EQUATOR

-10°

CETUS

-20°

PISCIS AUSTRINUS

CAPRICORNUS

August

AQUILA

SCULPTOR

-30°

-40°

PHOENIX

GRUS

-50°

ERI

MICROSCOPIUM

SAGITTARIUS

SERPENS CAUDA

HOR

-60°

INDUS

HYDRUS

-70°

TUCANA

PAVO

CORONA AUSTRALIS

SCUTUM

RET

CETUS

-80°

TELESCOPIUM

HOR

MENSA

-90°

VOLANS

OCTANS

APUS

ARA

July

CHAMAELEON

TRIANGULUM AUST.

SCORPIUS

OPHIUCHUS

CARINA

MUSCA

CIRCINUS

NORMA

VELA

CRUX

LUPUS

CENTAURUS

ANTLIA

HYDRA

LIBRA

June

SERPENS CAPUT

CRATER

CORVUS

VIRGO

May

April

	Galaxy
	Bright nebula
	Globular cluster
	Open cluster

Magnitudes:

| >0 | 0 | 1 | 2 | 3 | 4 | 5 | var. |

23

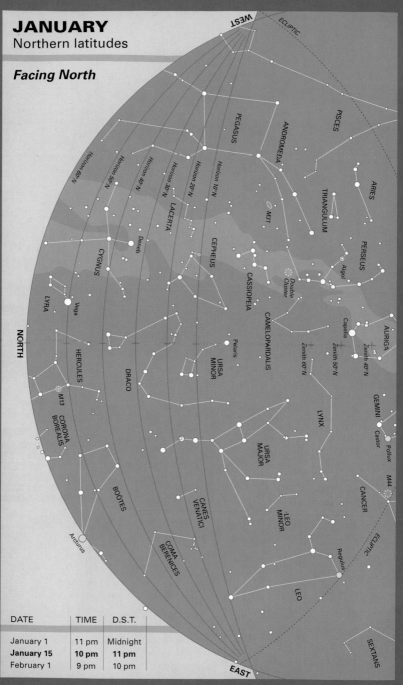

JANUARY
Northern latitudes

Facing North

DATE	TIME	D.S.T.
January 1	11 pm	Midnight
January 15	**10 pm**	**11 pm**
February 1	9 pm	10 pm

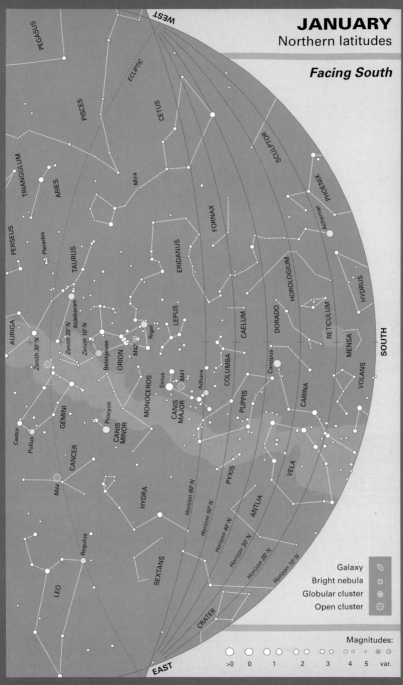

JANUARY
Northern latitudes

Facing South

WEST

PEGASUS
PISCES
ECLIPTIC
CETUS
SCULPTOR
TRIANGULUM
ARIES
Mira
PHOENIX
Achernar
PERSEUS
FORNAX
Pleiades
TAURUS
ERIDANUS
HOROLOGIUM
HYDRUS
Aldebaran
LEPUS
DORADO
RETICULUM
AURIGA
Zenith 30° N
Zenith 20° N
Zenith 10° N
CAELUM
MENSA
SOUTH
Rigel
M42
Betelgeuse
COLUMBA
Canopus
VOLANS
ORION
DORADO
CARINA
MONOCEROS
Sirius
M41
Adhara
GEMINI
Castor
Pollux
Procyon
CANIS MAJOR
PUPPIS
Pollux
CANCER
CANIS MINOR
M44
HYDRA
PYXIS
VELA
Horizon 60° N
Horizon 50° N
ANTLIA
Regulus
Horizon 40° N
Horizon 30° N
Horizon 20° N
Horizon 10° N
LEO
SEXTANS
CRATER

EAST

Galaxy
Bright nebula
Globular cluster
Open cluster

Magnitudes:
>0 0 1 2 3 4 5 var.

25

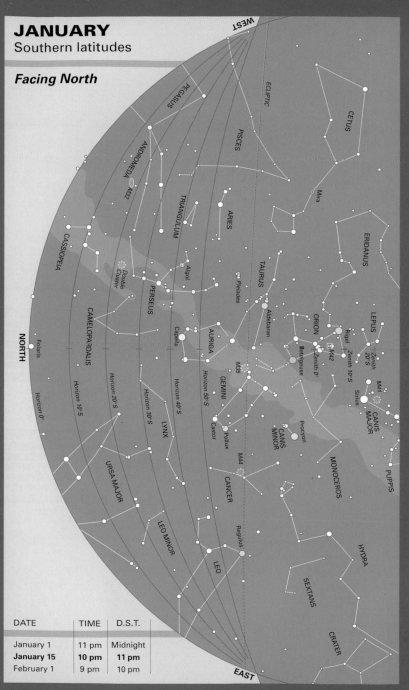

JANUARY
Southern latitudes

Facing North

WEST

ECLIPTIC

CETUS

PEGASUS

PISCES

ANDROMEDA

M32

TRIANGULUM

ARIES

Mira

ERIDANUS

CASSIOPEIA

Double Cluster

PERSEUS

Algol

Pleiades

TAURUS

LEPUS

Aldebaran

ORION

Zenith 20° S

Rigel

NORTH

Polaris

CAMELOPARDALIS

Capella

AURIGA

Betelgeuse

M42

Zenith 10° S

Zenith 0°

M41

Sirius

Horizon 0°

Horizon 10° S

Horizon 20° S

Horizon 30° S

Horizon 40° S

Horizon 50° S

M35

GEMINI

CANIS MAJOR

LYNX

Castor

Pollux

CANIS MINOR

Procyon

MONOCEROS

PUPPIS

URSA MAJOR

M44

CANCER

LEO MINOR

LEO

Regulus

HYDRA

SEXTANS

CRATER

EAST

DATE	TIME	D.S.T.
January 1	11 pm	Midnight
January 15	**10 pm**	**11 pm**
February 1	9 pm	10 pm

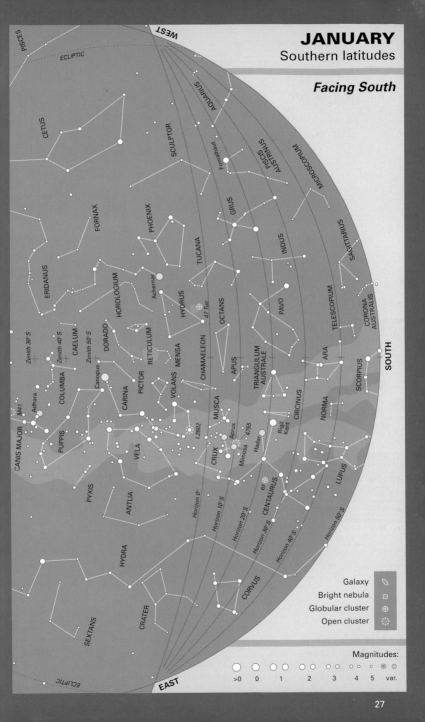

JANUARY
Southern latitudes

Facing South

WEST

PISCES
ECLIPTIC
CETUS
AQUARIUS
SCULPTOR
Fomalhaut
PISCIS AUSTRINUS
MICROSCOPIUM
FORNAX
PHOENIX
GRUS
TUCANA
SAGITTARIUS
ERIDANUS
HOROLOGIUM
Achernar
HYDRUS
47 Tuc
OCTANS
INDUS
PAVO
TELESCOPIUM
CORONA AUSTRALIS
Zenith 30°S
Zenith 40°S
Zenith 50°S
CAELUM
DORADO
RETICULUM
MENSA
CHAMAELEON
APUS
TRIANGULUM AUSTRALE
ARA
SOUTH
CANIS MAJOR
M41
Adhara
COLUMBA
Canopus
PICTOR
CARINA
VOLANS
MUSCA
SCORPIUS
Acrux
I2602
CRUX
Mimosa
4755
Hadar
Rigil Kent
CIRCINUS
NORMA
PUPPIS
VELA
ω
CENTAURUS
LUPUS
PYXIS
Horizon 0°
Horizon 10°S
Horizon 20°S
Horizon 30°S
Horizon 40°S
Horizon 50°S
ANTLIA
HYDRA
CORVUS
CRATER
SEXTANS
ECLIPTIC
EAST

Galaxy
Bright nebula
Globular cluster
Open cluster

Magnitudes:
>0 0 1 2 3 4 5 var.

27

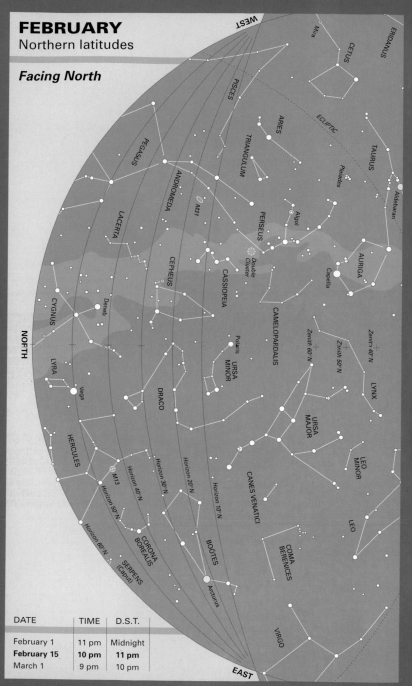

FEBRUARY
Northern latitudes

Facing North

WEST

Mira

CETUS

ERIDANUS

PISCES

ECLIPTIC

PEGASUS

ARIES

TRIANGULUM

TAURUS

Pleiades

Aldebaran

LACERTA

ANDROMEDA

M31

PERSEUS

Algol

AURIGA

Capella

CYGNUS

Deneb

CEPHEUS

CASSIOPEIA

Double Cluster

CAMELOPARDALIS

Zenith 40° N

NORTH

LYRA

Polaris

URSA MINOR

Zenith 60° N

Zenith 50° N

LYNX

Vega

DRACO

URSA MAJOR

LEO MINOR

HERCULES

M13

Horizon 40° N

Horizon 30° N

Horizon 20° N

Horizon 10° N

CANES VENATICI

LEO

Horizon 50° N

CORONA BOREALIS

BOOTES

COMA BERENICES

Horizon 60° N

SERPENS (Caput)

Arcturus

VIRGO

EAST

DATE	TIME	D.S.T.
February 1	11 pm	Midnight
February 15	**10 pm**	**11 pm**
March 1	9 pm	10 pm

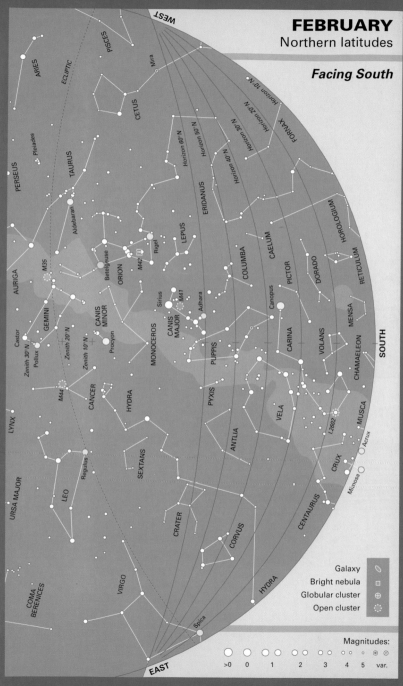

FEBRUARY
Northern latitudes

Facing South

29

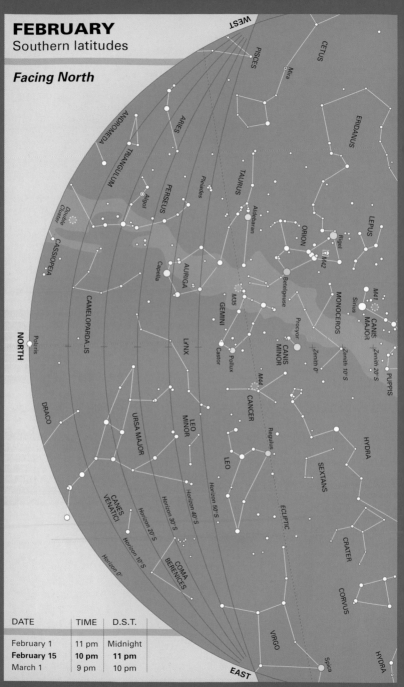

FEBRUARY
Southern latitudes

Facing North

DATE	TIME	D.S.T.
February 1	11 pm	Midnight
February 15	**10 pm**	**11 pm**
March 1	9 pm	10 pm

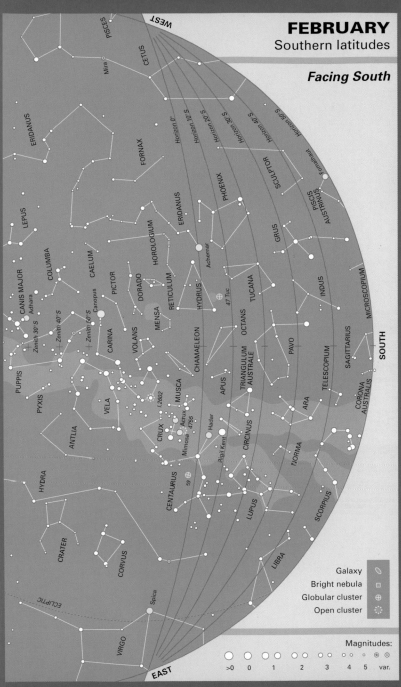

FEBRUARY
Southern latitudes

Facing South

Galaxy
Bright nebula
Globular cluster
Open cluster

Magnitudes:
>0 0 1 2 3 4 5 var.

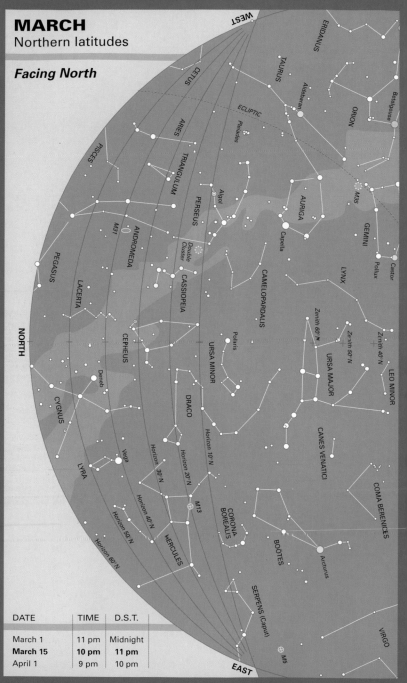

MARCH
Northern latitudes

Facing North

WEST

ERIDANUS

TAURUS

Aldebaran

Betelgeuse

ORION

CETUS

ECLIPTIC

Pleiades

ARIES

PISCES

TRIANGULUM

PERSEUS

Algol

M31

ANDROMEDA

Double Cluster

CASSIOPEIA

PEGASUS

LACERTA

CEPHEUS

URSA MINOR

Polaris

CYGNUS

Deneb

DRACO

LYRA

Vega

Horizon 10° N

Horizon 20° N

Horizon 30° N

Horizon 40° N

Horizon 50° N

Horizon 60° N

MERCULES

M13

SERPENS (Caput)

CORONA BOREALIS

BOÖTES

Arcturus

COMA BERENICES

VIRGO

M5

AURIGA

Capella

M35

GEMINI

Pollux

Castor

LYNX

CAMELOPARDALIS

Zenith 60° N

Zenith 50° N

Zenith 40° N

URSA MAJOR

LEO MINOR

CANES VENATICI

NORTH

EAST

DATE	TIME	D.S.T.
March 1	11 pm	Midnight
March 15	**10 pm**	**11 pm**
April 1	9 pm	10 pm

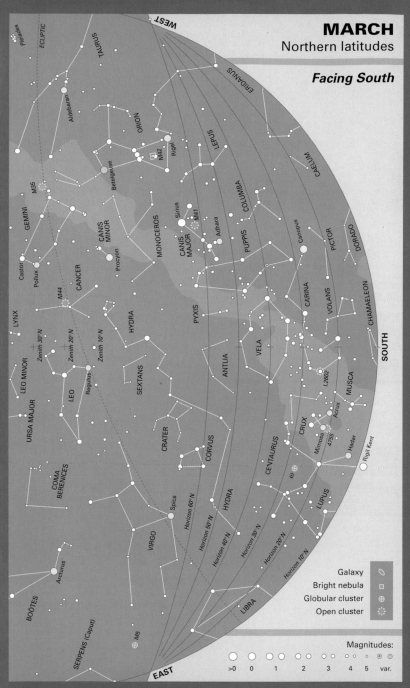

MARCH
Northern latitudes

Facing South

WEST

Pleiades
ECLIPTIC
TAURUS
Aldebaran
ORION
M42
Rigel
ERIDANUS
LEPUS
CAELUM
Betelgeuse
M35
GEMINI
CANIS MINOR
MONOCEROS
Sirius
M41
CANIS MAJOR
Adhara
COLUMBA
PUPPIS
Canopus
PICTOR
DORADO
Castor
Pollux
Procyon
CANCER
M44
Zenith 30° N
Zenith 20° N
Zenith 10° N
HYDRA
PYXIS
VELA
CARINA
VOLANS
CHAMAELEON
LYNX
LEO MINOR
URSA MAJOR
LEO
Regulus
SEXTANS
ANTLIA
I.2602
MUSCA
SOUTH
COMA BERENICES
CRATER
CORVUS
HYDRA
CENTAURUS
CRUX
Acrux
Mimosa
4755
Hadar
Rigil Kent
ω
VIRGO
Spica
Horizon 60° N
Horizon 50° N
Horizon 40° N
Horizon 30° N
Horizon 20° N
Horizon 10° N
LUPUS
BOÖTES
Arcturus
SERPENS (Caput)
M5
LIBRA

EAST

Galaxy
Bright nebula
Globular cluster
Open cluster

Magnitudes:
>0   0   1   2   3   4   5   var.

33

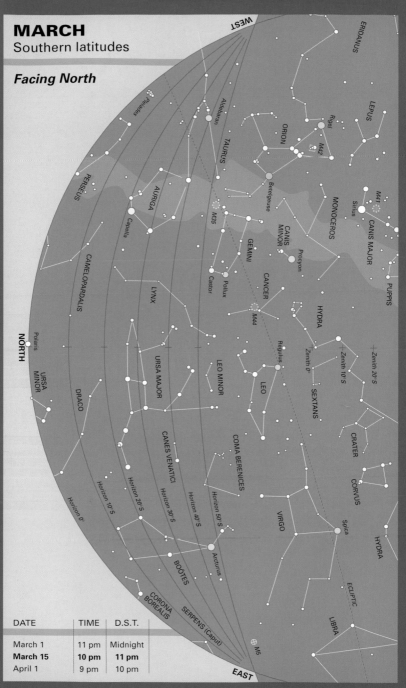

MARCH
Southern latitudes

Facing North

WEST

ERIDANUS

LEPUS

ORION

Rigel

M42

Pleiades

Aldebaran

TAURUS

Betelgeuse

MONOCEROS

Sirius

M41

CANIS MAJOR

PUPPIS

PERSEUS

AURIGA

Capella

M35

CANIS MINOR

Procyon

GEMINI

Castor

Pollux

CANCER

M44

HYDRA

CAMELOPARDALIS

LYNX

Zenith 0°

Zenith 10° S

Zenith 20° S

NORTH

Polaris

URSA MINOR

DRACO

URSA MAJOR

LEO MINOR

Regulus

LEO

SEXTANS

CRATER

CANES VENATICI

COMA BERENICES

CORVUS

Horizon 0°

Horizon 10° S

Horizon 20° S

Horizon 30° S

Horizon 40° S

Horizon 50° S

Arcturus

BOÖTES

VIRGO

Spica

HYDRA

CORONA BOREALIS

SERPENS (Caput)

ECLIPTIC

M5

LIBRA

EAST

DATE	TIME	D.S.T.
March 1	11 pm	Midnight
March 15	**10 pm**	**11 pm**
April 1	9 pm	10 pm

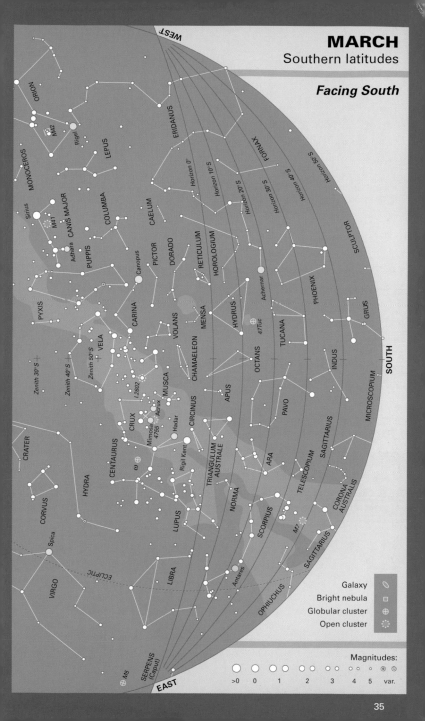

MARCH
Southern latitudes

Facing South

WEST

EAST

SOUTH

Constellations and objects labeled:

ORION, M42, Rigel, LEPUS, ERIDANUS, FORNAX, Horizon 0°, Horizon 10° S, Horizon 20° S, Horizon 30° S, Horizon 40° S, Horizon 50° S, S 50° S, SCULPTOR

MONOCEROS, Sirius, M41, CANIS MAJOR, Adhara, PUPPIS, COLUMBA, CAELUM, PICTOR, DORADO, RETICULUM, HOROLOGIUM, Achernar, PHOENIX, GRUS

PYXIS, Canopus, CARINA, VELA, VOLANS, MENSA, HYDRUS, 47Tuc, TUCANA, INDUS

Zenith 30° S, Zenith 40° S, Zenith 50° S, I.2602, MUSCA, CHAMAELEON, OCTANS, PAVO, SAGITTARIUS, MICROSCOPIUM

CRATER, CENTAURUS, CRUX, Acrux, Mimosa, 4755, Hadar, CIRCINUS, APUS, ARA, TELESCOPIUM, SAGITTARIUS, CORONA AUSTRALIS

CORVUS, HYDRA, ω, Rigil Kent, TRIANGULUM AUSTRALE, NORMA, SCORPIUS, M7

Spica, VIRGO, LIBRA, LUPUS, Antares, OPHIUCHUS, SAGITTARIUS

ECLIPTIC, M5, SERPENS (Caput)

Legend:

Galaxy	
Bright nebula	
Globular cluster	
Open cluster	

Magnitudes:

>0 0 1 2 3 4 5 var.

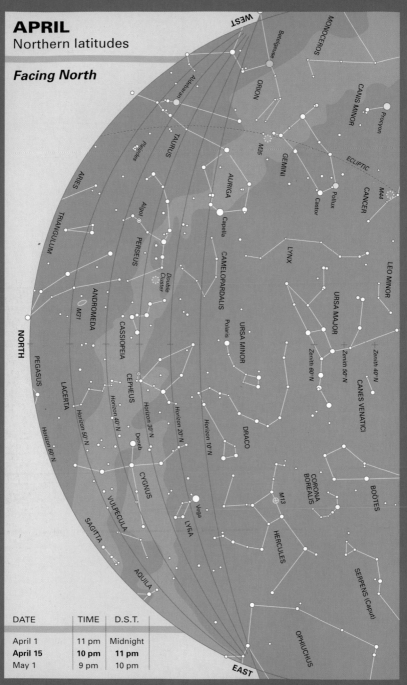

APRIL
Northern latitudes

Facing North

WEST

MONOCEROS

Betelgeuse

ORION

CANIS MINOR

Procyon

Aldebaran

Pleiades

TAURUS

M35

GEMINI

ECLIPTIC

AURIGA

Castor

Pollux

CANCER

M44

ARIES

Algol

Capella

LYNX

LEO MINOR

TRIANGULUM

PERSEUS

CAMELOPARDALIS

URSA MAJOR

Double Cluster

ANDROMEDA

M31

Zenith 60° N

Zenith 50° N

Zenith 40° N

NORTH

CASSIOPEIA

Polaris

URSA MINOR

CANES VENATICI

PEGASUS

CEPHEUS

Zenith 50° N

LACERTA

Horizon 50° N

Horizon 40° N

Horizon 30° N

Horizon 20° N

Horizon 10° N

DRACO

Horizon 60° N

Deneb

BOÖTES

CYGNUS

Vega

M13

CORONA BOREALIS

VULPECULA

LYRA

SAGITTA

HERCULES

SERPENS (Caput)

AQUILA

OPHIUCHUS

EAST

DATE	TIME	D.S.T.
April 1	11 pm	Midnight
April 15	**10 pm**	**11 pm**
May 1	9 pm	10 pm

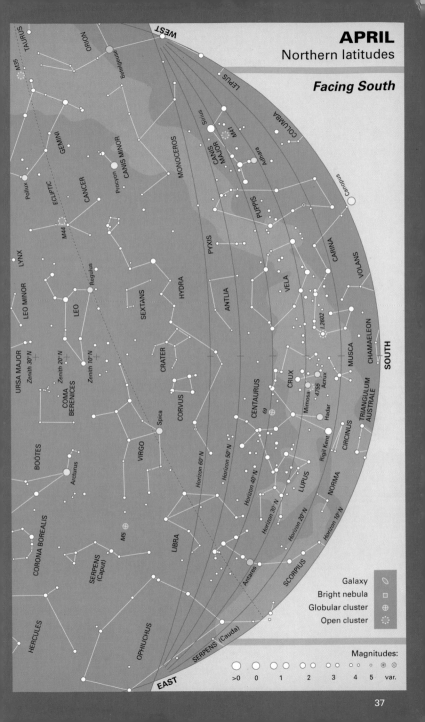

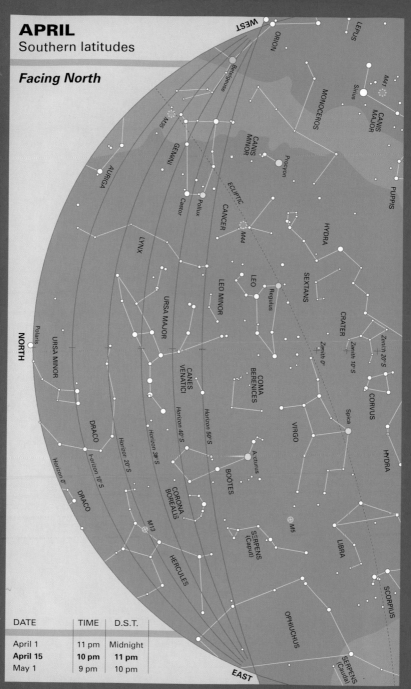

APRIL
Southern latitudes

Facing North

DATE	TIME	D.S.T.
April 1	11 pm	Midnight
April 15	**10 pm**	**11 pm**
May 1	9 pm	10 pm

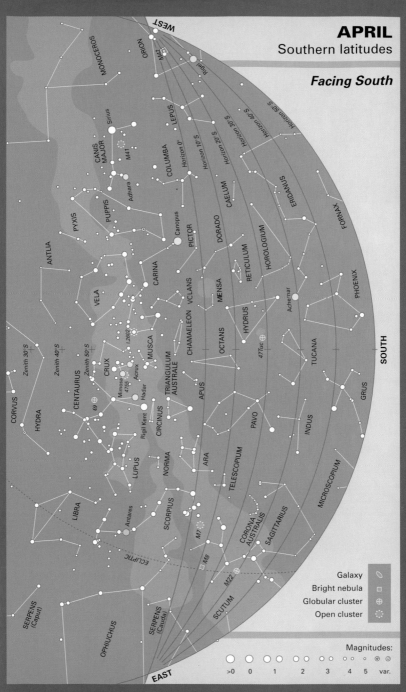

APRIL
Southern latitudes

Facing South

WEST

MONOCEROS
ORION
M42
Rigel
Sirius
CANIS MAJOR
M41
Adhara
LEPUS
COLUMBA
PUPPIS
PYXIS
Canopus
ANTLIA
PICTOR
CARINA
VELA
DORADO
CAELUM
ERIDANUS
FORNAX
PHOENIX
Achernar
RETICULUM
HOROLOGIUM
MENSA
V CLANS
CHAMAELEON
HYDRUS
OCTANS
47 Tuc
TUCANA
GRUS
Zenith 30° S
Zenith 40° S
Zenith 50° S
CENTAURUS
CRUX
Mimosa
4755
Acrux
Hadar
MUSCA
L2602
TRIANGULUM AUSTRALE
APUS
PAVO
INDUS
MICROSCOPIUM
CORVUS
HYDRA
ω
Rigil Kent
CIRCINUS
NORMA
ARA
TELESCOPIUM
LUPUS
LIBRA
Antares
SCORPIUS
M7
CORONA AUSTRALIS
SAGITTARIUS
M8
M22
SCUTUM
ECLIPTIC
SERPENS (Caput)
OPHIUCHUS
SERPENS (Cauda)

Horizon 0°
Horizon 10° S
Horizon 20° S
Horizon 30° S
Horizon 40° S
Horizon 50° S

SOUTH

EAST

Galaxy
Bright nebula
Globular cluster
Open cluster

Magnitudes:
>0 0 1 2 3 4 5 var.

39

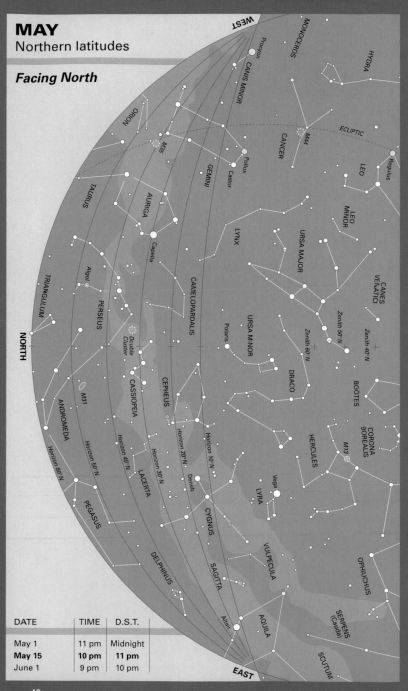

MAY
Northern latitudes

Facing North

WEST

NORTH

EAST

DATE	TIME	D.S.T.
May 1	11 pm	Midnight
May 15	**10 pm**	**11 pm**
June 1	9 pm	10 pm

40

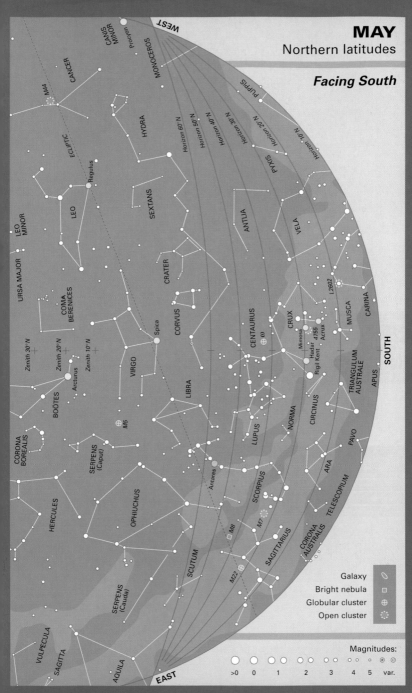

MAY
Northern latitudes

Facing South

WEST

CANIS MINOR
Procyon
CANCER
MONOCEROS
M44
ECLIPTIC
PUPPIS
Horizon 60° N
Horizon 50° N
Horizon 40° N
Horizon 30° N
Horizon 20° N
Horizon 10° N
HYDRA
LEO
Regulus
PYXIS
SEXTANS
ANTLIA
VELA
LEO MINOR
URSA MAJOR
CRATER
CARINA
CENTAURUS
CRUX
I.2602
MUSCA
COMA BERENICES
ω
CORVUS
Mimosa
4755
Acrux
Spica
Hadar
Rigil Kent
TRIANGULUM AUSTRALE
Zenith 30° N
Zenith 20° N
Zenith 10° N
Arcturus
VIRGO
APUS
SOUTH
BOOTES
LIBRA
NORMA
CIRCINUS
PAVO
CORONA BOREALIS
M5
LUPUS
ARA
SERPENS (Caput)
HERCULES
TELESCOPIUM
Antares
OPHIUCHUS
SCORPIUS
CORONA AUSTRALIS
M8
M7
SCUTUM
M22
SAGITTARIUS
VULPECULA
SERPENS (Cauda)
SAGITTA
AQUILA
EAST

Galaxy
Bright nebula
Globular cluster
Open cluster

Magnitudes:
>0   0   1   2   3   4   5   var.

41

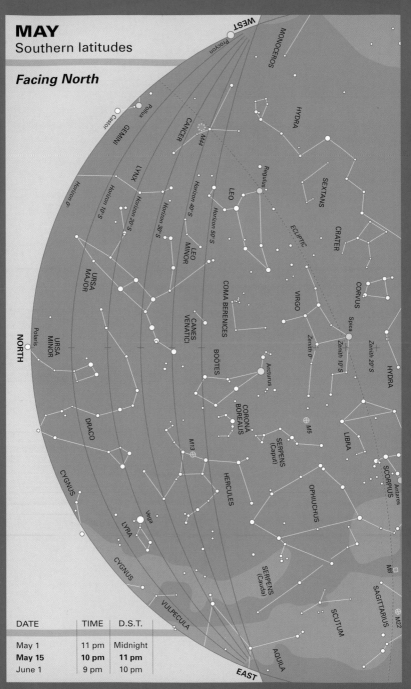

MAY
Southern latitudes

Facing North

DATE	TIME	D.S.T.
May 1	11 pm	Midnight
May 15	**10 pm**	**11 pm**
June 1	9 pm	10 pm

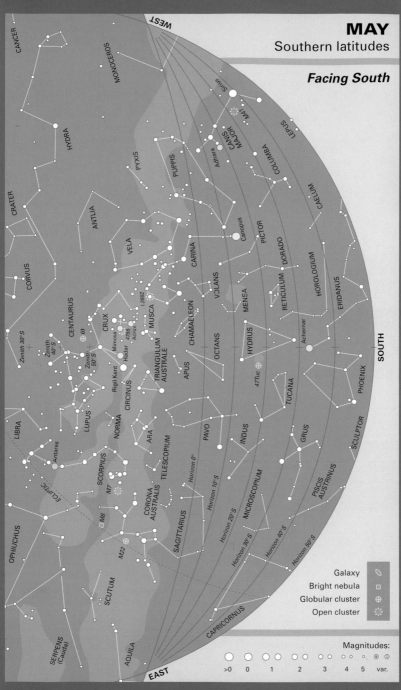

MAY
Southern latitudes

Facing South

WEST

EAST

SOUTH

CANCER

MONOCEROS

Sirius

M41

CANIS MAJOR

Adhara

LEPUS

COLUMBA

HYDRA

PYXIS

PUPPIS

CAELUM

CRATER

ANTLIA

VELA

CARINA

PICTOR

Canopus

DORADO

RETICULUM

HOROLOGIUM

ERIDANUS

CORVUS

CENTAURUS

VOLANS

CRUX

MENSA

MUSCA

4755

Acrux

Mimosa

Zenith 30° S

Zenith 40° S

ω

Zenith 50° S

Hadar

Rigil Kent

i 2602

CHAMAELEON

HYDRUS

Achernar

TRIANGULUM AUSTRALE

OCTANS

CIRCINUS

APUS

47 Tuc

PHOENIX

LIBRA

LUPUS

NORMA

ARA

TELESCOPIUM

PAVO

INDUS

TUCANA

GRUS

SCULPTOR

Antares

ECLIPTIC

SCORPIUS

M7

CORONA AUSTRALIS

Horizon 0° S

Horizon 10° S

MICROSCOPIUM

PISCIS AUSTRINUS

OPHIUCHUS

M8

SAGITTARIUS

Horizon 20° S

Horizon 30° S

Horizon 40° S

Horizon 50° S

M22

SCUTUM

SERPENS (Cauda)

AQUILA

CAPRICORNUS

⬭	Galaxy
◻	Bright nebula
⊕	Globular cluster
⊙	Open cluster

Magnitudes:

○ ○ ○ ○ ○ ∘ ∘ ∘ ◉ ◎

\>0 0 1 2 3 4 5 var.

43

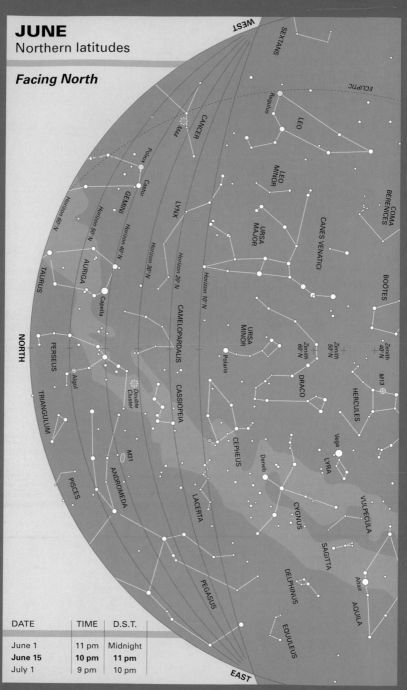

JUNE
Northern latitudes

Facing North

WEST

SEXTANS

ECLIPTIC

Regulus

LEO

CANCER

M44

LEO MINOR

COMA BERENICES

Polaris

Castor

GEMINI

LYNX

URSA MAJOR

CANES VENATICI

BOÖTES

Horizon 60° N

Horizon 50° N

Horizon 40° N

Horizon 30° N

Horizon 20° N

Horizon 10° N

TAURUS

AURIGA

Capella

CAMELOPARDALIS

URSA MINOR

Zenith 60° N

Zenith 50° N

Zenith 40° N

M13

HERCULES

NORTH

PERSEUS

Algol

Double Cluster

CASSIOPEIA

Polaris

DRACO

Vega

LYRA

TRIANGULUM

M31

ANDROMEDA

CEPHEUS

Deneb

CYGNUS

VULPECULA

PISCES

LACERTA

SAGITTA

DELPHINUS

AQUILA

Altair

PEGASUS

EQUULEUS

EAST

DATE	TIME	D.S.T.
June 1	11 pm	Midnight
June 15	**10 pm**	**11 pm**
July 1	9 pm	10 pm

44

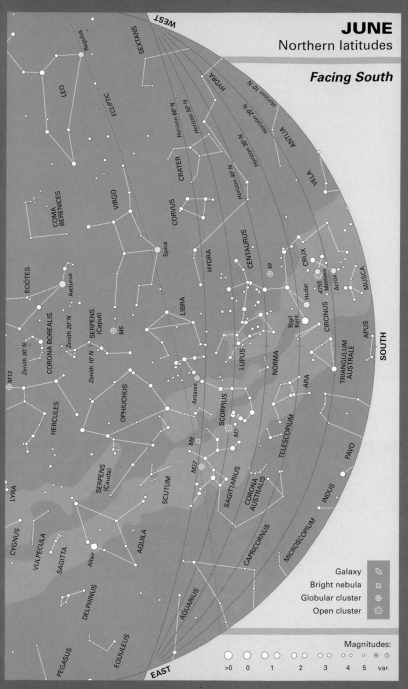

JUNE
Northern latitudes

Facing South

WEST

EAST

SOUTH

Galaxy
Bright nebula
Globular cluster
Open cluster

Magnitudes:
>0   0   1   2   3   4   5   var.

45

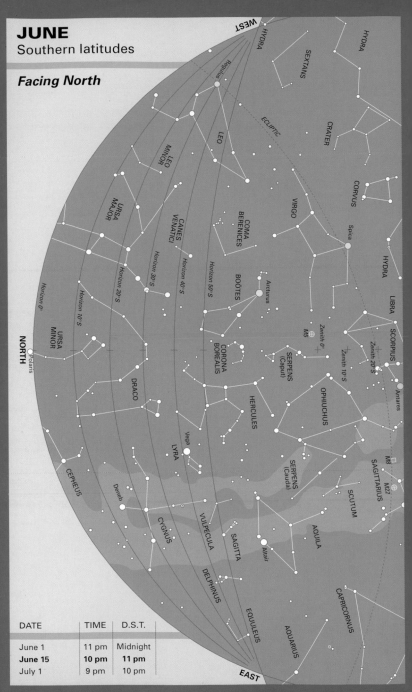

JUNE
Southern latitudes

Facing North

DATE	TIME	D.S.T.
June 1	11 pm	Midnight
June 15	**10 pm**	**11 pm**
July 1	9 pm	10 pm

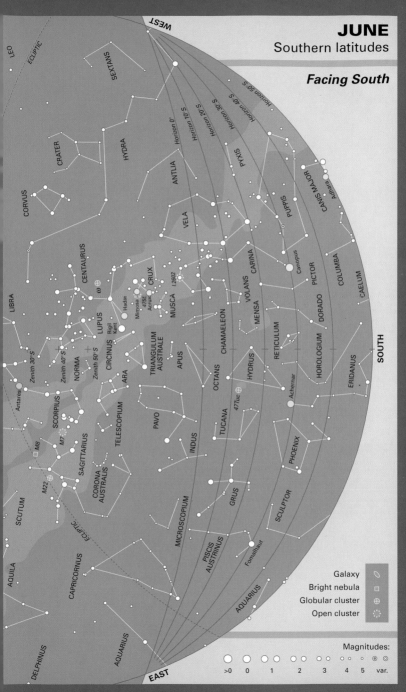

WEST

LEO

ECLIPTIC

SEXTANS

CRATER

HYDRA

Horizon 0°

Horizon 10° S

Horizon 20° S

Horizon 30° S

Horizon 40° S

Horizon 50° S

ANTLIA

PYXIS

CANIS MAJOR

Adhara

PUPPIS

CORVUS

CENTAURUS

VELA

CARINA

CRUX

ω

NGC 2602

LIBRA

LUPUS

Hadar

Mimosa

Acrux

MUSCA

VOLANS

CANOPUS

PICTOR

COLUMBA

CAELUM

SOUTH

Rigil
Kent

CIRCINUS

CHAMAELEON

MENSA

DORADO

Zenith 30° S

Zenith 40° S

NORMA

TRIANGULUM
AUSTRALE

APUS

OCTANS

HYDRUS

RETICULUM

ERIDANUS

Zenith 50° S

ARA

Antares

SCORPIUS

TELESCOPIUM

PAVO

47 Tuc

TUCANA

Achernar

M8

M7

SAGITTARIUS

INDUS

PHOENIX

SCUTUM

CORONA
AUSTRALIS

MICROSCOPIUM

GRUS

SCULPTOR

M22

AQUILA

CAPRICORNUS

PISCIS
AUSTRINUS

Fomalhaut

ECLIPTIC

AQUARIUS

DELPHINUS

AQUARIUS

EAST

Galaxy

Bright nebula

Globular cluster

Open cluster

Magnitudes:

>0   0   1   2   3   4   5   var.

47

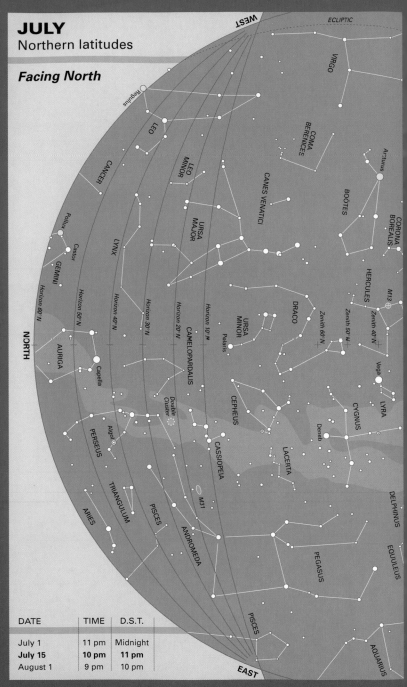

JULY
Northern latitudes

Facing North

WEST

ECLIPTIC

VIRGO

Regulus

LEO

COMA BERENICES

CANCER

CANES VENATICI

BOÖTES

Arcturus

CORONA BOREALIS

LEO MINOR

Pollux

URSA MAJOR

HERCULES

M13

Castor

LYNX

GEMINI

DRACO

Zenith 60° N

Zenith 50° N

Zenith 40° N

Vega

Horizon 60° N

Horizon 50° N

Horizon 40° N

Horizon 30° N

Horizon 20° N

Horizon 10° N

URSA MINOR

LYRA

NORTH

AURIGA

Polaris

CYGNUS

Capella

CAMELOPARDALIS

CEPHEUS

Deneb

Algol

Double Cluster

LACERTA

PERSEUS

CASSIOPEIA

DELPHINUS

TRIANGULUM

M31

EQUULEUS

ARIES

PISCES

ANDROMEDA

PEGASUS

PISCES

AQUARIUS

EAST

DATE	TIME	D.S.T.
July 1	11 pm	Midnight
July 15	**10 pm**	**11 pm**
August 1	9 pm	10 pm

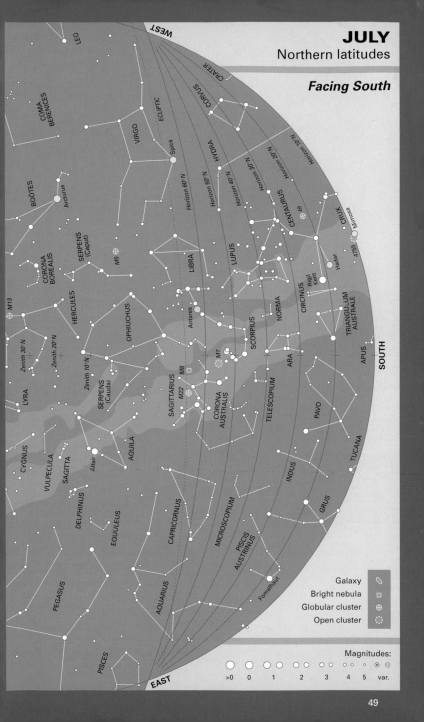

JULY
Northern latitudes

Facing South

49

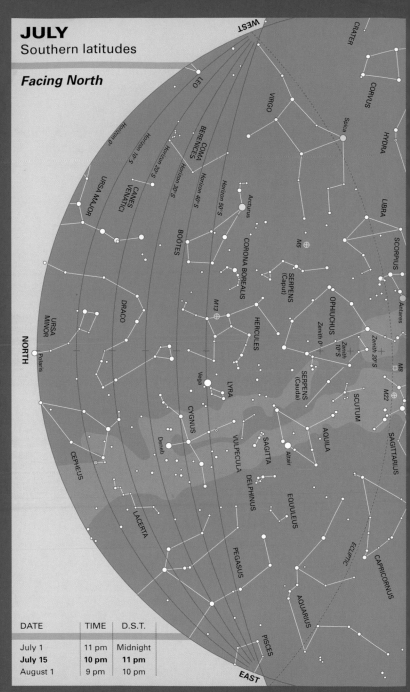

JULY
Southern latitudes

Facing North

50

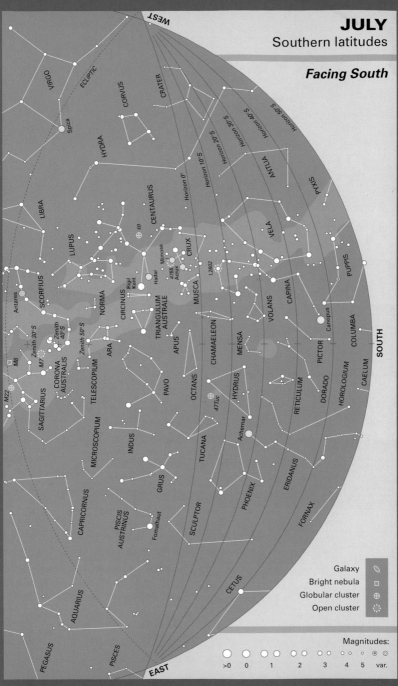

JULY
Southern latitudes

Facing South

WEST

EAST

SOUTH

VIRGO
Spica
ECLIPTIC
CORVUS
CRATER
HYDRA
CENTAURUS
ANTLIA
PYXIS
VELA
LIBRA
LUPUS
ω
Mimosa
CRUX
4755 Acrux
MUSCA
I 2602
PUPPIS
SCORPIUS
Antares
NORMA
CIRCINUS
Rigil Kent
Hadar
TRIANGULUM AUSTRALE
APUS
CHAMAELEON
VOLANS
CARINA
Canopus
Zenith 30° S
Zenith 40° S
Zenith 50° S
M8
M7
ARA
PAVO
OCTANS
MENSA
PICTOR
COLUMBA
M22
CORONA AUSTRALIS
TELESCOPIUM
47 Tuc
HYDRUS
RETICULUM
DORADO
HOROLOGIUM
CAELUM
SAGITTARIUS
MICROSCOPIUM
INDUS
TUCANA
Achernar
ERIDANUS
FORNAX
GRUS
PHOENIX
CAPRICORNUS
PISCIS AUSTRINUS
Fomalhaut
SCULPTOR
CETUS
AQUARIUS
PEGASUS
PISCES

Horizon 0°
Horizon 10° S
Horizon 20° S
Horizon 30° S
Horizon 40° S
Horizon 50° S

Galaxy
Bright nebula
Globular cluster
Open cluster

Magnitudes:
>0   0   1   2   3   4   5   var.

51

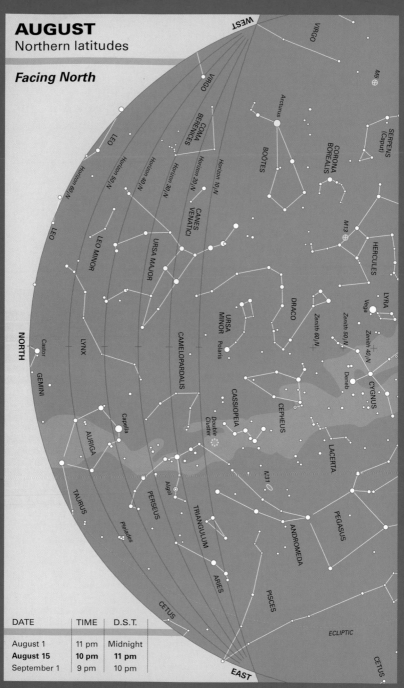

AUGUST
Northern latitudes

Facing North

DATE	TIME	D.S.T.
August 1	11 pm	Midnight
August 15	**10 pm**	**11 pm**
September 1	9 pm	10 pm

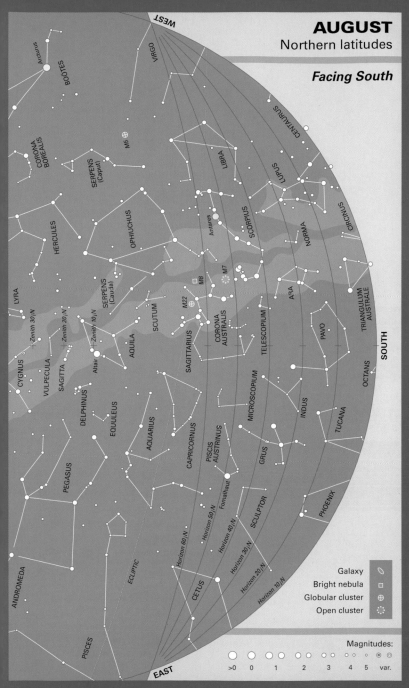

AUGUST
Northern latitudes

Facing South

WEST

VIRGO

Arcturus
BOÖTES

CORONA
BOREALIS

CENTAURUS

LIBRA

SERPENS
(Caput)

M5

LUPUS

HERCULES

OPHIUCHUS

Antares

SCORPIUS

NORMA

CIRCINUS

LYRA

SERPENS
(Cauda)

M8

M7

ARA

TRIANGULUM
AUSTRALE

Zenith 30°N

Zenith 20°N

Zenith 10°N

SCUTUM

M22

CORONA
AUSTRALIS

TELESCOPIUM

CYGNUS

VULPECULA

SAGITTA

Altair

AQUILA

SAGITTARIUS

PAVO

SOUTH

DELPHINUS

MICROSCOPIUM

OCTANS

EQUULEUS

INDUS

AQUARIUS

CAPRICORNUS

PISCIS
AUSTRINUS

GRUS

TUCANA

PEGASUS

Fomalhaut

PHOENIX

Horizon 60°N

Horizon 50°N

Horizon 40°N

SCULPTOR

ANDROMEDA

Horizon 30°N

ECLIPTIC

Horizon 20°N

CETUS

Horizon 10°N

PISCES

EAST

| | | | | | | | | Magnitudes: |
| Galaxy | | | | | | | | |

Galaxy

Bright nebula

Globular cluster

Open cluster

Magnitudes:

>0   0   1   2   3   4   5   var.

53

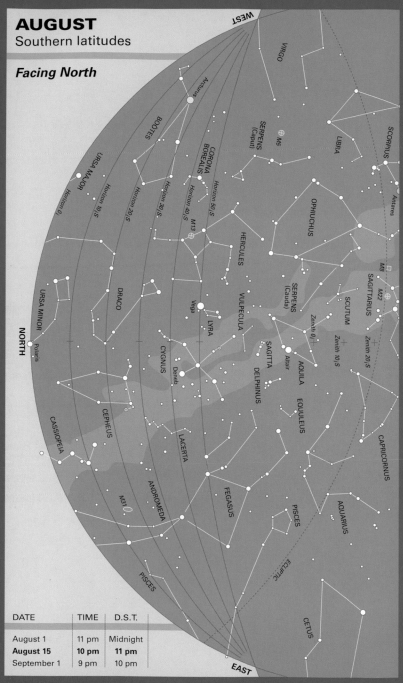

AUGUST
Southern latitudes

Facing North

WEST

VIRGO

SERPENS (Caput)
M5

LIBRA

SCORPIUS

BOOTES

Arcturus

CORONA BOREALIS

OPHIUCHUS

Antares

URSA MAJOR

Horizon 0

Horizon 10 /S

Horizon 20 /S

Horizon 30 /S

Horizon 40 /S

Horizon 50 /S

M13

HERCULES

SERPENS (Cauda)

M8

SAGITTARIUS

M22

DRACO

URSA MINOR

Vega

LYRA

VULPECULA

SCUTUM

Zenith 0 /

Zenith 10 /S

Zenith 20 /S

NORTH

Polaris

SAGITTA

Altair

AQUILA

CYGNUS

Deneb

DELPHINUS

EQUULEUS

CAPRICORNUS

CEPHEUS

LACERTA

PEGASUS

PISCES

AQUARIUS

CASSIOPEIA

ANDROMEDA

M31

PISCES

ECLIPTIC

CETUS

EAST

DATE	TIME	D.S.T.
August 1	11 pm	Midnight
August 15	**10 pm**	**11 pm**
September 1	9 pm	10 pm

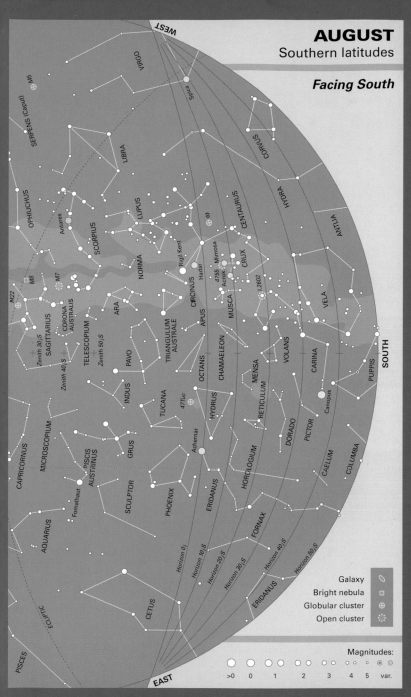

AUGUST
Southern latitudes

Facing South

55

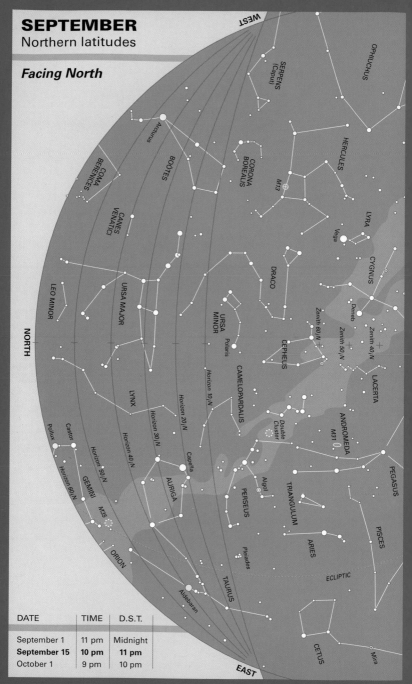

SEPTEMBER
Northern latitudes

Facing North

WEST

SERPENS (Caput)

OPHIUCHUS

HERCULES

Arcturus

BOÖTES

CORONA BOREALIS

M13

COMA BERENICES

LYRA

CANES VENATICI

Vega

CYGNUS

Deneb

DRACO

Zenith 60/N

Zenith 50/N

Zenith 40/N

LEO MINOR

URSA MAJOR

URSA MINOR

Polaris

CEPHEUS

LACERTA

NORTH

CAMELOPARDALIS

Horizon 10/N

ANDROMEDA

M31

LYNX

Horizon 20/N

Double Cluster

Horizon 30/N

Capella

PEGASUS

Pollux

Castor

Horizon 40/N

AURIGA

Algol

PERSEUS

TRIANGULUM

Horizon 50/N

PISCES

GEMINI

Horizon 60/N

M35

ARIES

Pleiades

ORION

ECLIPTIC

TAURUS

Aldebaran

CETUS

Mira

DATE	TIME	D.S.T.
September 1	11 pm	Midnight
September 15	**10 pm**	**11 pm**
October 1	9 pm	10 pm

EAST

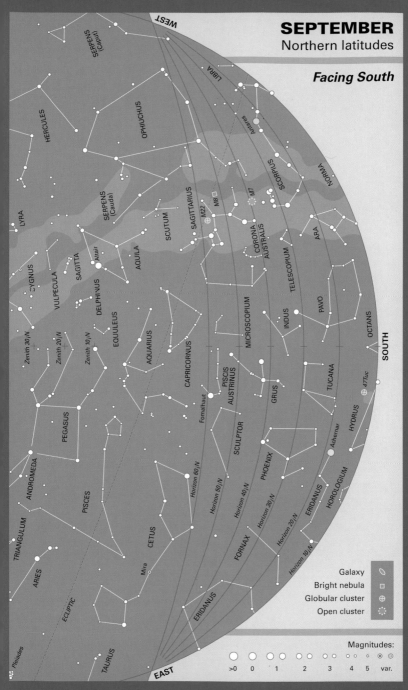

SEPTEMBER
Northern latitudes

Facing South

57

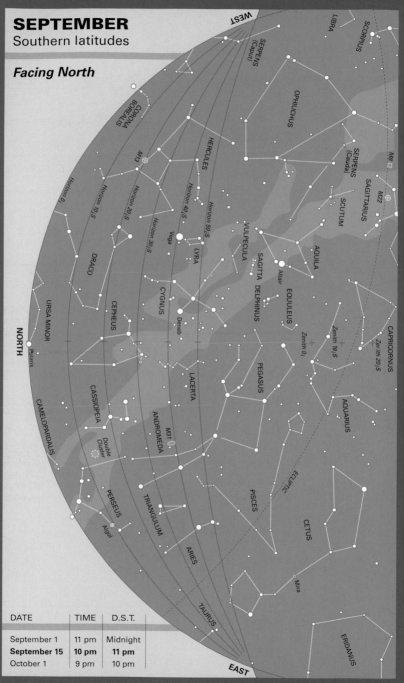

SEPTEMBER
Southern latitudes

Facing North

WEST

LIBRA

SCORPIUS

SERPENS (Caput)

CORONA BOREALIS

OPHIUCHUS

M13

HERCULES

SERPENS (Cauda)

M8

SAGITTARIUS

M22

SCUTUM

Horizon 0

Horizon 10 S

Horizon 20 S

Horizon 30 S

Horizon 40 S

Horizon 50 S

VULPECULA

DRACO

Vega

LYRA

SAGITTA

AQUILA

DELPHINUS

Altair

CEPHEUS

CYGNUS

EQUULEUS

CAPRICORNUS

URSA MINOR

NORTH

Polaris

Deneb

Zenith 10 S

Zenith 20 S

Zenith 0

LACERTA

PEGASUS

CASSIOPEIA

AQUARIUS

CAMELOPARDALIS

Double Cluster

ANDROMEDA

M31

PISCES

ECLIPTIC

CETUS

PERSEUS

TRIANGULUM

Algol

ARIES

Mira

TAURUS

ERIDANUS

EAST

DATE	TIME	D.S.T.
September 1	11 pm	Midnight
September 15	**10 pm**	**11 pm**
October 1	9 pm	10 pm

58

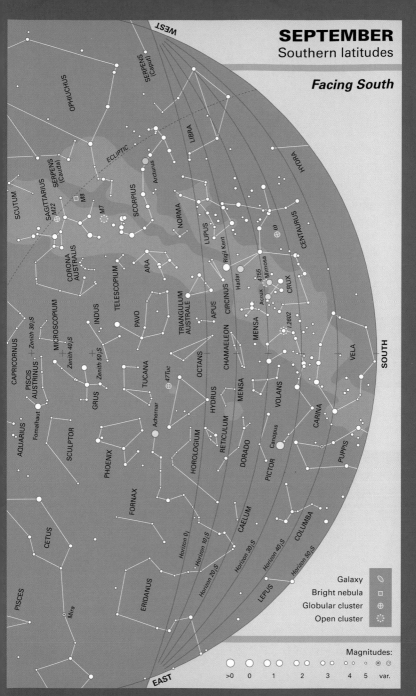

SEPTEMBER
Southern latitudes

Facing South

WEST

EAST

SOUTH

OPHIUCHUS
SERPENS (Caput)
SERPENS (Cauda)
SCUTUM
SAGITTARIUS
M22
M8
M7
CORONA AUSTRALIS
CAPRICORNUS
MICROSCOPIUM
PISCIS AUSTRINUS
Fomalhaut
AQUARIUS
SCULPTOR
INDUS
GRUS
TUCANA
47 Tuc
Achernar
PHOENIX
FORNAX
CETUS
PISCES
Mira
ERIDANUS
HOROLOGIUM
RETICULUM
CAELUM
DORADO
PICTOR
Canopus
COLUMBA
LEPUS
PUPPIS
CARINA
VELA
ECLIPTIC
LIBRA
SCORPIUS
Arcturus
NORMA
LUPUS
Rigil Kent
Hadar
λ765
Mimosa
Acrux
CRUX
CENTAURUS
HYDRA
ARA
TELESCOPIUM
PAVO
TRIANGULUM AUSTRALE
APUS
CIRCINUS
CHAMAELEON
MENSA
MENSA
VOLANS
OCTANS
HYDRUS
ω
λ2602

Zenith 30/S
Zenith 40/S
Zenith 50/S

Horizon 0/S
Horizon 10/S
Horizon 20/S
Horizon 30/S
Horizon 40/S
Horizon 50/S

Galaxy	⬭
Bright nebula	▫
Globular cluster	⊕
Open cluster	⬚

Magnitudes:

>0	0	1	2	3	4	5	var.

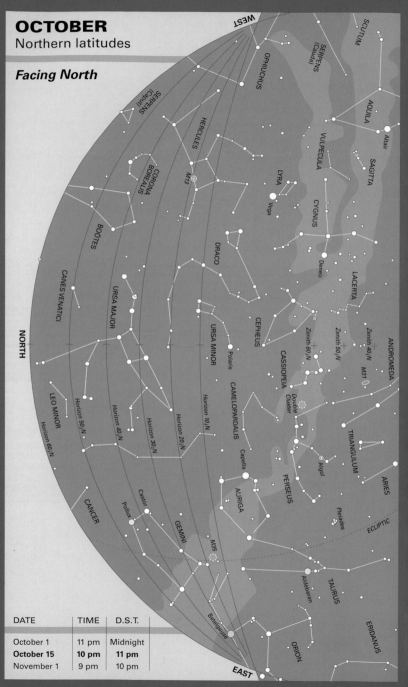

OCTOBER
Northern latitudes

Facing North

WEST

NORTH

EAST

SCUTUM
SERPENS (Cauda)
OPHIUCHUS
AQUILA
Altair
SAGITTA
HERCULES
VULPECULA
M13
SERPENS (Caput)
CYGNUS
LYRA
Vega
CORONA BOREALIS
Deneb
BOÖTES
LACERTA
DRACO
Zenith 40/N
CANES VENATICI
CEPHEUS
Zenith 50/N
ANDROMEDA
URSA MAJOR
URSA MINOR
Zenith 60/N
M31
Polaris
CASSIOPEIA
LEO MINOR
Double Cluster
CAMELOPARDALIS
TRIANGULUM
Horizon 60/N
Horizon 50/N
Horizon 40/N
Horizon 30/N
Horizon 20/N
Horizon 10/N
Capella
PERSEUS
Algol
ARIES
CANCER
AURIGA
Pollux
Castor
Pleiades
ECLIPTIC
GEMINI
M35
Aldebaran
TAURUS
Betelgeuse
ORION
ERIDANUS

DATE	TIME	D.S.T.
October 1	11 pm	Midnight
October 15	**10 pm**	**11 pm**
November 1	9 pm	10 pm

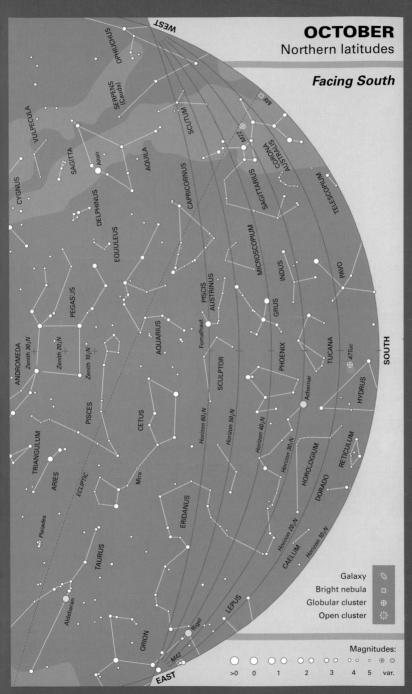

OCTOBER
Northern latitudes

Facing South

WEST

EAST

SOUTH

OPHIUCHUS
SERPENS (Cauda)
VULPECULA
CYGNUS
SAGITTA
DELPHINUS
AQUILA
Altair
SCUTUM
CAPRICORNUS
EQUULEUS
PEGASUS
SAGITTARIUS
CORONA AUSTRALIS
M8
M22
TELESCOPIUM
MICROSCOPIUM
INDUS
PAVO
PISCIS AUSTRINUS
ANDROMEDA
Zenith 30°N
Zenith 20°N
Zenith 10°N
AQUARIUS
Fomalhaut
GRUS
PHOENIX
TUCANA
47Tuc
PISCES
SCULPTOR
Achernar
HYDRUS
TRIANGULUM
CETUS
Horizon 60°N
Horizon 50°N
Horizon 40°N
Horizon 30°N
ARIES
ECLIPTIC
Mira
HOROLOGIUM
RETICULUM
Pleiades
ERIDANUS
Horizon 20°N
DORADO
CAELUM
Horizon 10°N
TAURUS
Aldebaran
LEPUS
Rigel
ORION
M42

Galaxy
Bright nebula
Globular cluster
Open cluster

Magnitudes:

>0   0   1   2   3   4   5   var.

61

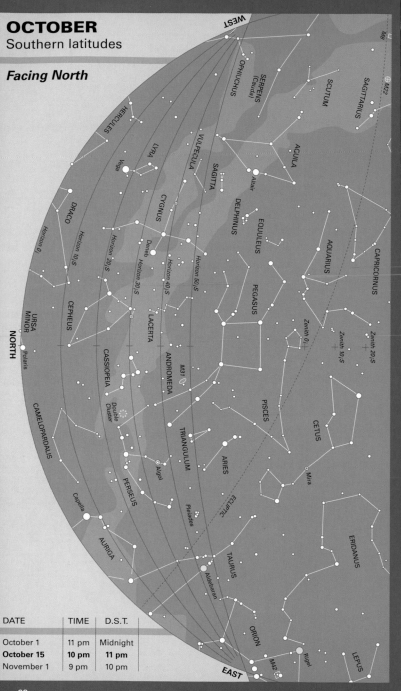

OCTOBER
Southern latitudes

Facing North

WEST

M8

OPHIUCHUS

SERPENS (Cauda)

SCUTUM

SAGITTARIUS

⊕ M22

HERCULES

Vega

LYRA

VULPECULA

SAGITTA

ACUILA

Altair

CYGNUS

DELPHINUS

EQUULEUS

DRACO

Deneb

Horizon 0/S

Horizon 10/S

Horizon 20/S

Horizon 30/S

Horizon 40/S

Horizon 50/S

PEGASUS

AQUARIUS

CAPRICORNUS

URSA MINOR

CEPHEUS

LACERTA

Zenith 0/S

Zenith 10/S

Zenith 20/S

NORTH

Polaris

CASSIOPEIA

ANDROMEDA

M31

PISCES

CETUS

Double Cluster

TRIANGULUM

CAMELOPARDALIS

Algol

ARIES

Mira

PERSEUS

Pleiades

ECLIPTIC

Capella

ERIDANUS

AURIGA

TAURUS

Aldebaran

ORION

LEPUS

EAST

M42

Rigel

DATE	TIME	D.S.T.
October 1	11 pm	Midnight
October 15	**10 pm**	**11 pm**
November 1	9 pm	10 pm

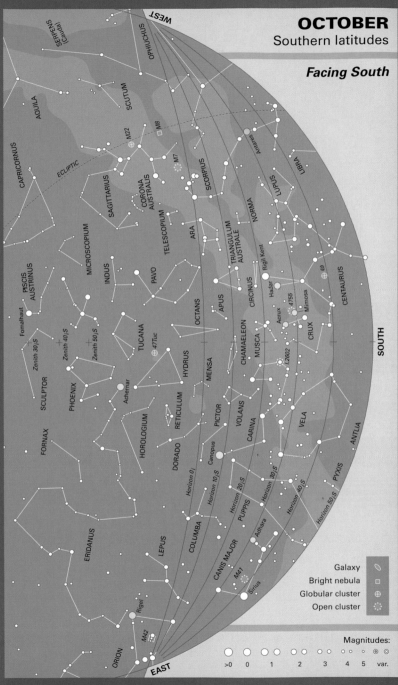

OCTOBER
Southern latitudes

Facing South

WEST

SERPENS (Cauda)
OPHIUCHUS
AQUILA
SCUTUM
CAPRICORNUS
M22
M8
ECLIPTIC
SAGITTARIUS
CORONA AUSTRALIS
M7
SCORPIUS
Antares
LIBRA
LUPUS
TELESCOPIUM
ARA
NORMA
MICROSCOPIUM
TRIANGULUM AUSTRALE
Rigil Kent
PISCIS AUSTRINUS
INDUS
PAVO
APUS
CIRCINUS
Hadar
4755
ω
CENTAURUS
Fomalhaut
Zenith 30°S
Zenith 40°S
Zenith 50°S
TUCANA
OCTANS
Acrux
Mimosa
47Tuc
MENSA
CHAMAELEON
CRUX
SOUTH
SCULPTOR
PHOENIX
Achernar
HYDRUS
MUSCA
2602
PHOENIX
RETICULUM
PICTOR
VOLANS
FORNAX
HOROLOGIUM
DORADO
Canopus
CARINA
VELA
ANTLIA
ERIDANUS
Horizon 0°S
Horizon 10°S
Horizon 20°S
Horizon 30°S
Horizon 40°S
Horizon 50°S
LEPUS
COLUMBA
PUPPIS
PYXIS
Rigel
M42
ORION
CANIS MAJOR
Adhara
M41
Sirius
EAST

Galaxy
Bright nebula
Globular cluster
Open cluster

Magnitudes:
>0 0 1 2 3 4 5 var.

63

NOVEMBER
Northern latitudes

Facing North

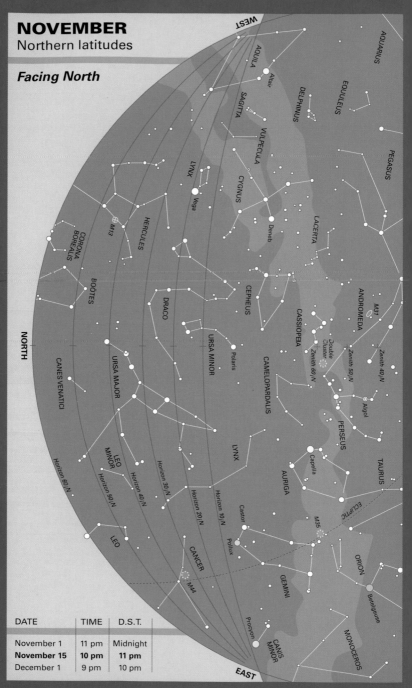

WEST

AQUARIUS
AQUILA
Altair
DELPHINUS
EQUULEUS
SAGITTA
PEGASUS
VULPECULA
LYNX
CYGNUS
Vega
M13
Deneb
LACERTA
HERCULES
ANDROMEDA
M31
CEPHEUS
CASSIOPEIA
CORONA BOREALIS
BOÖTES
DRACO
Zenith 40/N
Zenith 50/N
Zenith 60/N
Double Cluster
Algol
URSA MINOR
CAMELOPARDALIS
PERSEUS
Polaris
NORTH
CANES VENATICI
URSA MAJOR
Capella
LYNX
TAURUS
AURIGA
LEO MINOR
Horizon 60/N
Horizon 50/N
Horizon 40/N
Horizon 30/N
Horizon 20/N
Horizon 10/N
Castor
M35
Pollux
LEO
CANCER
GEMINI
ORION
M44
Betelgeuse
Procyon
MONOCEROS
CANIS MINOR
ECLIPTIC
EAST

DATE	TIME	D.S.T.
November 1	11 pm	Midnight
November 15	**10 pm**	**11 pm**
December 1	9 pm	10 pm

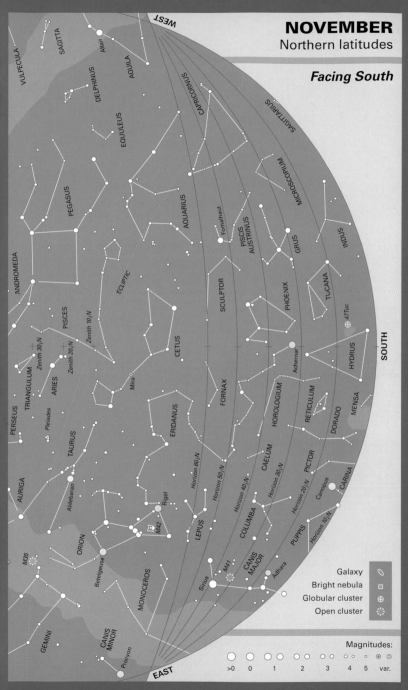

NOVEMBER
Northern latitudes

Facing South

WEST

VULPECULA
SAGITTA
Altair
AQUILA
DELPHINUS
EQUUELEUS
PEGASUS
ANDROMEDA
CAPRICORNUS
SAGITTARIUS
AQUARIUS
MICROSCOPIUM
Fomalhaut
PISCIS AUSTRINUS
INDUS
GRUS
SCULPTOR
PHOENIX
TUCANA
ECLIPTIC
PISCES
Zenith 10/N
Zenith 20/N
Zenith 30/N
TRIANGULUM
ARIES
Mira
CETUS
FORNAX
Achernar
⊕ 47 Tuc
HYDRUS
SOUTH
PERSEUS
TAURUS
HOROLOGIUM
RETICULUM
DORADO
MENSA
Aldebaran
Pleiades
ERIDANUS
CAELUM
Horizon 60/N
Horizon 50/N
Horizon 40/N
Horizon 30/N
PICTOR
AURIGA
Rigel
LEPUS
COLUMBA
Horizon 20/N
Canopus
CARINA
ORION
M42
Betelgeuse
Horizon 10/N
M35
MONOCEROS
Sirius
M41
CANIS MAJOR
Adhara
PUPPIS
GEMINI
CANIS MINOR
Procyon

EAST

Galaxy	⬭
Bright nebula	☐
Globular cluster	⊕
Open cluster	⁂

Magnitudes:

| ◯ | ◯ | ◯ | ◯ | ◯ | ◯ ◯ | ◦ ◦ | ◦ | ◉ | ◎ |
| >0 | 0 | 1 | 2 | 3 | 4 | 5 | var. |

65

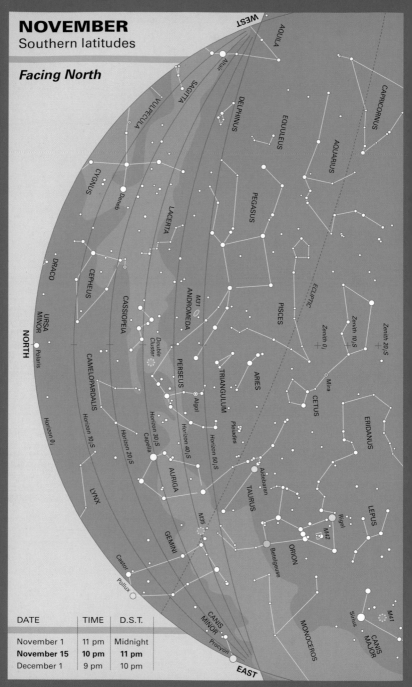

NOVEMBER
Southern latitudes

Facing North

WEST

EAST

NORTH

Aquila
Capricornius
Sagitta
Vulpecula
Delphinus
Equuleus
Aquarius
Cygnus
Deneb
Lacerta
Pegasus
Draco
Cepheus
Pisces
Ecliptic
Cassiopeia
M31
Andromeda
Zenith 0 J
Zenith 10 /S
Zenith 20 /S
Ursa Minor
Polaris
Double Cluster
Perseus
Triangulum
Aries
Mira
Cetus
Eridanus
Camelopardalis
Algol
Pleiades
Horizon 0 J
Horizon 10 /S
Horizon 20 /S
Horizon 30 /S
Horizon 40 /S
Horizon 50 /S
Capella
Aldebaran
Lynx
Auriga
Taurus
Lepus
M35
M42
Rigel
Gemini
Orion
Betelgeuse
Castor
Pollux
Monoceros
Canis Minor
M41
Procyon
Sirius
Canis Major

DATE	TIME	D.S.T.
November 1	11 pm	Midnight
November 15	**10 pm**	**11 pm**
December 1	9 pm	10 pm

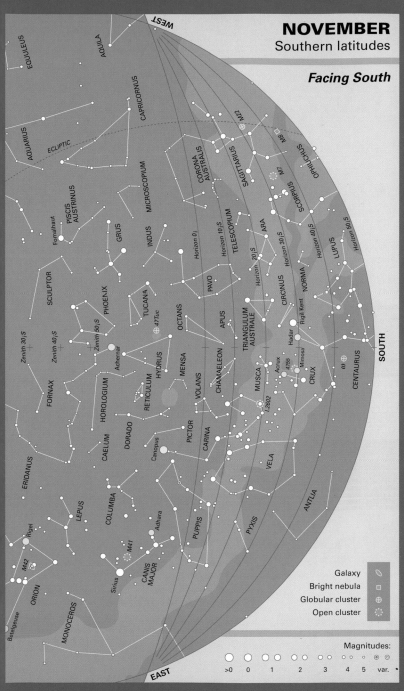

NOVEMBER
Southern latitudes

Facing South

WEST

AQUILA
EQUULEUS
AQUARIUS
ECLIPTIC
CAPRICORNUS
MICROSCOPIUM
PISCIS AUSTRINUS
Fomalhaut
GRUS
INDUS
SCULPTOR
PHOENIX
TUCANA
47 Tuc
Zenith 50/S
Zenith 30/S
Zenith 40/S
FORNAX
Achernar
HYDRUS
OCTANS
PAVO
APUS
MENSA
CHAMAELEON
RETICULUM
HOROLOGIUM
CAELUM
DORADO
PICTOR
VOLANS
CARINA
Canopus
ERIDANUS
LEPUS
COLUMBA
M41
Adhara
CANIS MAJOR
Sirius
PUPPIS
PYXIS
VELA
ANTLIA
Rigel
M42
ORION
Betelgeuse
MONOCEROS

CORONA AUSTRALIS
SAGITTARIUS
M22
M8
OPHIUCHUS
M7
SCORPIUS
ARA
TELESCOPIUM
Horizon 0/S
Horizon 10/S
Horizon 20/S
Horizon 30/S
Horizon 40/S
Horizon 50/S
CIRCINUS
NORMA
LUPUS
TRIANGULUM AUSTRALE
Rigil Kent
Hadar
Acrux
4755
Mimosa
CRUX
ω
CENTAURUS
MUSCA
I.2602

SOUTH

EAST

Galaxy ⊘
Bright nebula ▢
Globular cluster ⊕
Open cluster ⊛

Magnitudes:
>0   0   1   2   3   4   5   var.

67

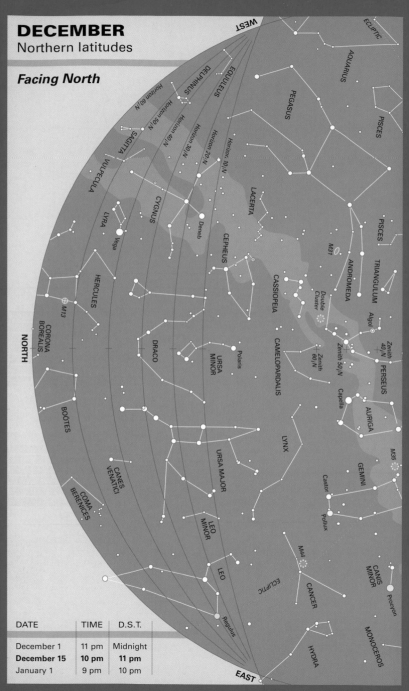

DECEMBER
Northern latitudes

Facing North

WEST

ECLIPTIC

AQUARIUS

PISCES

DELPHINUS

EQUULEUS

PEGASUS

Horizon 60/N
Horizon 50/N
Horizon 40/N
Horizon 30/N
Horizon 20/N
Horizon 10/N

SAGITTA

VULPECULA

LACERTA

PISCES

LYRA

Vega

CYGNUS

Deneb

ANDROMEDA

M31

TRIANGULUM

HERCULES

CEPHEUS

CASSIOPEIA

Double
Cluster

Algol

Zenith
40/N

PERSEUS

M13

CORONA
BOREALIS

NORTH

DRACO

URSA
MINOR

Polaris

CAMELOPARDALIS

Zenith
60/N

Zenith
50/N

Capella

AURIGA

BOÖTES

LYNX

M35

CANES
VENATICI

URSA MAJOR

GEMINI

Castor

COMA
BERENICES

LEO
MINOR

Pollux

LEO

M44

CANCER

CANIS
MINOR

Procyon

Regulus

ECLIPTIC

MONOCEROS

HYDRA

EAST

DATE	TIME	D.S.T.
December 1	11 pm	Midnight
December 15	**10 pm**	**11 pm**
January 1	9 pm	10 pm

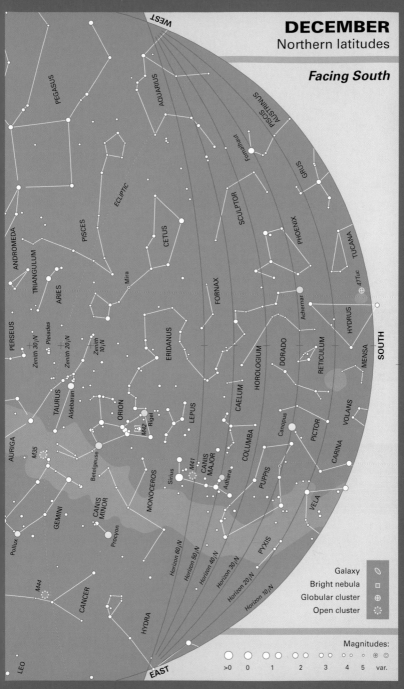

DECEMBER
Northern latitudes

Facing South

WEST

PEGASUS
AQUARIUS
PISCIS AUSTRINUS
ANDROMEDA
TRIANGULUM
ARIES
PISCES
ECLIPTIC
CETUS
Mira
GRUS
Fomalhaut
SCULPTOR
PHOENIX
TUCANA
47 Tuc
PERSEUS
FORNAX
Achernar
HYDRUS
SOUTH
Zenith 30/N
Pleiades
Zenith 20/N
Zenith 10/N
ERIDANUS
HOROLOGIUM
DORADO
RETICULUM
MENSA
TAURUS
Aldebaran
ORION
LEPUS
CAELUM
COLUMBA
Canopus
PICTOR
VOLANS
AURIGA
M42
[3]
Rigel
M35
Betelgeuse
CARINA
Sirius
M41
CANIS MAJOR
Adhara
PUPPIS
VELA
MONOCEROS
CANIS MINOR
Procyon
GEMINI
Pollux
Horizon 60/N
Horizon 50/N
Horizon 40/N
Horizon 30/N
Horizon 20/N
PYXIS
M44
CANCER
Horizon 10/N
Galaxy
Bright nebula
Globular cluster
Open cluster
LEO
HYDRA

Magnitudes:
>0 0 1 2 3 4 5 var.

EAST

69

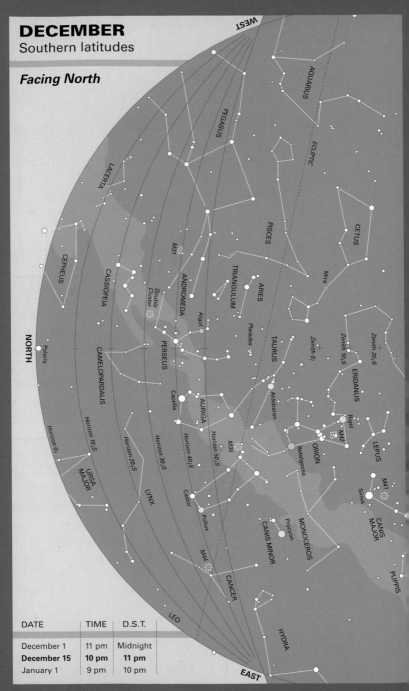

DECEMBER
Southern latitudes

Facing North

WEST

AQUARIUS
ECLIPTIC
PEGASUS
LACERTA
PISCES
CETUS
CEPHEUS
CASSIOPEIA
M31
ANDROMEDA
TRIANGULUM
ARIES
Mira
Double Cluster
Algol
Pleiades
TAURUS
Zenith 0J
Zenith 10/S
Zenith 20/S
ERIDANUS
NORTH
Polaris
CAMELOPARDALIS
PERSEUS
Capella
AURIGA
Aldebaran
Rigel
M42
ORION
LEPUS
Horizon 0J
Horizon 10/S
Horizon 20/S
Horizon 30/S
Horizon 40/S
Horizon 50/S
M38
Betelgeuse
URSA MAJOR
LYNX
Castor
Pollux
Procyon
MONOCEROS
CANIS MINOR
M41
Sirius
CANIS MAJOR
M44
CANCER
PUPPIS
LEO
HYDRA
EAST

DATE	TIME	D.S.T.
December 1	11 pm	Midnight
December 15	**10 pm**	**11 pm**
January 1	9 pm	10 pm

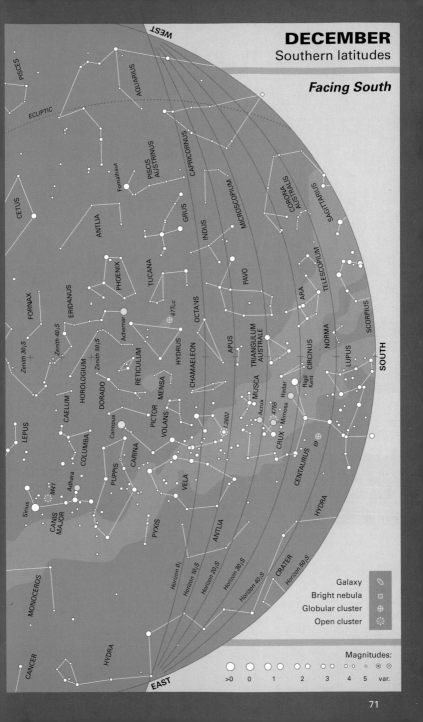

DECEMBER
Southern latitudes

Facing South

WEST

PISCES

AQUARIUS

ECLIPTIC

CETUS

PISCIS AUSTRINUS

Fomalhaut

CAPRICORNUS

MICROSCOPIUM

CORONA AUSTRALIS

SAGITTARIUS

ANTLIA

GRUS

INDUS

TELESCOPIUM

FORNAX

ERIDANUS

PHOENIX

TUCANA

PAVO

ARA

SCORPIUS

⊕ 47Tuc

OCTANS

NORMA

Zenith 30/S

Zenith 40/S

Achernar

RETICULUM

HYDRUS

CHAMAELEON

APUS

TRIANGULUM AUSTRALE

CIRCINUS

LUPUS

SOUTH

Zenith 50/S

HOROLOGIUM

CAELUM

DORADO

MENSA

MUSCA

Rigil Kent

Acrux

Hadar

LEPUS

PICTOR

VOLANS

η 2602

4755

Mimosa

COLUMBA

Canopus

CARINA

CRUX

ω

CENTAURUS

M41

Adhara

PUPPIS

VELA

Sirius

CANIS MAJOR

PYXIS

ANTLIA

HYDRA

MONOCEROS

Horizon 0/S

Horizon 10/S

Horizon 20/S

Horizon 30/S

Horizon 40/S

CRATER

Horizon 50/S

Galaxy

Bright nebula

Globular cluster

Open cluster

CANCER

HYDRA

EAST

Magnitudes:

>0   0   1   2   3   4   5   var.

71

ANDROMEDA

Andromeda represents the daughter of Queen Cassiopeia who was chained to a rock as a sacrifice to the sea monster Cetus until saved by Perseus, whom she subsequently married. The constellation originated in ancient times. Despite its fame, Andromeda is not particularly striking: its brightest stars are only 2nd magnitude. Its most prominent feature is a crooked line of four stars extending from the Square of Pegasus; the first of these stars marks a corner of the Square, although it is actually part of Andromeda. This star, known both as Alpheratz and Sirrah, marks the head of the chained Andromeda; another star in the line, Mirach, represents her waist and a third, Almaak, is her chained foot. The most celebrated object in the constellation is the Andromeda Galaxy, M31, a spiral galaxy like our own Milky Way; it is the most distant object visible to the naked eye. Two stars leading from Mirach, β (beta) Andromedae, act as a guide to it.

α (alpha) Andromedae, 0h 08m +29°.1, (Alpheratz or Sirrah), mag. 2.1, is a blue-white star 97 l.y. away.

β (beta) And, 1h 10m +35°.6, (Mirach), mag. 2.1, is a red giant 199 l.y. away.

γ (gamma) And, 2h 04m +42°.3, (Almaak or Almach), 355 l.y. away, is an outstanding triple star. Its two brightest components, of mags. 2.3 and 4.8, form one of the finest pairs for small telescopes: their colours are orange and blue. The fainter, blue star also has a close 6th-mag. blue-white companion that orbits it every 61 years. This fainter pair is closest around the year 2013 and is unresolvable by amateur telescopes for several years either side.

δ (delta) And, 0h 39m +30°.9, mag. 3.3, is an orange giant 101 l.y. away.

μ (mu) And, 0h 57m +38°.5, mag. 3.9, is a white star 136 l.y. away.

π (pi) And, 0h 37m +33°.7, 660 l.y. away, is a blue-white star of mag. 4.3 with a mag. 8.9 companion visible in small telescopes. ▶

M31, the Andromeda Galaxy, is a magnificent spiral visible to the naked eye, with two smaller companion galaxies that can be seen in small telescopes: M32 just below centre, and the larger but fainter M110 to its upper right. (Bill Schoening, Vanessa Harvey/ REU program/ AURA/NOAO/NSF)

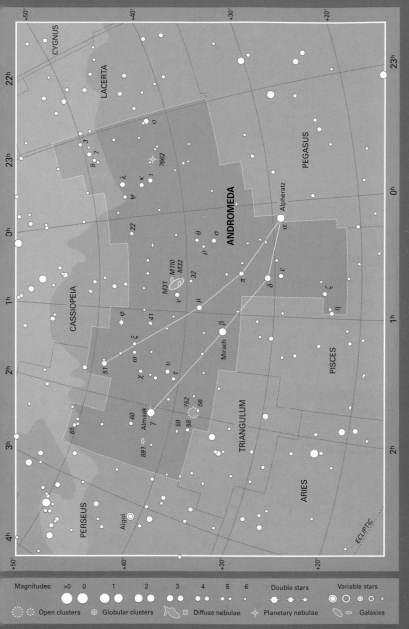

Magnitudes: >0 0 1 2 3 4 5 6 Double stars Variable stars

Open clusters Globular clusters Diffuse nebulae Planetary nebulae Galaxies

υ (upsilon) And, 1h 37m, +41°.4, mag. 4.1, is a yellow-white main-sequence star 44 l.y. away. Astronomers have found that it is accompanied by three planets, the first known multi-planet system around a star other than our own Sun.

56 And, 1h 56m +37°.3, is a yellow giant star of mag. 5.7, 320 l.y. away, with an orange giant companion, mag. 5.9 and 990 l.y. away, easily split in binoculars. The two stars are found near the star cluster NGC 752 but are closer to us.

M31 (NGC 224), 0h 43m +41°.3, the Andromeda Galaxy, is a spiral galaxy normally reckoned to lie 2.4 million l.y. away, although Hipparcos results suggest a distance closer to 3 million l.y. It is visible to the naked eye as an elliptical fuzzy patch, and becomes more prominent in binoculars or a telescope with low magnification (too high a power reduces the contrast and renders the fainter parts of the galaxy less visible). Dark lanes can be seen in the spiral arms surrounding the nucleus. But the full extent of the galaxy becomes apparent only on long-exposure photographs – visual observers see just its brightest, central portion. If the entire Andromeda Galaxy were bright enough to be seen by the naked eye, it would appear five or six times the diameter of the full Moon. M31 is accompanied by two small satellite galaxies, the equivalents of our Magellanic Clouds but both elliptical rather than irregular in shape. The brighter of these, M32 (NGC 221), is visible in small telescopes as a fuzzy, 8th-mag. star-like glow ½° south of M31's core. The second companion, M110 (also known as NGC 205), is larger but more elusive, and over 1° northwest of M31.

NGC 752, 1h 58m +37°.7, is a widely spread cluster of about 60 stars of 9th mag. and fainter, 1200 l.y. away, visible in binoculars and easily resolved in telescopes.

NGC 7662, 23h 26m +42°.6, is one of the brightest and easiest planetary nebulae to see with a small telescope. At low powers it appears as a fuzzy, 9th-mag. blue-green star, but magnifications of ×100 or so reveal its slightly elliptical disk. Larger apertures show a central hole; the central star is a difficult object for amateur telescopes. NGC 7662 lies about 4000 l.y. away.

ANTLIA   The Air Pump

An obscure southern constellation introduced on a map published in 1756 by the French astronomer Nicolas Louis de Lacaille to commemorate the air pump invented by the French physicist Denis Papin. Lacaille, the first person to map the southern skies completely (which he did from an observatory at the Cape of Good Hope), introduced 14 new constellations to fill gaps between existing ones. Most of these new figures are unremarkable, as is Antlia.

α (alpha) Antliae, 10h 27m −31°.1, mag. 4.3, the constellation's brightest star, is an orange giant 366 l.y. distant.

δ (delta) Ant, 10h 30m −30°.6, 481 l.y. away, is a blue-white star of mag. 5.6 with a mag. 9.6 companion visible in small telescopes.

ζ¹ ζ² (zeta¹ zeta²) Ant, 9h 31m −31°.9, 373 l.y. away, is a wide pair of stars of mags. 5.8 and 5.9, visible in binoculars. Small telescopes show that ζ¹ Ant is itself double, with components of mags. 6.2 and 7.0. ▶

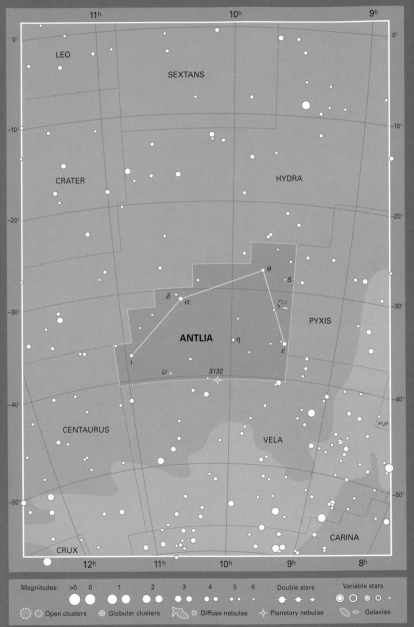

Knots of stars and gas as well as lanes of dark dust are visible in the curling arms of NGC 2997 in Antlia, a handsome 9th-magnitude spiral galaxy presented virtually face-on to us. Moderate-sized amateur telescopes are required to see it. (European Southern Observatory)

APUS   The Bird of Paradise

A faint constellation near the south celestial pole, introduced in the 1590s by the Dutch navigators Pieter Dirkszoon Keyser and Frederick de Houtman during voyages to the southern hemisphere. They introduced a total of 12 constellations in all. This one represents the bird of paradise, native to New Guinea.

α (alpha) Apodis, 14h 48m −79°.0, mag. 3.8, is an orange giant star 411 l.y. away.

β (beta) Aps, 16h 43m −77°.5, mag. 4.2, is an orange giant 158 l.y. away.

γ (gamma) Aps, 16h 33m −78°.9, mag. 3.9, is an orange star 160 l.y. away.

δ¹ δ² (delta¹ delta²) Aps, 16h 20m −78°.7, is a naked-eye or binocular pair of red and orange giant stars of mags. 4.7 and 5.3, 765 and 663 l.y. away.

θ (theta) Aps, 14h 05m −76°.8, 328 l.y. away, is a red giant that varies semi-regularly between 5th and 7th mags. every 4 months or so.

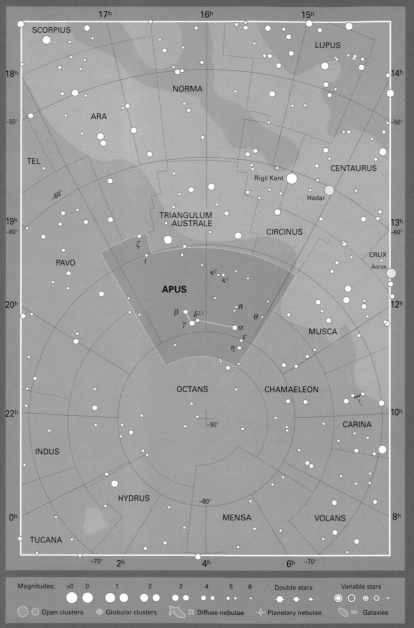

AQUARIUS   The Water Carrier

Aquarius is one of the most ancient constellations. The Babylonians saw in this area of sky the figure of a man pouring water from a jar. In Greek mythology the constellation represents Ganymede, a shepherd boy carried off by Zeus to Mount Olympus, where he became wine-waiter to the gods. The most prominent part of Aquarius is the Y-shaped asterism of four stars representing the Water Jar itself, centred on the star ζ (zeta) Aquarii. Aquarius is in an area of 'watery' constellations that includes Pisces, Cetus and Capricornus. The Sun is in the constellation Aquarius from late February to early March. Aquarius will one day contain the vernal equinox, the point at which the Sun crosses into the northern celestial hemisphere each year. This astronomically important point, from which the coordinate of right ascension is measured, will move into Aquarius from neighbouring Pisces in AD 2597 because of the effect of precession. Hence the so-called Age of Aquarius is a long way off yet.

Three main meteor showers radiate from Aquarius each year. The first, the Eta Aquarids, is the richest, reaching a maximum of 35 meteors per hour around May 5–6. The Delta Aquarids have a double radiant: the southerly one produces about 20 meteors per hour around July 29, while the northerly radiant (which is actually near the Circlet of Pisces) peaks at about 10 an hour on August 6. A weaker stream, the Iota Aquarids, produces a maximum of 8 meteors per hour, also on August 6. Each shower is named after the bright star closest to its radiant.

α (alpha) Aquarii, 22h 06m −0°.3, (Sadalmelik, from the Arabic for 'the lucky stars of the king'), mag. 2.9, is a yellow supergiant 760 l.y. away.

β (beta) Aqr, 21h 32m −5°.6, (Sadalsuud, from the Arabic for 'luckiest of the lucky stars'), mag. 2.9, is a yellow supergiant 610 l.y. away.

γ (gamma) Aqr, 22h 22m −1°.4, (Sadachbia), mag. 3.9, is a blue-white star 158 l.y. away.

δ (delta) Aqr, 22h 55m −15°.8, (Skat), mag. 3.3, is a blue-white star 160 l.y. away.

ε (epsilon) Aqr, 20h 48m −9°.5, (Albali), mag. 3.8, is a blue-white star 102 l.y. away.

ζ (zeta) Aqr, 22h 29m −0°.0, 103 l.y. away, is a celebrated binary consisting of twin white stars of mags. 4.3 and 4.5 orbiting each other every 760 years. The two stars are gradually moving apart as seen from Earth and hence are becoming increasingly easy to divide in small telescopes.

M2 (NGC 7089), 21h 34m −0°.8, is a mag. 6.5 globular cluster easily visible in binoculars or small telescopes, but requiring 100 mm aperture to resolve the brightest individual stars. This rich and highly concentrated globular lies 37,000 l.y. away.

M72 (NGC 6981), 20h 54m −12°.5, is a 9th-mag. globular cluster 56,000 l.y. away, much smaller and less impressive than M2.

NGC 7009, 21h 04m −11°.4, is a famous planetary nebula, 3000 l.y. away, known as the Saturn Nebula because of its resemblance to that planet when seen in ▶

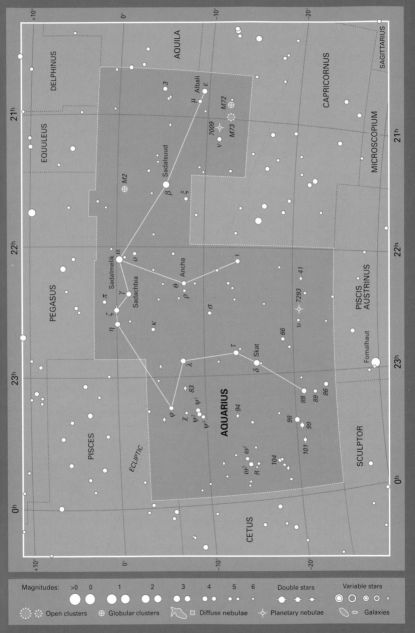

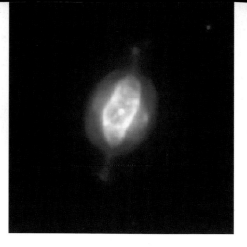

NGC 7009 in Aquarius is popularly known as the Saturn Nebula because of the faint 'handles' either side of it, like Saturn's rings, that can be seen through large telescopes and in photographs. The name Saturn Nebula was given to it in the 19th century by Lord Rosse. (SAAO)

large telescopes (see photograph above). But in most amateur telescopes, of 75 mm aperture or more, it appears as merely an 8th-mag. blue-green ellipse of similar apparent size to the globe of Saturn. Its central star is of mag. 11.5.

NGC 7293, 22h 30m −20°.8, is the nearest planetary nebula to the Sun, only about 300 l.y. away, and is commonly known as the Helix Nebula. It is the largest planetary nebula in apparent size, covering ¼° of sky, half the apparent size of the Moon. Despite its size the Helix Nebula appears quite faint and is best found with binoculars or very low power on a telescope, when it appears as a circular misty patch, not as impressive as its large size would suggest.

AQUILA   The Eagle

A constellation dating from ancient times, representing the bird that in Greek mythology carried the thunderbolts of Zeus. According to Greek myth, Zeus sent an eagle (or, in a variant of the tale, turned himself into an eagle) to abduct the shepherd boy Ganymede, represented by neighbouring Aquarius. Aquila's brightest star, Altair, forms one corner of the Summer Triangle that is completed by Deneb in Cygnus and Vega in Lyra. The name Altair comes from the Arabic *al-nasr al-tair*, 'the flying eagle'. Two fainter stars, β (beta) and γ (gamma) Aquilae, stand like sentinels either side of it; they are called Alshain and Tarazed, both from the Persian *shahin-i tarazu*, a translation of an Arabic name meaning 'the balance' which was jointly applied to these two stars and Altair. Aquila lies in the Milky Way and contains rich starfields, particularly towards Scutum in the south. It is an abundant area for novae.

α (alpha) Aquilae, 19h 51m +8°.9, (Altair, 'the flying eagle'), mag. 0.76, is a white star 17 l.y. away, among the closest naked-eye stars.

β (beta) Aql, 19h 55m +6°.4, (Alshain), mag. 3.7, is a yellow star 45 l.y. away.   ▶

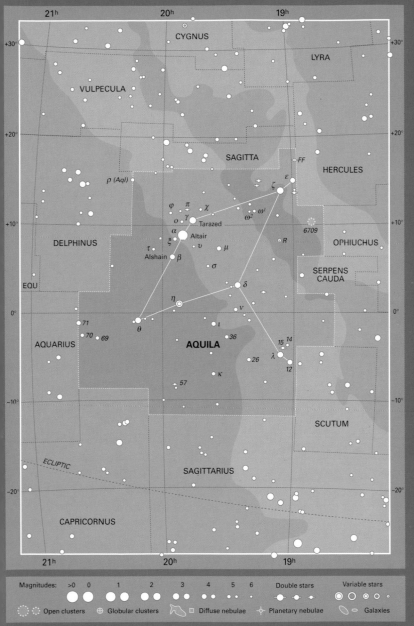

γ (gamma) Aql, 19h 46m +10°.6, (Tarazed), mag. 2.7, is an orange giant star 460 l.y. away.

ζ (zeta) Aql, 19h 05m +13°.9, mag. 3.0, is a blue-white star 83 l.y. away.

η (eta) Aql, 19h 52m +1°.0, a yellow-white supergiant 1200 l.y. away, is one of the brightest Cepheid variable stars. Its brightness ranges from mag. 3.5 to 4.4 with a period of 7.2 days.

15 Aql, 19h 05m −4°.0, is an orange giant star of mag. 5.4, 325 l.y. away, with a purplish mag. 7.0 companion, 550 l.y. away, easily visible in small telescopes.

57 Aql, 19h 55m −8°.2, is an easy double for small telescopes, consisting of a bluish star of mag. 5.7 with a mag. 6.5 companion, both about 350 l.y. away.

R Aql, 19h 06m +8°.2, 690 l.y. away, is a red giant variable of Mira type, about 400 times the diameter of the Sun, ranging from 6th to 12th mag. every 9 months or so.

FF Aql, 18h 58m +17°.4, is a yellow-white supergiant Cepheid variable, ranging from mag. 5.2 to 5.7 every 4.5 days. It lies about 2500 l.y. away.

NGC 6709, 18h 52m +10°.3, is a loosely scattered cluster of some 40 stars of mags. 9 to 11, about 3000 l.y. away.

ARA   The Altar

This constellation, although faint and relatively little-known, originated with the Greeks, who visualized it as the altar on which the gods of Olympus swore an oath of allegiance before their defeat of the Titans. Ara lies in a rich part of the Milky Way, south of Scorpius.

α (alpha) Arae, 17h 32m −49°.9, mag. 2.8, is a blue-white star 242 l.y. away.

β (beta) Ara, 17h 25m −55°.5, mag. 2.8, is an orange supergiant 600 l.y. away.

γ (gamma) Ara, 17h 25m −56°.4, mag. 3.3, is a blue supergiant 1140 l.y. away.

δ (delta) Ara, 17h 31m −60°.7, mag. 3.6, is a blue-white star 187 l.y. away.

ζ (zeta) Ara, 16h 59m −56°.0, mag. 3.1, is an orange giant star 574 l.y. away.

NGC 6193, 16h 41m −48°.8, is a 5th-mag. open cluster of about 30 stars 4200 l.y. away, about half the apparent size of the full Moon. The brightest member is a blue-white star of mag. 5.6 which small telescopes show has a companion of mag. 6.9. Around the cluster is an irregularly shaped patch of faint nebulosity, NGC 6188, which shows up well only on photographs.

NGC 6397, 17h 41m −53°.7, is a 6th-mag. globular cluster that appears as a fuzzy star through binoculars; under good conditions it is visible to the naked eye. Small telescopes resolve the brightest stars in its outer regions, which extend across at least half the apparent diameter of the full Moon. It is one of the closer globulars to us, 10,500 l.y. away.

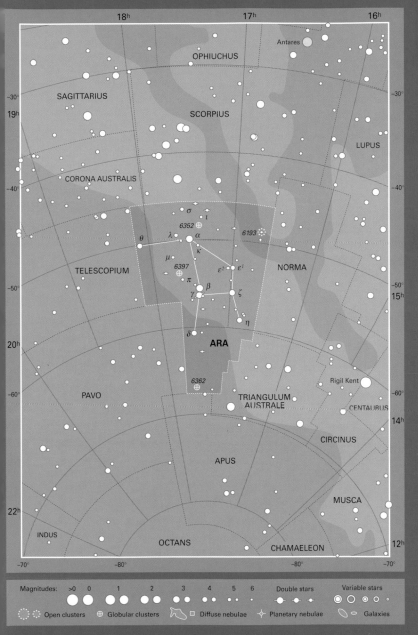

ARIES   The Ram

A constellation lying between Taurus and Andromeda, the origin of which dates back to ancient times. Aries represents the ram of Greek legend whose golden fleece was sought by Jason and the Argonauts. Despite its faintness, Aries has assumed great importance in astronomy because, over 2000 years ago, it contained the point where the Sun passes from south to north across the celestial equator each year. This point, the vernal equinox, marked the start of northern hemisphere spring, and from it the celestial coordinate known as right ascension is measured; this point is also known as the First Point of Aries, although it no longer lies in Aries, having moved into neighbouring Pisces (see diagram below) as a result of the slight wobble of the Earth in space known as precession. Currently, the Sun is in Aries from late April to mid-May.

α (alpha) Arietis, 2h 07m +23°.5, (Hamal, from the Arabic for 'lamb'), mag. 2.0, is an orange giant star 66 l.y. away.

β (beta) Ari, 1h 55m +20°.8, (Sheratan, from the Arabic for 'two'), mag. 2.6, is a blue-white star 60 l.y. away.

γ (gamma) Ari, 1h 54m +19°.3, (Mesartim), 204 l.y. away, is a striking double consisting of twin white stars of mags. 4.7 and 4.6, clearly visible through small telescopes even under low magnification.

ε (epsilon) Ari, 2h 59m +21°.3, 290 l.y. away, is a challenging double star for apertures of 100 mm or more. High magnification reveals a tight pair of white stars of mags. 5.2 and 5.5.

λ (lambda) Ari, 1h 58m +23°.6, 133 l.y. away, is a white star of mag. 4.8 with a yellow mag. 7.3 companion visible in small telescopes or even good binoculars.

π (pi) Ari, 2h 49m +17°.5, 600 l.y. away, is a blue-white star of mag. 5.3, with a close mag. 8.5 companion, difficult to distinguish in the smallest telescopes.

Movement of the so-called First Point of Aries over 800 years. At present it lies in Pisces and is approaching Aquarius. (Wil Tirion)

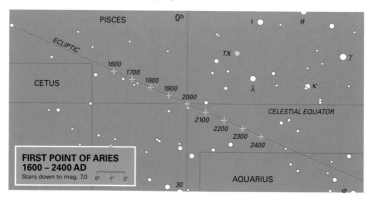

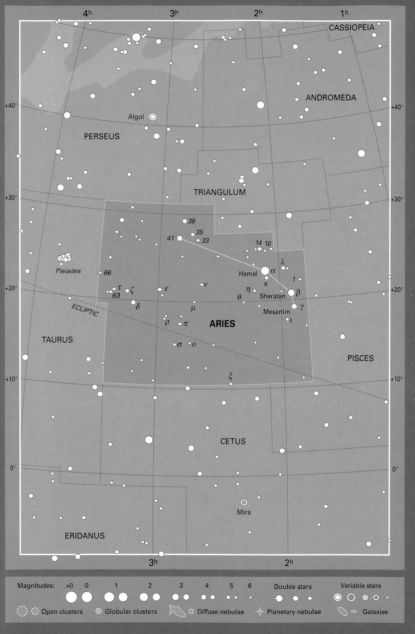

AURIGA   The Charioteer

A large and prominent constellation, usually identified by the ancient Greeks as Erichthonius, a legendary king of Athens and a skilled charioteer. Auriga's leading star is Capella, sixth-brightest in the whole sky and the most northerly first-magnitude star. In legend this star represented the she-goat Amaltheia which suckled the infant Zeus; the stars ζ (zeta) and η (eta) Aurigae are supposedly her kids. The star marking the charioteer's foot, once known as γ (gamma) Aurigae, was originally regarded as being shared with Taurus the Bull, but it is now assigned exclusively to Taurus as β (beta) Tauri.

α (alpha) Aurigae, 5h 17m +46°.0, (Capella, 'she-goat'), mag. 0.08, lies 42 l.y. away. It is a spectroscopic binary, consisting of two yellow giant stars orbiting every 104 days, although they do not eclipse each other.

β (beta) Aur, 6h 00m +44°.9, (Menkalinan, 'shoulder of the charioteer'), 82 l.y. away, is an eclipsing variable of mag. 1.9 consisting of two blue-white stars that orbit every 3.96 days, causing two dips in brightness of 0.1 mag. on each orbit.

ε (epsilon) Aur, 5h 02m +43°.6, a white supergiant about 2000 l.y. away, is an eclipsing binary of exceptionally long period. Normally it shines at mag. 3.0, but every 27 years it sinks to mag. 3.8 as it is eclipsed by a dark companion, remaining at minimum for a year. One theory is that ε Aurigae's companion is a binary star enveloped in a disk of matter. Its next eclipse starts in late 2009.

ζ (zeta) Aur, 5h 02m +41°.1, 790 l.y. away, is a famous eclipsing binary of contrasting stars: an orange giant orbited every 972 days by a smaller blue companion. During eclipses, ζ Aurigae's brightness drops from mag. 3.7 to 4.0.

θ (theta) Aur, 6h 00m +37°.2, 173 l.y. away, is a blue-white star of mag. 2.6. It has a yellowish companion of mag. 7.1 which, because of its closeness and relative faintness, needs at least 100 mm aperture and high magnification to distinguish. This is a tough double for steady nights.

ψ¹ (psi¹) Aur, 6h 25m +49°.3, is a variable orange supergiant that ranges between mags. 4.8 and 5.7 with no set period. Its distance is uncertain.

4 Aur, 4h 59m +37°.9, 159 l.y. away, is a double star of mags. 5.0 and 8.1, visible in small telescopes.

14 Aur, 5h 15m +32°.7, 270 l.y. away, is a white star of mag. 5.0 with a mag. 7.9 companion, 82 l.y. away, split in small telescopes.

RT Aurigae, 6h 29m +30°.5, a yellow-white supergiant, is a Cepheid variable ranging between mag. 5.0 and 5.8 every 3.7 days. It lies about 1600 l.y. away.

UU Aur, 6h 37m +38°.4, is a semi-regular variable, ranging between 5th and 7th mags. with a rough period of 234 days. It is a giant star appearing deep red in colour, around 2000 l.y. away.

M36 (NGC 1960), 5h 36m +34°.1, is a small, bright open cluster of about 60 stars, visible in binoculars and resolvable into stars in small telescopes. M36 lies 3900 l.y. away.

M37 (NGC 2099), 5h 52m +33°.5, is the largest and richest of the clusters in Auriga, containing about 150 stars. In binoculars the cluster appears as a hazy, ▶

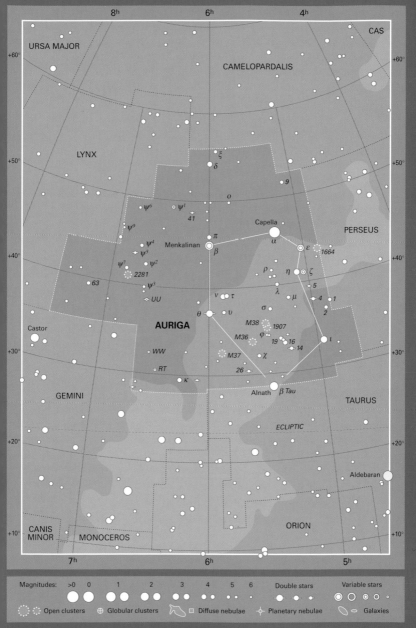

unresolved patch, but a 100-mm telescope resolves it into a sparkling field of faint stardust, with a brighter orange star at the centre. It lies 4200 l.y. away.

M38 (NGC 1912), 5h 29m +35°.8, is a large, scattered cluster of about 100 faint stars, visible in binoculars, with a noticeable cross-shape when seen through a telescope. Its distance is 3900 l.y. Half a degree south of it lies the small fuzzy blob of NGC 1907, a much smaller and fainter cluster 4200 l.y. away.

NGC 2281, 6h 49m +41°.1, is a binocular cluster of about 30 stars, 1500 l.y. away. Through a telescope the stars appear to be arranged in a crescent, with four brighter stars forming a diamond shape.

BOÖTES   The Herdsman

An ancient constellation, representing a herdsman driving a bear (Ursa Major) around the sky; the herdsman is often depicted holding the leash of the hunting dogs, Canes Venatici. The name of the constellation's brightest star, Arcturus, actually means 'bear-keeper' in Greek. In Greek mythology Boötes represents Arcas, the son of Zeus and the nymph Callisto (neighbouring Ursa Major represents Callisto herself, who was turned into a bear by Zeus's jealous wife Hera). Arcturus is the brightest star in the northern half of the sky, and is easily identified: the curving handle of the Big Dipper or Plough acts as a pointer to it. Arcturus forms the base of a large 'Y' shape completed by ε (epsilon) Boötis, γ (gamma) Boötis and α (alpha) Coronae Borealis. The other stars of the constellation are much fainter than Arcturus, but include many doubles of interest. The year's most abundant meteor shower, the Quadrantids, radiates from the northern part of Boötes, an area of sky that was once occupied by the now-abandoned constellation of Quadrans Muralis, the Mural Quadrant (hence the shower's name). The Quadrantid meteors reach a peak of about 100 per hour on January 3–4 each year, although they are not as bright as other rich showers such as the Perseids and Geminids.

α (alpha) Boötis, 14h 16m +19°.2, (Arcturus), mag. −0.05, is the fourth-brightest star in the entire sky. It is an orange giant, about 27 times the diameter of the Sun, lying 37 l.y. away; its ruddy colour is noticeable to the naked eye, and is more striking with optical aid. Arcturus has a mass similar to our Sun's, and it is believed that the Sun will swell up to become a red giant like Arcturus 5000 million years from now.

β (beta) Boo, 15h 02m +40°.4, (Nekkar, corrupted from the Arabic for 'ox-driver', referring to the whole constellation), mag. 3.5, is a yellow giant 219 l.y. away.

γ (gamma) Boo, 14h 32m +38°.3, (Seginus), mag. 3.0, is a white star 85 l.y. away.

δ (delta) Boo, 15h 16m +33°.3, mag. 3.5, is a yellow giant 117 l.y. away. It has a wide binocular companion of mag. 7.8.   ▶

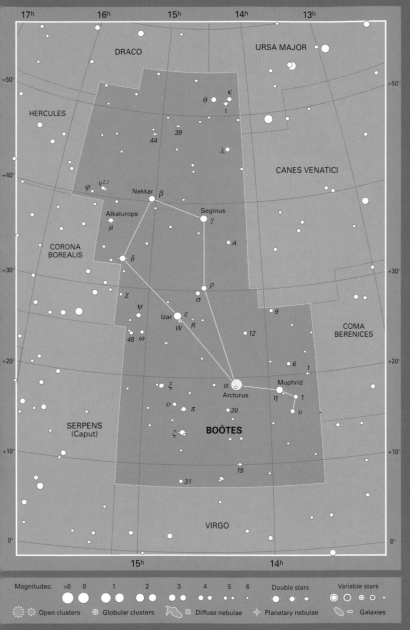

DRACO

URSA MAJOR

HERCULES

θ κ
    ι

39

44

λ

CANES VENATICI

φ v²·¹

Nekkar β

Alkalurops
μ

Seginus
γ

A

CORONA
BOREALIS

δ

χ

ρ

σ

9

ψ

Izar ε
   W R

12

COMA
BERENICES

45 ω

6 1

ξ

α
Arcturus

Muphrid

η τ

υ

o π

20

ζ

BOÖTES

SERPENS
(Caput)

15

31

VIRGO

Magnitudes: >0 0 1 2 3 4 5 6 Double stars Variable stars

Open clusters ⊕ Globular clusters Diffuse nebulae Planetary nebulae Galaxies

ε (epsilon) Boo, 14h 45m +27°.1, (Izar, 'girdle' or 'loincloth'), 210 l.y. away, is a celebrated double star: an orange giant primary of mag. 2.5 with a blue main-sequence companion of mag. 4.6. This close double of contrasting colours requires a telescope of at least 75 mm at ×100 power or more, because the bright primary tends to overwhelm its fainter companion; but its appearance when split has led to the alternative name Pulcherrima, meaning 'most beautiful'.

ι (iota) Boo, 14h 16m +51°.4, 97 l.y. away, is a wide double star of mags. 4.8 and 7.5.

κ (kappa) Boo, 14h 13m +51°.8, is an easy double star for small telescopes, con-sisting of unrelated components of mags. 4.5 and 6.6, distances 155 and 196 l.y.

μ (mu) Boo, 15h 24m +37°.4, (Alkalurops, 'club' or 'staff'), 121 l.y. distant, is an attractive triple star. To the naked eye it appears as a blue-white star of mag. 4.3, but binoculars reveal a wide companion of mag. 6.5. Telescopes of 75 mm aper-ture with high magnification show that this companion actually consists of two close stars of mags. 7.0 and 7.6; they orbit each other every 260 years.

ν^1 ν^2 (nu^1 nu^2) Boo, 15h 31m +40°.8, is an unrelated binocular duo: ν^1 is an orange giant of mag. 5.0, 870 l.y. distant, while ν^2 is a white star also of mag. 5.0, 430 l.y. away.

π (pi) Boo, 14h 41m +16°.4, 317 l.y. away, is a double star with blue-white components of mags. 4.9 and 5.8, visible in small telescopes.

ξ (xi) Boo, 14h 51m +19°.1, 22 l.y. away, is a showpiece double for small tele-scopes, consisting of yellow and orange stars of mags. 4.7 and 7.0, orbiting each other every 150 years.

44 Boo, 15h 04m +47°.7, 42 l.y. away, is a complex double–variable star. To the naked eye it appears as a yellow star of mag. 4.8. In fact, it is a binary of mags. 5.3 and 6.1 orbiting every 220 years. Until the year 2015 the two can be split in apertures of 75 mm, but thereafter they will become more difficult as they move together, reaching their closest in 2028 when they will be indivisible in amateur telescopes. The fainter star is itself an eclipsing binary with a period of 6.4 hours and a range of 0.6 mag.

CAELUM   The Chisel

An obscure, almost irrelevant constellation at the foot of Eridanus, representing an engraving tool. It was introduced in the 1750s by Nicolas Louis de Lacaille during his mapping of the southern sky.

α (alpha) Caeli, 4h 41m −41°.9, mag. 4.4, is a white main-sequence star 66 l.y. away.

β (beta) Cae, 4h 42m −37°.1, mag. 5.0, is a white star 90 l.y. away.

γ (gamma) Cae, 5h 04m −35°.5, mag. 4.6, is an orange giant 185 l.y. away. It has a close mag. 8.1 companion, difficult to see in the smallest apertures because of the brightness contrast.

δ (delta) Cae, 4h 31m −45°.0, mag. 5.1, is a blue-white star 710 l.y. away.

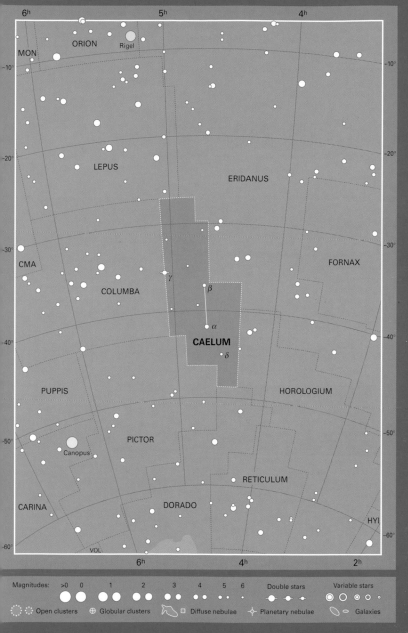

CAMELOPARDALIS   The Giraffe

A faint and obscure constellation of the north polar region of the sky, sometimes written as Camelopardus, an obsolete variant of the name. It was invented in 1613 by the Dutch theologian and astronomer Petrus Plancius, and supposedly represents the animal on which Rebecca rode into Canaan to marry Isaac.

α (alpha) Camelopardalis, 4h 54m +66°.3, mag. 4.3, is a highly luminous blue supergiant star approximately 5000 l.y. away, exceptionally distant for a naked-eye star.

β (beta) Cam, 5h 03m +60°.4, at mag. 4.0 the brightest star in the constellation, is a yellow supergiant 1000 l.y. distant. It has a wide mag. 8.6 companion star visible in small telescopes or even good binoculars.

11 Cam, 5h 06m +59°.0, mag. 5.2, forms an easy binocular pairing with 12 Cam, mag. 6.1. Both lie at a similar distance, about 650 l.y., but are too widely separated to be a genuine double.

Σ 1694 (Struve 1694), 12h 49m +83°.4, 300 l.y. away, is a pair of blue-white stars, mags. 5.4 and 5.9, easily split in small telescopes. In some old catalogues the star was listed as 32 Cam.

NGC 1502, 4h 08m +62°.3, is a small 6th-mag. open star cluster with about 45 members visible in binoculars or a small telescope, somewhat triangular in shape and with an easy 7th-mag. double star at its centre. It lies 3000 l.y. away. Note a chain of stars called Kemble's Cascade which runs for 2½° from NGC 1502 towards Cassiopeia, parallel to the Milky Way (see illustration below).

NGC 2403, 7h 37m +65°.6, is an 8th-mag. spiral galaxy ¼° long, visible as an elliptical glow in a 100-mm telescope. It lies about 12 million l.y. away.

Kemble's Cascade is a chain of faint stars near NGC 1502. (Wil Tirion)

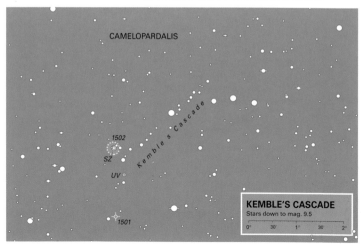

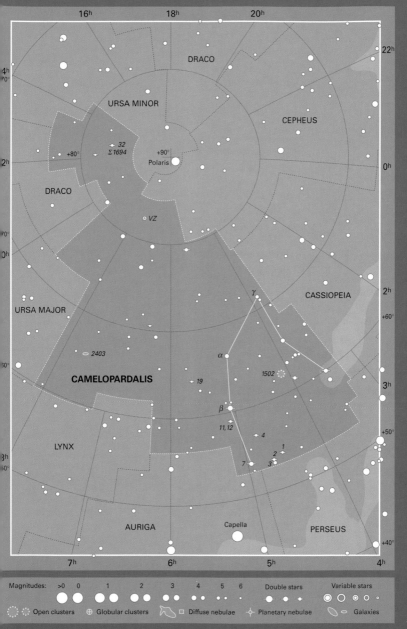

Magnitudes: >0 0 1 2 3 4 5 6   Double stars   Variable stars

Open clusters   Globular clusters   Diffuse nebulae   Planetary nebulae   Galaxies

CANCER   The Crab

Cancer represents the crab that attacked Hercules when he was fighting the multi-headed Hydra; the luckless crab was crushed underfoot by mighty Hercules but was subsequently elevated to the heavens. In ancient times, the Sun reached its most northerly point in the sky each year while it was in Cancer. The date on which the Sun is farthest north of the Earth's equator, on or around June 21, is the northern summer solstice; on this day the Sun appears overhead at noon at latitude 23½° north on Earth. This latitude came to be known as the Tropic of Cancer, a name it retains today even though, because of the effect of precession, the Sun now lies in the constellation of Taurus on that date. With only two stars brighter than mag. 4.0, Cancer is the faintest of the 12 constellations of the zodiac, but it nevertheless contains much of interest, notably the star cluster Praesepe, the Manger. Praesepe is flanked by two stars, Asellus Borealis and Asellus Australis, the northern and southern donkeys, visualized as feeding at the stellar manger.

α (alpha) Cancri, 8h 58m +11°.9, (Acubens, 'the claw'), mag. 4.3, is a white star 174 l.y. away. It has a 12th-mag. companion visible with telescopes of 75 mm aperture and over.

β (beta) Cnc, 8h 17m +9°.2, mag. 3.5, an orange giant 290 l.y. away, is the brightest star in the constellation.

γ (gamma) Cnc, 8h 43m +21°.5, (Asellus Borealis, 'northern donkey'), mag. 4.7, is a white star 158 l.y. away.

δ (delta) Cnc, 8h 45m +18°.2, (Asellus Australis, 'southern donkey'), mag. 3.9, is an orange giant star 136 l.y. away.

ζ (zeta) Cnc, 8h 12m +17°.6, 83 l.y. away, is an interesting multiple star. A small telescope reveals two yellow stars of mags. 5.0 and 6.2; they form a genuine binary with an estimated orbital period of 1100 years. Larger telescopes split the brighter component into a tight binary of mags. 5.6 and 6.0, orbital period 59.5 years. These two stars are currently moving apart; at their widest, around the year 2018, 100 mm should be sufficient to split them, but 150 mm will be needed before about 2010.

ι (iota) Cnc, 8h 47m +28°.8, 298 l.y. away, is a yellow giant of mag. 4.0 with a blue-white mag. 6.6 companion just visible in binoculars, or easily seen through a small telescope.

M44 (NGC 2632), 8h 40m +20°.0, Praesepe ('manger'), commonly called the Beehive Cluster, is a swarm of about 50 stars of 6th mag. and fainter, visible as a misty patch to the naked eye and best seen through binoculars. The brightest member of this open cluster, ε (epsilon) Cancri, is mag. 6.3. Praesepe sprawls over 1½° of sky, three times the apparent diameter of the Moon. The distance to the cluster's centre is 577 l.y. according to the Hipparcos measurements.

M67 (NGC 2682), 8h 50m +11°.8, is a smaller and denser open cluster than M44, visible as a Moon-sized misty ellipse in binoculars or small telescopes and needing apertures of at least 75 mm to resolve the brightest of its 200 or so individual stars of 10th mag. and fainter. It lies 2500 l.y. away.

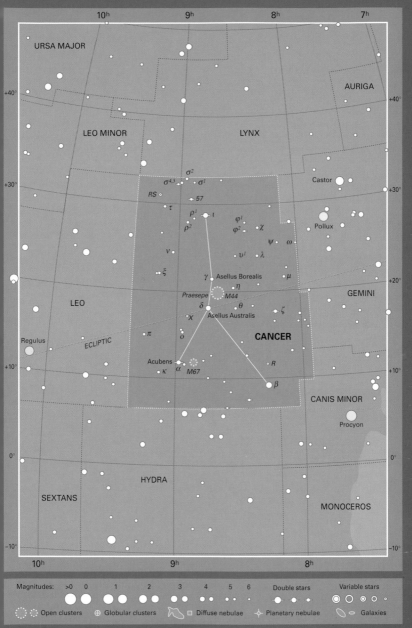

CANES VENATICI   The Hunting Dogs

A constellation introduced in 1687 by the Polish astronomer Johannes Hevelius, consisting of a sprinkling of faint stars below Ursa Major. It represents two dogs, Asterion ('little star') and Chara ('joy'), held on a leash by neighbouring Boötes as they pursue the Great Bear around the pole. Canes Venatici contains numerous galaxies, the most famous being M51, the Whirlpool, a beautiful face-on spiral (pictured on page 284). It was the first galaxy in which spiral form was detected, by Lord Rosse in 1845 with his 72-inch (1.8-m) reflector at Birr Castle, Ireland.

α (alpha) Canum Venaticorum, 12h 56m +38°.3, is popularly called Cor Caroli, meaning 'Charles's heart', a reference to the executed King Charles I of England; it is reputed, doubtless apocryphally, to have shone particularly brightly in 1660 on the arrival of Charles II in England at the Restoration of the monarchy. It is a double star of mags. 2.9 and 5.6, easily split in small telescopes. Both stars are white, but various observers have reported subtle shades of colour when viewed through a telescope. The brighter star is the standard example of a rare class of stars with strong and variable magnetic fields; its brightness fluctuates slightly but not enough to be noticeable to the eye. Cor Caroli is 110 l.y. away.

β (beta) CVn, 12h 34m +41°.4, (Chara, 'joy'), mag. 4.2, is the only other star of any prominence in the constellation. It is a yellow main-sequence star similar to the Sun, 27 l.y. away.

Y CVn, 12h 45m +45°.4, 710 l.y. away, is a semi-regular variable supergiant of deep red colour sometimes known as La Superba. Its range is about mag. 5.0 to 6.5 and its period is approximately 160 days.

M3 (NGC 5272), 13h 42m +28°.4, is a rich globular cluster located midway between Cor Caroli and Arcturus, regarded as one of the finest globulars in the northern sky. At 6th mag. it is on the naked-eye limit, but can be picked up easily as a hazy star in binoculars or a small telescope; a 5th-mag. star nearby acts as a guide. In small telescopes the cluster appears as a condensed ball of light with a faint outer halo. Apertures of 100 mm or more are needed to resolve individual stars in its outer regions. M3 is 32,000 l.y. away.

M51 (NGC 5194), 13h 30m +47°.2, the Whirlpool Galaxy, is an 8th-mag. spiral galaxy about 20 million l.y. away with a smaller satellite galaxy, NGC 5195, apparently lying at the end of one of its arms; in reality, this companion lies slightly behind M51, having brushed past it some time in the last 100 million years or so. The Whirlpool can be seen in binoculars, appearing elongated. It is disappointing in small telescopes, which show a faint milky radiance around the starlike nuclei of the galaxy and its satellite; apertures of at least 250 mm are needed to see the arms well. Nevertheless, M51 is well worth hunting for on clear, dark nights. (For a photograph see page 284.)

M63 (NGC 5055), 13h 16m +42°.0, is a 9th-mag. spiral galaxy visible in small telescopes as an elliptical haze with a somewhat mottled texture. It is popularly known as the Sunflower Galaxy because of its appearance in large instruments.

M94 (NGC 4736), 12h 51m +41°.1, is a compact spiral galaxy presented nearly face-on. In amateur telescopes it looks like an 8th-mag. comet, with a fuzzy star-like nucleus surrounded by an elliptical halo. M94 is about 15 million l.y. away.

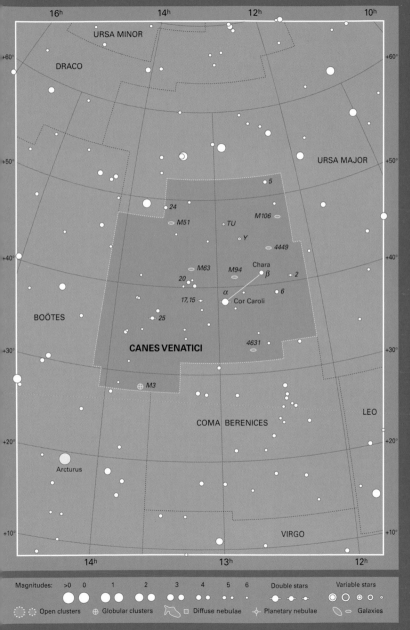

CANIS MAJOR   The Greater Dog

An ancient constellation, representing one of the two dogs (the other being Canis Minor) following at the heels of Orion. Canis Major contains many brilliant stars, making it one of the most prominent constellations; its leading star, Sirius, is the brightest in the entire sky. Sirius features in many legends, and the ancient Egyptians based their calendar on its yearly motion around the sky.

α (alpha) Canis Majoris, 6h 45m −16°.7, (Sirius, from the Greek meaning 'searing' or 'scorching'), mag. −1.44, is a brilliant white star 8.6 l.y. away, one of the Sun's closest neighbours; the Hipparcos satellite detected variability of 0.1 mag. or so in its brightness. It has a white dwarf companion of mag. 8.4 that orbits it every 50 years. The brilliance of Sirius overpowers this white dwarf so that even when the two stars are at their greatest separation, as between the years 2020 and 2025, telescopes of 200 mm aperture or more and steady atmospheric conditions are required to see it. The two were last at their closest in 1993.

β (beta) CMa, 6h 23m −18°.0, (Mirzam, 'the announcer', i.e. of Sirius), mag. 2.0, is a blue giant 500 l.y. away. It is a pulsating star whose variations, of a few hundredths of a magnitude every 6 hours, are undetectable to the naked eye.

δ (delta) CMa, 7h 08m −26°.4, (Wezen, 'the weight'), mag. 1.8, is a white supergiant star 1800 l.y. away.

ε (epsilon) CMa, 6h 59m −29°.0, (Adhara, 'the virgins'), mag. 1.5, is a blue giant 430 l.y. away. It has a mag. 7.4 companion which is difficult to see in small telescopes because of the glare from the primary.

η (eta) CMa, 7h 24m −29°.3, (Aludra), mag. 2.4, is a blue supergiant about 3200 l.y. away.

μ (mu) CMa, 6h 56m −14°.0, mag. 5.0, is a yellow giant about 900 l.y. away with a close 7th-mag. blue-white companion difficult to pick up in the smallest apertures because of the magnitude contrast.

ν^1 (nu^1) CMa, 6h 36m −18°.7, mag. 5.7, is a yellow giant 278 l.y. away with a mag. 8.1 companion, visible in small telescopes.

UW CMa, 7h 19m −24°.6, is an eclipsing binary of β (beta) Lyrae type that varies between mags. 4.8 and 5.3 in 4.4 days. It lies about 3000 l.y. away.

M41 (NGC 2287), 6h 47m −20°.7, is a large and bright open cluster of about 80 stars, easily visible through binoculars or a small telescope and, with a total magnitude of 4.5, detectable by the naked eye under good conditions – it was known to the ancient Greeks. A low-power view in a small telescope shows the individual stars grouped in bunches and curves, covering an area of sky equivalent to the apparent diameter of the Moon. The brightest stars in the cluster are 7th-mag. orange giants. M41 is 2100 l.y. away.

NGC 2362, 7h 19m −25°.0, is a compact cluster surrounding the mag. 4.4 blue supergiant star τ (tau) Canis Majoris, which is a genuine member. Small telescopes show about 60 stars in the cluster, which lies some 5200 l.y. away.

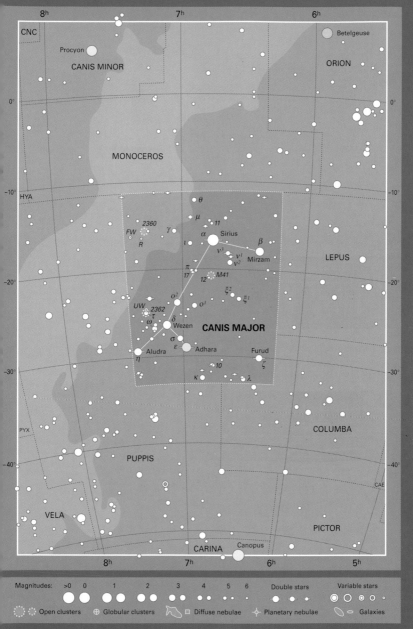

Magnitudes: >0 0 1 2 3 4 5 6   Double stars   Variable stars

Open clusters   Globular clusters   Diffuse nebulae   Planetary nebulae   Galaxies

CANIS MINOR   The Lesser Dog

The second of the two dogs of Orion, the other being Canis Major. Apart from its leading star Procyon, the eighth-brightest star in the sky, there are few objects of importance in Canis Minor. Procyon forms a prominent equilateral triangle with the bright stars Sirius (in Canis Major) and Betelgeuse (in Orion).

α (alpha) Canis Minoris, 7h 39m +5°.2, (Procyon, from the Greek meaning 'preceding the dog', referring to its rising before Sirius), mag. 0.40, is a yellow-white star 11.4 l.y. away, and therefore among the nearest stars to the Sun. Like Sirius, Procyon has a white dwarf companion, but this star, of mag. 10.7, is even more difficult to see than the companion of Sirius, requiring the use of large professional telescopes. Procyon's companion orbits it every 41 years.

β (beta) CMi, 7h 27m +8°.3, (Gomeisa), mag. 2.9, is a blue-white main-sequence star 170 l.y. away.

White dwarfs

By a remarkable coincidence both Sirius and Procyon, the brightest stars in Canis Major and Canis Minor respectively, are accompanied by tiny, faint stars known as white dwarfs. The existence of these companion stars was predicted in 1844 by the German astronomer Friedrich Wilhelm Bessel, who detected a wobble in the proper motions (page 14) of Sirius and Procyon. Bessel realized that this wobble was most probably caused by the presence of unseen companions orbiting around the visible stars. The companion of Sirius, called Sirius B, was first seen in 1862 by the American astronomer Alvan G. Clark using a 47-cm (18½-inch) refractor, while the companion to Procyon (Procyon B) was first seen in 1896 by John M. Schaeberle with the 91-cm (36-inch) refractor at Lick Observatory. But not until 1915 did astronomers realize the truly extraordinary nature of these stars. Observations showed that Sirius B was very hot, very small and very dense. In fact Sirius B has the mass of the Sun packed into a sphere less than 1 per cent of the Sun's diameter. The resulting density of Sirius B is over 100,000 times that of water. A white dwarf is a star at the end of its life; it is the shrunken remnant of a once-proud star like the Sun whose central nuclear fires have burnt out. The cause of the immense densities of white dwarfs is the inexorable pull of gravity, which squeezes the electrons of the dying star as closely together as is physically possible.

Magnitudes: >0 0 1 2 3 4 5 6   Double stars   Variable stars

Open clusters   Globular clusters   Diffuse nebulae   Planetary nebulae   Galaxies

CAPRICORNUS The Sea Goat

Capricornus is depicted as a goat with a fish's tail. Amphibious creatures feature prominently in ancient legends, and the origin of Capricornus certainly dates back to ancient times. In Greek legend, the constellation represented the goat-headed god Pan who jumped into a river to escape the approach of the monster Typhon, turning his lower half into a fish. Before about 130 BC the Sun lay in Capricornus when it reached its farthest point south of the equator each year; this point, known as the northern winter solstice, currently occurs on December 21 or 22 (the date can vary from year to year). The latitude on Earth at which the Sun appears overhead at noon on that date, $23\frac{1}{2}°$ south, became known as the Tropic of Capricorn. Because of precession, the northern winter solstice has since moved from Capricornus into the neighbouring constellation Sagittarius (and will reach Ophiuchus in the year 2269), but the Tropic of Capricorn retains its name. Capricornus is the smallest constellation of the zodiac. The Sun is within its boundaries from late January to mid-February.

α^1 α^2 (alpha^1 alpha^2) Capricorni, 20h 18m −12°.5, (Algedi or Giedi, both from the Arabic meaning 'the kid' in reference to the constellation as a whole) is a multiple star, consisting of an unrelated yellow supergiant of mag. 4.3 and a yellow giant of mag. 3.6, 690 and 109 l.y. away respectively, visible separately with the naked eye or binoculars. Telescopes reveal that each star is itself double. The fainter of the pair, α^1, has a wide mag. 9.2 companion visible in small telescopes; α^2 has its own companion of mag. 11. Telescopes of at least 100 mm aperture show that this faint companion is itself composed of two 11th-mag. stars. α Capricorni is therefore a fascinating hybrid system.

β (beta) Cap, 20h 21m −14°.8, (Dabih, from the Arabic meaning 'the lucky stars of the slaughterer'), 340 l.y. away, is a golden-yellow giant star of mag. 3.1 with a wide, blue-white companion of mag. 6.1, visible through binoculars or small telescopes.

γ (gamma) Cap, 21h 40m −16°.7, (Nashira), mag. 3.7, is a white giant star 139 l.y. away.

δ (delta) Cap, 21h 47m −16°.1, (Deneb Algedi, 'the kid's tail'), mag. 2.9, is the brightest star in the constellation. It is an eclipsing binary of β (beta) Lyrae type, varying by a barely perceptible 0.2 mag. over 24.5 hours. It lies 39 l.y. away.

π (pi) Cap, 20h 27m −18°.2, 670 l.y. away, is a blue-white star of mag. 5.1 with a close mag. 8.3 companion visible with a small telescope.

M30 (NGC 7099), 21h 40m −23°.2, is a mag. 7.5 globular cluster 30,000 l.y. away, visible in small telescopes and resolvable in 100 mm aperture, notably centrally condensed with finger-like chains of stars extending northwards.

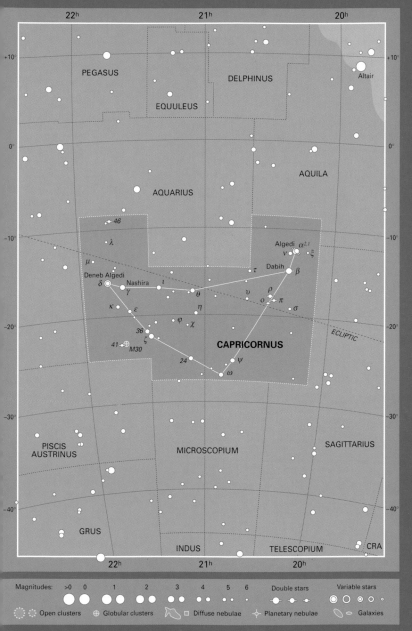

CARINA   The Keel

This constellation was originally part of the extensive Argo Navis, the Ship of the Argonauts, until subdivided in 1763 by the French celestial cartographer Nicolas Louis de Lacaille. As a part of Argo Navis, Carina originated in ancient Greek times and is associated with the legend of Jason and the Argonauts and their quest for the Golden Fleece. Carina lies in the Milky Way, providing rich starfields and clusters for binoculars. The stars ι (iota) and ε (epsilon) Carinae, together with κ (kappa) and δ (delta) Velorum, form the False Cross, sometimes confused with the real Southern Cross.

α (alpha) Carinae, 6h 24m −52°.7, (Canopus), mag. −0.62, the second-brightest star in the sky, is a white supergiant 313 l.y. away; the Hipparcos satellite found it to be variable by 0.1 mag. or so. It is named after the helmsman of the Greek King Menelaus, and appropriately enough is now used by spacecraft as a guide for navigation.

β (beta) Car, 9h 13m −69°.7, (Miaplacidus), mag. 1.7, is a blue-white star 111 l.y. away.

ε (epsilon) Car, 8h 23m −59°.5, mag. 1.9, is an orange giant star 630 l.y. away.

η (eta) Car, 10h 45m −59°.7, 7500 l.y. away, is a peculiar nova-like variable star embedded in the nebula NGC 3372 (see page 106). In the past, η Carinae has fluctuated erratically in brightness, reaching a maximum of mag. −1 in 1843 when it was temporarily the second-brightest star in the sky; it subsequently settled at around 6th mag., but brightened to 5th mag. in 1998. The star is estimated to be over 100 times more massive and 4 million times brighter than the Sun, with an unseen companion orbiting it every 5½ years. The binary pair is surrounded by a shell of dust and gas thrown off in the 1843 outburst.

θ (theta) Car, 10h 43m −64°.4, mag. 2.7, is a blue-white star 440 l.y. away, a member of the sparkling cluster IC 2602 (see page 106).

ι (iota) Car, 9h 17m −59°.3, mag. 2.2, is a white supergiant 690 l.y. away.

υ (upsilon) Car, 9h 47m −65°.1, 1600 l.y. away, is a double star consisting of two blue-white giants of mags. 3.0 and 6.0, divisible in small telescopes.

l Car, 9h 45m −62°.5, a yellow supergiant 1500 l.y. away, is the brightest Cepheid whose variations are large enough to be obvious to the naked eye. It rises and falls between mags. 3.3 and 4.2 every 35.5 days.

R Car, 9h 32m −62°.8, 416 l.y. away, is a red giant variable star of Mira type that ranges between 4th and 10th mags. with a period of 309 days.

S Car, 10h 09m −61°.5, about 1300 l.y. away, is a red giant variable similar to nearby R Car, varying from 5th to 10th mag. over 150 days.

NGC 2516, 7h 58m −60°.9, is a naked-eye open cluster of some 80 stars, 1100 l.y. away, as large as the full Moon. It is a sparkling sight in binoculars, which show it to be cross-shaped. Its brightest member is a mag. 5.2 red giant.

NGC 3114, 10h 03m −60°.1, is a widely scattered open cluster the same apparent size as the full Moon, containing stars of 6th mag. and fainter, about 3000 l.y. away. It is best seen in binoculars and small telescopes with low power.   ▶

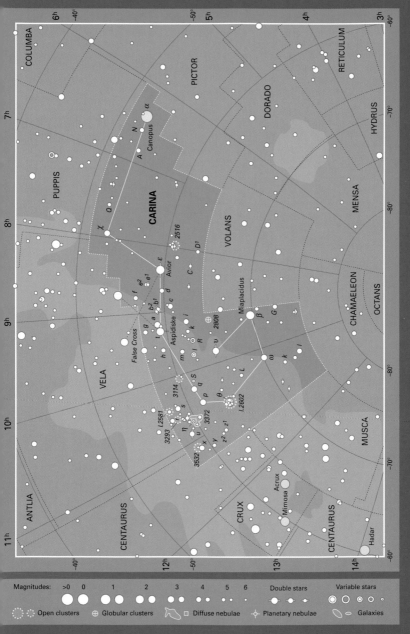

Magnitudes: >0 0 1 2 3 4 5 6   Double stars   Variable stars

Open clusters   ⊕ Globular clusters   Diffuse nebulae   Planetary nebulae   Galaxies

NGC 3372, 10h 44m −59°.9, is a celebrated diffuse nebula easily visible to the naked eye as a brilliant patch of the Milky Way four Moon diameters wide, surrounding the erratic variable star η (eta) Carinae. The nebula shines from the light of brilliant young stars born within it. Binoculars and small telescopes show jewelled star clusters and swirls of glowing gas alternating with dark lanes. Most famous is a dark notch, called the Keyhole because of its distinctive shape, silhouetted against the nebula's brightest central portion near η Carinae itself. The whole NGC 3372 nebula lies about 7500 l.y. from us, the same distance as η Carinae. (For a photograph see page 275.)

NGC 3532, 11h 06m −58°.7, 1350 l.y. away, is an outstanding open cluster, visible to the naked eye as a brighter patch nearly 1° wide among rich Milky Way starfields and glorious in binoculars. It contains 150 or so stars of 7th mag. and fainter, including several orange giants, arranged in an elliptical shape with a star-free lane across its centre. The mag. 3.9 yellow-white supergiant x Carinae at the cluster's edge is not a member but a background object five times farther off.

IC 2602, 10h 43m −64°.4, is a large and brilliant open cluster of 60 stars, popularly known as the Southern Pleiades, lying 480 l.y. away. Its brightest members are visible to the naked eye, notably the 3rd-mag. θ Carinae (see page 104). The whole cluster appears twice as wide as the full Moon.

CASSIOPEIA

In Greek legend, Cassiopeia was the beautiful but boastful Queen of Ethiopia, wife of King Cepheus and mother of Andromeda. In the sky she is depicted sitting in a chair. The constellation is easily identifiable by the distinctive W-shape of its five brightest stars. Cassiopeia lies on the opposite side of the Pole Star from Ursa Major, in a rich part of the Milky Way. Near the star κ (kappa) Cassiopeiae, at 0h 25.3m, +64° 09′, occurred the famous supernova outburst of 1572, Tycho's Star, observed by Tycho Brahe. The remains of this supernova are now a strong radio source. The remains of another supernova, which erupted around 1660 but which went unseen at the time, form the strongest radio source in the sky, Cassiopeia A; it lies at 23h 23.4m, +58° 50′, and is some 10,000 l.y. away.

α (alpha) Cassiopeiae, 0h 41m +56°.5, (Shedir, 'the breast'), mag. 2.2, is an orange giant star 229 l.y. away. It has a wide mag. 8.9 companion, unrelated.

β (beta) Cas, 0h 09m +59°.1, (Caph), mag. 2.3, is a white star 54 l.y. away.

γ (gamma) Cas, 0h 57m +60°.7, 613 l.y. away, is a remarkable blue-white variable of the type known as a shell star. It throws off rings of gas at irregular intervals, apparently because its high-speed rotation makes it unstable, causing it to vary unpredictably between mags. 3.0 and 1.6. Currently it hovers around mag. 2.2.

δ (delta) Cas, 1h 26m +60°.2, (Ruchbah, 'the knee'), mag. 2.7, is a blue-white star 99 l.y. away. It is an eclipsing binary of the Algol type that varies by about 0.1 mag. with a period of 2 years 1 month.

ε (epsilon) Cas, 1h 54m +63°.7, mag. 3.3, is a blue-white star 442 l.y. away. ▶

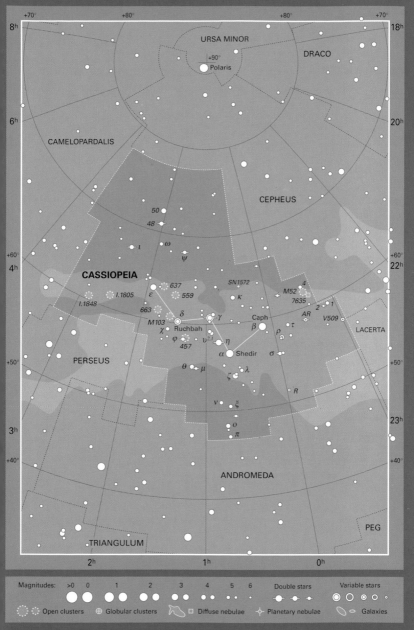

η (eta) Cas, 0h 49m +57°.8, 19 l.y. away, is a beautiful double star with yellow and red components of mags. 3.5 and 7.5, visible in small telescopes. They form a true binary with a period of 480 years.

ι (iota) Cas, 2h 29m +67°.4, 142 l.y. away, is a mag. 4.5 white star with a wide mag. 8.4 companion visible through a 60-mm telescope. With an aperture of 100 mm and high magnification, the brighter star is seen to have a closer mag. 6.9 yellow companion, making this an impressive triple.

ρ (rho) Cas, 23h 54m +57°.5, is a yellow-white supergiant, one of the most luminous stars known, giving out as much light as half a million Suns. It is a semi-regular pulsating variable, ranging between mag. 4.1 and 6.2 every 320 days or so. Its distance is not known accurately, but is probably over 10,000 l.y.

σ (sigma) Cas, 23h 59m +55°.8, 1500 l.y. away, is a close pair of mags. 5.0 and 7.3 appearing green and blue, in striking contrast to the warmer hues of η Cas. An aperture of 75 mm and high power will split the pair.

ψ (psi) Cas, 1h 26m +68°.1, 193 l.y. away, is a mag. 4.7 orange giant with a wide 9th-mag. companion visible in a small telescope. High powers reveal that this companion is itself a close binary.

M52 (NGC 7654), 23h 24m +61°.6, is an open cluster of about 100 stars, 5200 l.y. away, visible as a misty patch in binoculars. It is somewhat kidney-shaped, with an 8th-mag. orange star embedded at one edge, like a poorer version of the celebrated Wild Duck Cluster (M11 in Scutum). M52 can be resolved into stars with 75 mm aperture.

M103 (NGC 581), 1h 33m +60°.7, is a small, elongated group of about 25 faint stars, 8200 l.y. away. The brightest apparent member, a double of 7th and 10th mags. near the cluster's northern tip, is in fact a foreground object.

NGC 457, 1h 19m +58°.3, is a loose open cluster of about 80 stars, 10,000 l.y. away, seemingly arranged in chains. The mag. 5.0 white supergiant φ (phi) Cas on its southern outskirts is probably a true member.

NGC 663, 1h 46m +61°.2, is a prominent binocular cluster of about 80 stars appearing half the size of the full Moon, 8200 l.y. away.

CENTAURUS   The Centaur

A large and rich constellation representing a centaur, the mythical beast that was half man, half horse. Reputedly Centaurus depicts the scholarly centaur Chiron, the tutor of many Greek gods and heroes, who was raised to the sky after being accidentally struck by a poisoned arrow from Hercules. A line from α (alpha) through β (beta) Centauri points to Crux, the Southern Cross (see the photograph on page 118). The rapid proper motion of α Centauri is taking it towards β, and about 4000 years from now the two will be little more than ½° (one Moon diameter) apart, making a stunning naked-eye pair. Centaurus is of particular interest because it contains the closest star to the Sun, Proxima Centauri, which is a member of the α Centauri family of three stars. One of the strongest ▶

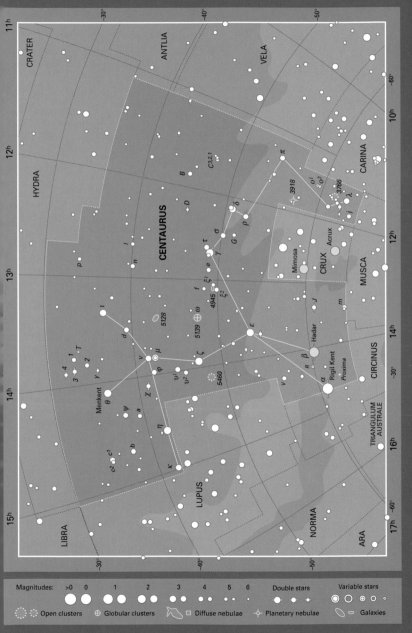

Magnitudes: >0 0 1 2 3 4 5 6   Double stars   Variable stars

Open clusters   ⊕ Globular clusters   Diffuse nebulae   ✧ Planetary nebulae   Galaxies

radio sources in the sky, Centaurus A, is associated with the unusual galaxy NGC 5128. Centaurus lies in a prominent part of the Milky Way and contains more naked-eye stars than any other constellation: 281 brighter than magnitude 6.5 according to the Hipparcos catalogue.

α (alpha) Centauri, 14h 40m −60°.8, (Rigil Kentaurus, abbreviated as Rigil Kent, 'foot of the centaur', or Toliman) lies 4.4 l.y. away. To the naked eye it shines at mag. −0.28, the third-brightest star in the sky. But the smallest of telescopes reveals that it consists of twin yellow stars of mags. −0.01 and 1.35; the brighter of these is very similar to the Sun. They orbit each other every 80 years and are always divisible in amateur telescopes, although at their closest, around the years 2037–8, they will need 75 mm aperture. Also associated with α Centauri is an 11th-mag. red dwarf called Proxima Centauri, lying 2° away and therefore not even in the same telescopic field of view (see the finder chart below.) This star is estimated to take as long as a million years to orbit its two brilliant companions. At present, Proxima Centauri is about 0.2 l.y. closer to us than the two other members of α Centauri. Proxima Centauri is a flare star, suddenly increasing in brightness by as much as one magnitude for several minutes at a time.

β (beta) Cen, 14h 04m −60°.4, (Hadar or Agena), mag. 0.6, is a blue giant 525 l.y. away. It is in fact a very close double with a 4th-mag. companion, divisible only in larger apertures.

γ (gamma) Cen, 12h 42m −49°.0, 130 l.y. away, is a close double with blue-white components each of mag. 2.9, orbiting each other every 85 years. Together they shine as a star of mag. 2.2. As seen from Earth they are currently moving together, requiring 220 mm aperture to separate by 2005. At their closest, between 2010 and 2020, amateur telescopes will be unable to split them, but they will become divisible in 220 mm again by 2025, and in 150 mm from 2030.

3 Cen, 13h 52m −33°.0, 298 l.y. away, is a blue-white star of mag. 4.6 with a companion of mag. 6.1, forming a striking pair in small telescopes.

Finder chart for Proxima Centauri, showing its proper motion over 200 years. (Wil Tirion)

ω (omega) Centauri, the finest of all globular star clusters, appears noticeably elliptical in shape. (AURA/NOAO/NSF)

R Cen, 14h 17m −59°.9, is a red giant variable of Mira type, which varies between mags. 5.3 and 11.8 in about 18 months. It lies about 2100 l.y. away.

ω (omega) Cen (NGC 5139), 13h 27m −47°.5, is the largest and brightest globular cluster in the sky – so prominent that it was labelled as a star on early charts. It appears as a hazy star of mag. 3.7 to the naked eye, noticeably elliptical in shape and as large as the full Moon. It is inherently the most luminous of all globulars, with the light output of a million Suns. Small telescopes or even binoculars begin to resolve its outer regions into stars, and it is a showpiece for all apertures. Its brilliance and large apparent size are due in part to its relative closeness, 17,000 l.y., which places it among the nearest globular clusters to us.

NGC 3766, 11h 36m −61°.6, is a naked-eye open cluster of about 100 stars of mag. 7 and fainter, resolvable in binoculars, 6300 l.y. away.

NGC 3918, 11h 50m −57°.2, is an 8th-mag. planetary nebula 2600 l.y. away, discovered by John Herschel and called by him the Blue Planetary. It is similar in appearance to the planet Uranus, but three times the apparent diameter. Its central star, of 11th mag., should be detectable in modest amateur instruments.

NGC 5128, 13h 25m −43°.0, is a peculiar 7th-mag. galaxy known to radio astronomers as Centaurus A. In long-exposure photographs it appears as a giant elliptical galaxy with an encircling band of dust, and is apparently the result of a merger between an elliptical and a spiral galaxy. Under good skies it is visible in binoculars, but at least 100 mm aperture is necessary to trace its outline and the dark bisecting lane of dust. NGC 5128 is 13 million l.y. away.

NGC 5460, 14h 08m −48°.3, is a large, 6th-mag. open cluster of about 40 stars visible in binoculars or small telescopes. It lies 2500 l.y. away.

CEPHEUS

An ancient constellation representing the mythological King Cepheus of Ethiopia, husband of Cassiopeia and father of Andromeda, themselves depicted by nearby constellations. Cepheus is replete with double and variable stars, including the celebrated δ (delta) Cephei, prototype of the Cepheid variables which are used as 'standard candles' for distance-finding in space. This star's fluctuations in light output were discovered in 1784 by the English amateur astronomer John Goodricke.

α (alpha) Cephei, 21h 19m +62°.6, (Alderamin), mag. 2.5, is a white star 49 l.y. away.

β (beta) Cep, 21h 29m +70°.6, (Alfirk, meaning 'flock', i.e. of sheep), 595 l.y. away, is both a double and a variable star. A small telescope shows that this blue giant, of mag. 3.2, has a mag. 7.9 companion. β Cephei is the prototype of a class of pulsating variable stars (also known as β Canis Majoris stars) with periods of a few hours and tiny brightness fluctuations. Over a period of 4.6 hours or so β Cephei varies by 0.1 mag., an amount indistinguishable to the naked eye but detectable by sensitive instruments.

γ (gamma) Cep, 23h 39m +77°.6, (Errai, 'the shepherd'), mag. 3.2, is an orange star 45 l.y. away.

δ (delta) Cep, 22h 29m +58°.4, 980 l.y. away, is a famous pulsating variable star, the prototype of the classic Cepheid variables. This yellow supergiant varies between mags. 3.5 and 4.4 in 5 days 9 hours, changing in size between about 40 and 46 times the Sun's diameter as it does so. Less well known is that δ Cephei is also an attractive double star for binoculars or the smallest telescopes, with a wide, bluish mag. 6.3 companion.

μ (mu) Cep, 21h 44m +58°.8, 2800 l.y. away, is a famous red star, called the Garnet Star by William Herschel because of its striking tint, which is notable in binoculars. μ Cephei is a red supergiant, a prominent example of a class of variable stars known as semi-regular variables. It varies between mags. 3.4 and 5.1 with a period of around 2 years.

ξ (xi) Cep, 22h 04m +64°.6, 102 l.y. away, is a double star of mags. 4.4 and 6.5 visible in small telescopes. The components are blue-white and yellow and form a true binary with an estimated orbital period of nearly 4000 years.

o (omicron) Cep, 23h 19m +68°.1, 211 l.y. away, is an orange giant of mag. 4.9 with a close mag. 7.1 companion for telescopes of 60 mm aperture and above. They orbit each other every 800 years.

T Cep, 21h 10m +68°.5, 685 l.y. away, is a red giant variable of the Mira type, about 500 times the diameter of the Sun, ranging between mags. 5.2 and 11.3 in around 13 months.

VV Cep, 21h 57m +63°.6, is an enormous red supergiant that varies semi-regularly between mags. 4.8 and 5.4. It is also an eclipsing binary with the un-usually long period of 20.3 years, but the light dip during eclipse is too slight to be noticeable to the naked eye. Its estimated distance is more than 2000 l.y. and its diameter is probably over 1000 Suns, one of the largest stars known.

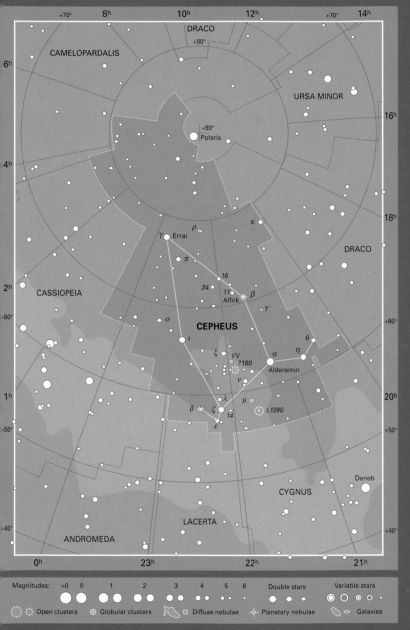

Magnitudes: >0   0   1   2   3   4   5   6   Double stars   Variable stars

Open clusters   Globular clusters   Diffuse nebulae   Planetary nebulae   Galaxies

CETUS   The Whale

An ancient constellation depicting the sea monster that threatened to devour Andromeda before she was rescued by Perseus. In the sky, Cetus is found basking on the banks of Eridanus, the River. The constellation is large but not prominent; nevertheless it contains several stars of particular interest, notably o (omicron) Ceti and τ (tau) Ceti. One faint but famous star is UV Ceti, position 1h 38.8m, −17° 57′, which consists of a pair of 13th-mag. red dwarfs 8.7 l.y. away, one of which is the prototype of a class of erratic variables known as flare stars; these are red dwarfs that undergo sudden increases in light output lasting only a few minutes. The outbursts of the flare star component of UV Ceti can take it from its normal level of 13th mag. to as bright as 7th mag.

▶

Finder chart for the variable star Mira, also known as o (omicron) Ceti. The numbers against the surrounding stars are their magnitudes with the decimal points omitted. This convention prevents confusion between faint stars and decimal points. The magnitude of Mira may be estimated by comparison with these stars. (Wil Tirion)

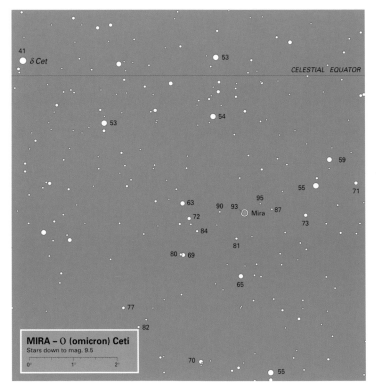

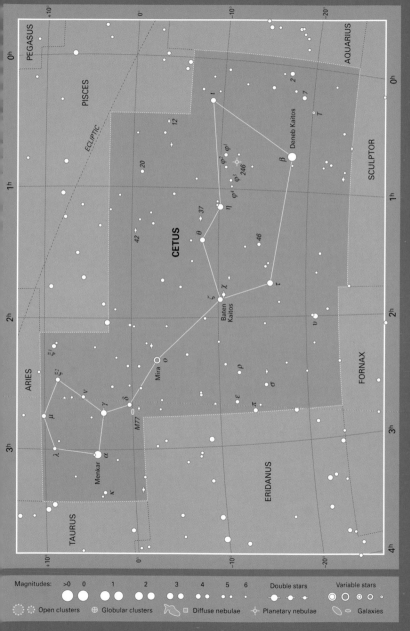

Magnitudes:	>0	0	1	2	3	4	5	6	Double stars	Variable stars

Open clusters ⊕ Globular clusters Diffuse nebulae ✧ Planetary nebulae Galaxies

α (alpha) Ceti, 3h 02m +4°.1, (Menkar, 'nose'), mag. 2.5, is a red giant star 220 l.y. away. Binoculars show a wide mag. 5.6 blue-white companion, 93 Ceti, which is unrelated, lying twice as far away.

β (beta) Cet, 0h 44m −18°.0, (Deneb Kaitos, 'tail of the whale', or Diphda), mag. 2.0, the brightest star of the constellation, is an orange giant 96 l.y. away.

γ (gamma) Cet, 2h 43m +3°.2, 82 l.y. away, is a close double star needing telescopes of at least 60 mm aperture and high power to split. The stars are of mags. 3.5 and 6.6, the colours yellow and bluish.

o (omicron) Cet, 2h 19m −3°.0, (Mira, 'the amazing one'), 420 l.y. away, is the prototype of a famous class of red giant long-period variable stars. Mira itself varies between about 3rd and 9th mags. (although it can become as bright as 2nd mag.) in an average of 332 days, changing in diameter from about 400 to 500 times the size of the Sun as it does so. Mira's light variations were first noted in 1596 by the Dutch astronomer David Fabricius, making it the first variable star (other than novae) to be discovered. (See the finder chart on page 114.)

τ (tau) Cet, 1h 44m −15°.9, mag. 3.5, a yellow main-sequence star, is one of the nearest stars to us, lying 11.9 l.y. away. Its main claim to fame is that, of all the nearby single stars, it is the one most like the Sun, although we do not yet know whether it also has planets.

M77 (NGC 1068), 2h 43m −0°.0, is a small, softly glowing 9th-mag. face-on spiral galaxy with a 10th-mag. starlike nucleus. Apertures of 100 mm show the brighter patches in the spiral arms. M77 is the brightest example of a Seyfert galaxy, a close relative of the quasars, and it is a radio source. It lies roughly 50 million l.y. away. (See photograph on page 288.)

CHAMAELEON   The Chameleon

A faint and unremarkable constellation, introduced into the southern skies by the Dutch navigators Pieter Dirkszoon Keyser and Frederick de Houtman at the end of the 16th century.

α (alpha) Chamaeleontis, 8h 19m −76°.9, mag. 4.1, is a white star 63 l.y. away.

β (beta) Cha, 12h 18m −79°.3, mag. 4.2, is a blue-white star 271 l.y. away.

γ (gamma) Cha, 10h 35m −78°.6, mag. 4.1, is a red giant 413 l.y. away.

δ¹ δ² (delta¹ delta²) Cha, 10h 45m −80°.5, consists of a wide pair of unrelated stars, both clearly seen in binoculars: δ¹ Cha, mag. 5.5, is an orange giant 354 l.y. away and δ² Cha, mag. 4.4, is a blue star 364 l.y. away.

NGC 3195, 10h 09m −80°.9, is a faint planetary nebula of similar apparent size to the planet Jupiter, needing at least 100 mm aperture to be seen well.

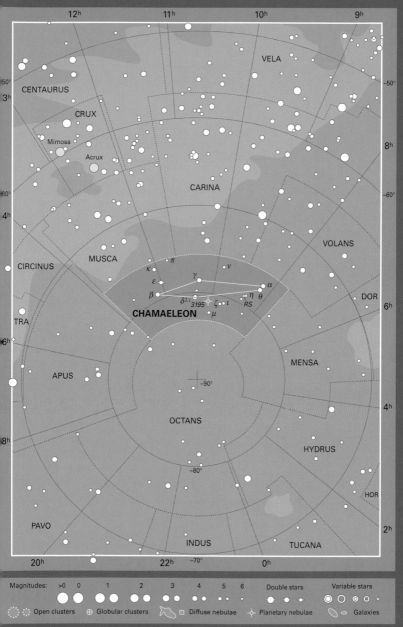

CIRCINUS   The Compasses

Another of the small and obscure southern constellations introduced in 1756 by the French astronomer Nicolas Louis de Lacaille. It represents a pair of compasses as used by surveyors, and is appropriately placed in the sky next to Norma, the Set Square. It is overshadowed by the brilliance of neighbouring Centaurus.

α (alpha) Circini, 14h 43m −65°.0, 53 l.y. away, is a white main-sequence star of mag. 3.2 with a mag. 8.5 companion easily visible in small telescopes.

γ (gamma) Cir, 15h 23m −59°.3, 500 l.y. away, consists of a very close pair of blue and yellow stars of mags. 5.1 and 5.5, requiring an aperture of at least 150 mm and high magnification to see separately. Their orbital period is calculated to be about 2000 years.

Circinus is most easily found by reference to the prominent stars Alpha and Beta Centauri (seen here left of centre), which point towards the familiar figure of Crux, the Southern Cross, at right. Alpha Circini is the star below and to the right of Alpha Centauri. (Robin Scagell)

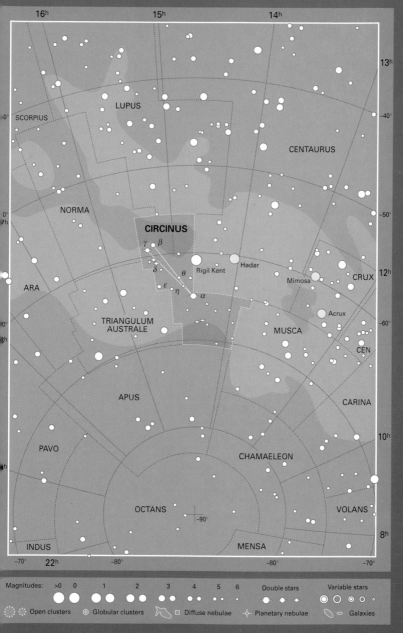

COLUMBA   The Dove

A constellation representing the dove that followed Noah's Ark, or possibly the dove that the Argonauts sent ahead to help them pass safely between the Symplegades, the Clashing Rocks, at the mouth of the Black Sea; fittingly, Columba is placed next to Puppis, the stern of the ship Argo. Columba originated in 1592 when the Dutchman Petrus Plancius formed it from some stars adjacent to Canis Major that had not previously been part of any constellation. It contains little of interest for amateur telescopes.

α (alpha) Columbae, 5h 40m −34°.1, (Phact, 'ring dove'), mag. 2.7, is a blue-white star 268 l.y. away.

β (beta) Col, 5h 51m −35°.8, (Wazn), mag. 3.1, is an orange giant 86 l.y. away.

NGC 1851, 5h 14m −40°.1, is a 7th-mag. globular cluster, visible in small telescopes but requiring moderate apertures to resolve its brightest stars. It lies 35,000 l.y. away.

In Coma Berenices is the classically elegant NGC 4565, a spiral galaxy seen exactly edge-on (see page 124) Note the central bulge of stars and the dark lane of dust in the galaxy's plane. (AURA/NOAO/NSF)

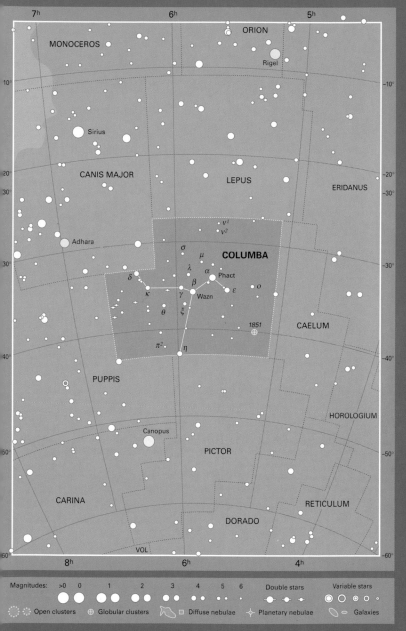

Magnitudes: >0 0 1 2 3 4 5 6   Double stars   Variable stars

Open clusters   Globular clusters   Diffuse nebulae   Planetary nebulae   Galaxies

COMA BERENICES   Berenice's Hair

This faint constellation represents the flowing locks of Queen Berenice of Egypt, who cut off her hair in gratitude to the gods for the safe return of her husband Ptolemy III Euergetes from battle. Although the legend dates from Greek times, this group of stars was regarded as part of Leo until 1551 when the Dutch cartographer Gerardus Mercator made them into a separate constellation. The main part of the queen's severed tresses is represented by the extensive Coma Star Cluster. Coma Berenices also contains another type of cluster – a cluster of galaxies. The Coma Cluster of galaxies (not to be confused with the Coma Star Cluster) lies about 280 million l.y. away, so its members are too faint for all but the largest amateur telescopes. But the constellation also contains some brighter galaxies, members of the nearer Virgo Cluster, the brightest of which are visible in amateur telescopes. The north pole of our Galaxy lies in Coma Berenices.

α (alpha) Comae Berenices, 13h 10m +17°.5, (Diadem), mag. 4.3, is a tight binary 47 l.y. away, consisting of twin yellow-white stars of mag. 5.1 that orbit each other every 26 years. Even at their widest, around the year 2010, they are at the limit of resolution of a 220-mm telescope.

β (beta) Com, 13h 12m +27°.9, mag. 4.2, is a yellow main-sequence star 30 l.y. away.

γ (gamma) Com, 12h 27m +28°.3, mag. 4.4, is an orange giant 170 l.y. away, which appears to be a member of the Coma Star Cluster but is actually a foreground star.

24 Com, 12h 35m +18°.4, is a beautiful coloured double star for small telescopes, consisting of an orange giant of mag. 5.0, 610 l.y. away, and an unrelated blue-white companion of mag 6.6.

35 Com, 12h 53m +21°.2, 324 l.y. away, is a tight binary star with yellow and white components of mags. 5.1 and 7.2 orbiting every 360 years and divisible in 150-mm apertures. Small telescopes show a wider 9th-mag. companion.

FS Com, 13h 06m +22°.6, 572 l.y. away, is a red giant that varies semi-regularly between mags. 5.3 and 6.1 every two months or so.

Coma Star Cluster (Melotte 111), 12h 25m +26°, is a scattered group of about 50 stars best seen in binoculars. The cluster's brightest members, of 5th mag., form a noticeable V-shape extending for several degrees south of γ (gamma) Com, which is not actually a member but a foreground star. The brightest true member seems to be 12 Com, mag. 4.8. The distance to the cluster's centre is 288 light years.

M53 (NGC 5024), 13h 13m +18°.2, is an 8th-mag. globular cluster 56,000 l.y. away, visible in small telescopes as a rounded, hazy patch.

M64 (NGC 4826), 12h 57m +21°.7, is a famous spiral galaxy, called the Black Eye Galaxy because of a dark cloud of dust silhouetted against its nucleus. This dark dust lane shows up well in apertures above 150 mm; observers with smaller instruments must content themselves simply with locating this 9th-mag. galaxy some 15 million l.y. away, closer than the Virgo Cluster and not a member of it. ▶

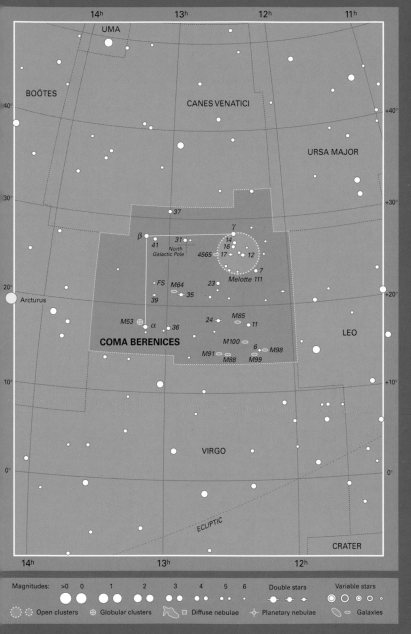

M85 (NGC 4382), 12h 25m +18°.2, is a 9th-mag. elliptical galaxy in the Virgo Cluster, 55 million l.y. away. Small telescopes show a brighter, starlike centre.

M88 (NGC 4501), 12h 32m +14°.4, is a 10th-mag. spiral galaxy in the Virgo Cluster, 55 million l.y. away. It is presented at an angle to us, so that it appears elliptical.

M99 (NGC 4254), 12h 19m +14°.6, is a 10th-mag. spiral galaxy 55 million l.y. away in the Virgo Cluster, presented face-on so that it appears almost circular.

M100 (NGC 4321), 12h 23m +15°.8, is a 9th-mag. Virgo Cluster spiral seen face-on, similar to M99 but larger. The Hubble Space Telescope has measured its distance accurately at 56 million l.y.

NGC 4565, 12h 36m +26°.0, is a 10th-mag. spiral galaxy seen edge-on. It is the most famous of the edge-on spirals, and is pictured on page 120. Apertures of 100 mm show its cigar-shaped body with a central bulge and starlike core, but larger instruments are needed to trace the dark band (actually a dust lane) that splits it lengthwise. NGC 4565 is not a member of the Virgo Cluster but closer, about 20 million l.y. away.

CORONA AUSTRALIS   The Southern Crown

The southern counterpart of the Northern Crown (Corona Borealis). Corona Australis has been known since the time of the Greek astronomer Ptolemy in the 2nd century AD who visualized it not as a crown but as a wreath. In one legend it represents the crown placed in the sky by Bacchus when he rescued his dead mother from the Underworld; alternatively, it could simply have slipped from the head of the centaur Sagittarius, at whose feet it lies. Although faint, it is a distinctive figure, and is situated on the edge of the Milky Way.

α (alpha) Coronae Australis, 19h 09m −37°.9, mag. 4.1, is a blue-white main-sequence star 130 l.y. away.

β (beta) CrA, 19h 10m −39°.3, mag. 4.1, is an orange giant 510 l.y. away.

γ (gamma) CrA, 19h 06m −37°.1, 58 l.y. away, consists of a pair of near-identical yellow-white stars, mags. 4.9 and 5.0, orbiting every 122 years, forming a tight double for small telescopes. The stars were closest together during the 1990s, but since 2000 have become progressively easier to split with 100 mm aperture.

κ (kappa) CrA, 18h 33m −38°.6, is a pair of unrelated blue-white stars of mags. 5.7 and 6.3, distances about 1700 and 490 l.y., easily divisible in small telescopes.

λ (lambda) CrA, 18h 44m −38°.3, 202 l.y. away, is a mag. 5.1 blue-white star with a wide mag. 9.7 companion, visible in small telescopes.

NGC 6541, 18h 08m −43°.7, is a 7th-mag. globular cluster 22,000 l.y. away, visible in binoculars or small telescopes.

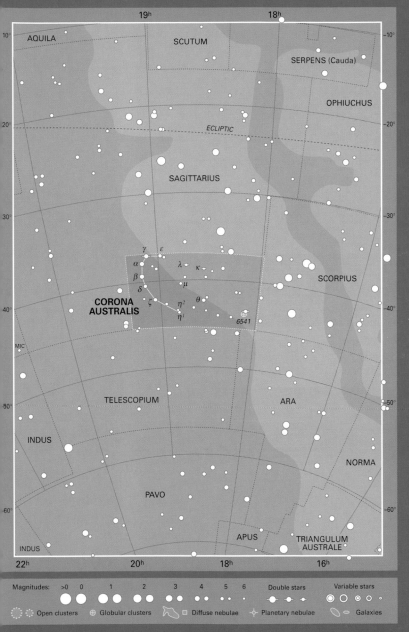

AQUILA

SCUTUM

SERPENS (Cauda)

OPHIUCHUS

ECLIPTIC

SAGITTARIUS

SCORPIUS

γ ε
α λ ← κ →
β μ
δ θ
CORONA
AUSTRALIS
ζ η²
η¹
6541

MIC

TELESCOPIUM

ARA

INDUS

NORMA

PAVO

INDUS

APUS

TRIANGULUM
AUSTRALE

Magnitudes: >0 0 1 2 3 4 5 6   Double stars   Variable stars

Open clusters ⊕ Globular clusters ⬡ Diffuse nebulae ✧ Planetary nebulae ⬯ Galaxies

CORONA BOREALIS   The Northern Crown

An ancient constellation, representing the jewelled crown worn by Ariadne when she married Bacchus and cast by him into the sky to mark the happy event. It consists of an arc of seven stars, all but one of 4th magnitude. The exception is 2nd-magnitude Alphekka (derived from the Arabic name for the constellation), which is set in the crown like a central gem, as its alternative name Gemma implies. Corona Borealis contains a famous cluster of about 400 galaxies more than 1000 million light years away. Being so very distant, the galaxies are no brighter than 16th magnitude and are thus far beyond the reach of amateur telescopes.

α (alpha) Coronae Borealis, 15h 35m +26°.7, (Alphekka or Gemma), mag. 2.2, is a blue-white main-sequence star 75 l.y. distant. It is an eclipsing binary of the Algol type, but its variation every 17.4 days is only 0.1 mag., too slight to be noticeable to the naked eye.

ζ (zeta) CrB, 15h 39m +36°.6, 470 l.y. away, is a pair of blue-white stars, mags. 5.0 and 6.0, visible in small telescopes.

ν¹ ν² (nu¹ nu²) CrB, 16h 22m +33°.8, is a wide binocular pair of red and orange giants of mags. 5.2 and 5.4, both about 550 l.y. from us but moving in different directions, so probably not a true binary.

σ (sigma) CrB, 16h 15m +33°.9, 71 l.y. away, is a pair of yellow stars for small telescopes, of mags. 5.6 and 6.6. They form a genuine binary with an estimated orbital period of about 1000 years.

R CrB, 15h 49m +28°.2, is a remarkable yellow supergiant star lying within the arc of the Crown, halfway between the stars α (alpha) and ι (iota). It usually appears about 6th mag., but occasionally and unpredictably drops in a matter of weeks to as faint as mag. 15, from where it may take many months to regain its former brightness. Recent catastrophic declines in the star's brilliance were seen in 1962, 1972 and 1977, although less extreme fades are more frequent and another could occur at any time. These sudden dips in the light of R Coronae Borealis are believed to be due to the accumulation of carbon particles (i.e. soot) in its atmosphere. R Coronae Borealis is estimated to lie over 7000 l.y. away.

T CrB, 16h 00m +25°.9, is another spectacular variable star, known as the Blaze Star, which performs in almost the opposite way to R Coronae Borealis. It is a recurrent nova, usually slumbering at around mag. 11 but which can suddenly and unpredictably brighten to mag. 2. Its last recorded outburst was in 1946, and the previous one was 80 years before that. It is not known when it may erupt again.

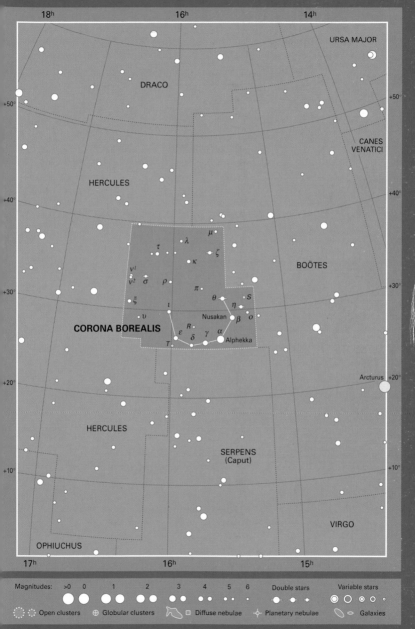

CORVUS   The Crow

In Greek legend, Corvus is associated with the neighbouring constellations Crater, the Cup, and Hydra, the Water Snake. The crow is said to have been sent by Apollo to fetch water in a cup, but along the way it dallied to eat figs. When the crow returned to Apollo it carried the water snake in its claws, claiming that this creature had been blocking the spring and was the cause of its delay. Apollo, seeing through the lie, banished the the trio to the sky, where the Crow and the Cup lie on the back of Hydra. For its misdeed the crow was condemned to suffer from eternal thirst, which is why crows croak so harshly; in the sky, the cup is just out of the thirsty crow's reach. In another legend, a snow-white crow brought Apollo the bad news that his lover Coronis had been unfaithful to him. In his anger, Apollo turned the crow black. Apollo and crows are closely linked in legend, for during the war waged by the giants on the gods, Apollo turned himself into a crow.

α (alpha) Corvi, 12h 08m −24°.7, (Alchiba), mag. 4.0, is a white star 48 l.y. away.

β (beta) Crv, 12h 34m −23°.4, mag. 2.7, is a yellow giant star 140 l.y. away.

γ (gamma) Crv, 12h 16m −17°.5, (Gienah, 'wing'), mag. 2.6, the brightest star in the constellation, is a blue-white giant 165 l.y. away.

δ (delta) Crv, 12h 30m −16°.5, (Algorab, 'the raven'), is a wide double star for small telescopes. The brighter component, visible to the naked eye, is a mag. 2.9 blue-white star 88 l.y. away, which is accompanied by a 9th-mag. star often described as purplish in colour.   ▶

Just over the northern border of Corvus into Virgo lies the Sombrero Galaxy, M104, a spiral galaxy with a dark central lane of dust. See page 258. (ESO)

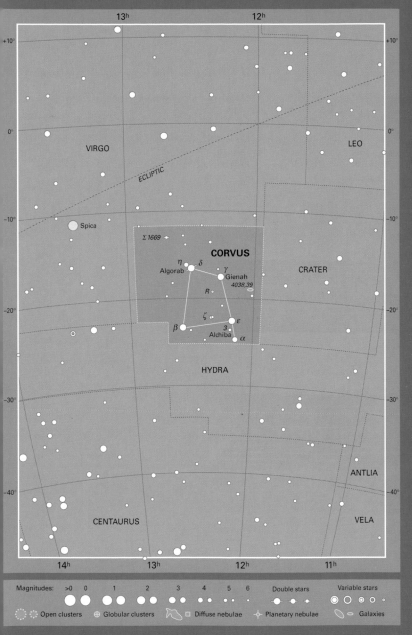

Σ 1669 (Struve 1669), 12h 41m −13°.0, is a neat pair of white stars 280 l.y. away, appearing to the naked eye as a single star of mag. 5.2 but divisible by small telescopes into near-identical components of mags. 5.9 and 6.0.

NGC 4038–9, 12h 02m −18°.9, the Antennae, are a pair of spiral galaxies of 10th mag. that have recently collided. Streams of gas and stars thrown off in the collision, visible on long-exposure photographs like that below, form the antennae that give the galaxies their popular name. They lie 63 million l.y. away.

NGC 4038–9, the peculiar pair of interacting galaxies known as the Antennae, lie in Corvus near the border with Crater. (François Schweizer)

CRATER   The Cup

An ancient constellation representing the chalice of Apollo; in legend, it is associated with neighbouring Corvus (see page 128). Crater contains no objects of particular interest.

α (alpha) Crateris, 11h 00m −18°.3, (Alkes, 'the cup'), mag. 4.1, is an orange giant star 174 l.y. away.

β (beta) Crt, 11h 12m −22°.8, mag. 4.5, is a blue-white star 266 l.y. away.

γ (gamma) Crt, 11h 25m −17°.7, mag. 4.1, is a white star 84 l.y. away. It has a mag. 9.6 companion visible in small telescopes.

δ (delta) Crt, 11h 19m −14°.8, mag. 3.6, the constellation's brightest star, is an orange giant lying 195 l.y. away.

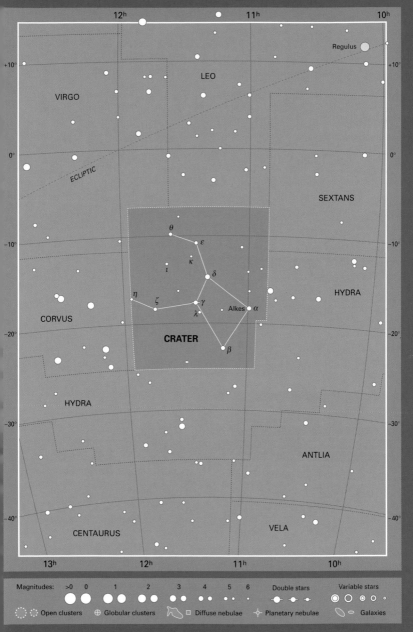

CRUX   The Southern Cross

The smallest constellation in the sky, but one of the most celebrated and distinctive. It was formed from some of the stars of Centaurus by various seamen and astronomers in the 16th century. Its long axis, from γ (gamma) via α (alpha) Crucis, points towards the south celestial pole. Crux lies in a dense and brilliant part of the Milky Way, and the famous dark nebula known as the Coalsack is particularly striking in silhouette against the starry background. Crux, like Centaurus, was visible from the Mediterranean area in ancient times, so its stars were known to Greek astronomers; the effects of precession have since carried it below the horizon from such northerly latitudes. α (alpha) Crucis is the most southerly first-magnitude star. α (alpha) and β (beta) Crucis are, by a small margin, the bluest first-magnitude stars.

α (alpha) Crucis, 12h 27m −63°.1, (Acrux), 321 l.y. away, appears to the naked eye as a bluish star of mag. 0.8, but a small telescope divides it into a sparkling double with components of mags. 1.3 and 1.8. There is also a wider 5th-mag. companion that can be seen with binoculars.

β (beta) Cru, 12h 48m −59°.7, (Mimosa), mag. 1.3, is a blue giant 353 l.y. away. It is a variable star of the β Cephei type, fluctuating by less than 0.1 mag. every 5 hours, too small to be noticeable to the eye.

γ (gamma) Cru, 12h 31m −57°.1, (Gacrux), mag. 1.6, is a red giant star 88 l.y. away. There is a very wide mag. 6.5 unrelated companion, three times as distant, visible in binoculars.

δ (delta) Cru, 12h 15m −58°.7, mag. 2.8, the faintest of the four main stars in the Cross, is a blue-white star 364 l.y. away.

ε (epsilon) Cru, 12h 21m −60°.4, mag. 3.6, is an orange giant 228 l.y. away.

ι (iota) Cru, 12h 46m −61°.0, 125 l.y. away, is an orange giant of mag. 4.7 with a mag. 9.5 companion visible in small telescopes.

μ (mu) Cru, 12h 55m −57°.2, is a wide pair of blue-white stars of mags. 4.0 and 5.1, splittable by the smallest telescopes or even good binoculars. They both lie about 370 l.y away.

NGC 4755, 12h 54m −60°.3, the Jewel Box or κ (kappa) Crucis Cluster, is one of the finest open clusters in the sky, visible to the naked eye as a hazy 4th-mag. star. Binoculars resolve the brightest individual members, which are blue supergiants of 6th and 7th mag.; three of these form a chain across the cluster like a mini Orion's belt, with an 8th-mag. red supergiant next to the middle one. The star κ Crucis itself, of mag. 5.9, is the most southerly member of this chain. A small telescope brings at least 50 stars into view. John Herschel gave this cluster its popular name when he likened it to a piece of multicoloured jewellery. The distance of the Jewel Box is just under 5000 l.y.

The Coalsack Nebula is a dark, pear-shaped dust cloud silhouetted against the Milky Way, covering nearly 7° by 5° of sky and spilling over into neighbouring Centaurus and Musca. It lies an estimated 600 l.y. away. The Coalsack has no NGC or other identification number.

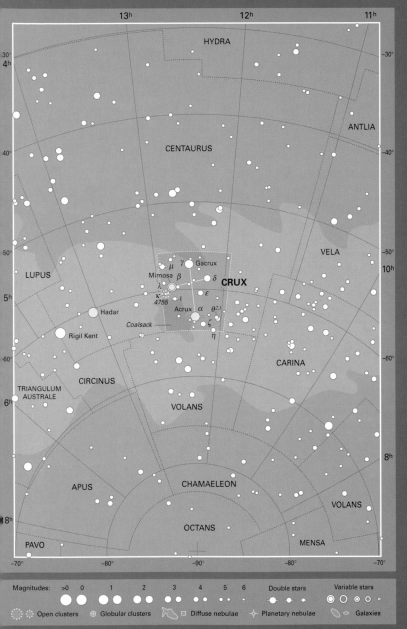

| Magnitudes: | >0 | 0 | 1 | 2 | 3 | 4 | 5 | 6 | Double stars | Variable stars |

- ◌ ◌ Open clusters
- ⊕ Globular clusters
- ◌ ▢ Diffuse nebulae
- ✦ Planetary nebulae
- ◌ ◌ Galaxies

CYGNUS The Swan

Cygnus represents a swan flying down the Milky Way. In Greek mythology, the swan was the guise in which the god Zeus visited Leda, wife of King Tyndareus of Sparta; the result of their union was Pollux, one of the heavenly twins. The flying swan's tail is marked by the star Deneb, its beak by Albireo, and its wings by δ (delta) and ε (epsilon) Cygni. These stars form a distinctive cross-shape, so the constellation is also referred to as the Northern Cross; it is far larger than the Southern Cross. Cygnus lies in a rich part of the Milky Way, which is split here by a dark lane of dust known as the Cygnus Rift or the Northern Coalsack. Deneb, the constellation's brightest star, forms one corner of the Summer Triangle, completed by Altair and Vega.

Among the fascinating objects in Cygnus is an X-ray source called Cygnus X-1, believed to be a black hole orbiting a 9th-mag. blue supergiant star; it lies at 19h 58.4m, +35° 12′, near η (eta) Cygni. Near γ (gamma) Cygni, at 19h 59.5m, +40° 44′, is Cygnus A, a powerful radio source, believed to be two distant galaxies in collision. Long-exposure photographs of the region between ε (epsilon) Cygni and the border with Vulpecula reveal beautiful swirls of gas known as the Veil Nebula, the brightest part of which, NGC 6992, is detectable in amateur instruments (see page 137).

α (alpha) Cygni, 20h 41m +45°.3, (Deneb, 'tail'), mag. 1.2, is a blue-white supergiant star about 3200 l.y. away.

β (beta) Cyg, 19h 31m +28°.0, (Albireo), 380 l.y. away, is one of the sky's showpiece doubles. It consists of gloriously contrasting amber and blue-green stars, like a celestial traffic light. The brighter star, of mag. 3.1, is an orange giant, and its blue-green companion is mag. 5.1. They can be separated through good binoculars and are a beautiful sight in any amateur telescope.

γ (gamma) Cyg, 20h 22m +40°.3, (Sadr, 'breast'), mag. 2.2, is a yellow-white supergiant about 1500 l.y. away.

δ (delta) Cyg, 19h 45m +45°.1, 171 l.y. away, is a blue-white giant of mag. 2.9 with a close mag. 6.6 companion, visible in a telescope of 100 mm aperture or above at high magnification. The stars have an orbital period of over 800 years.

ε (epsilon) Cyg, 20h 46m +34°.0, (Gienah, 'wing'), mag. 2.5, is an orange giant 72 l.y. away.

μ (mu) Cyg, 21h 44m +28°.7, 73 l.y. away, is a pair of white stars of mags. 4.8 and 6.2 orbiting each other about every 790 years. They are currently closing slowly and should remain divisible in 100 mm apertures until about 2020, although 150 mm will probably be necessary when they are closest, between 2043 and 2050. A wide mag. 6.9 binocular companion is an unrelated background star.

o[1] (omicron[1]) Cyg, 20h 14m +46°.7, also known as 31 Cygni, forms perhaps the most beautiful binocular double in the heavens with 30 Cygni. The stars are orange and turquoise, mags. 3.8 and 4.8, 1400 l.y. and 720 l.y. away, like a wider version of Albireo. A small telescope, or binoculars held steadily, shows a closer blue companion of mag. 7.0 to the brighter (orange giant) star. ▶

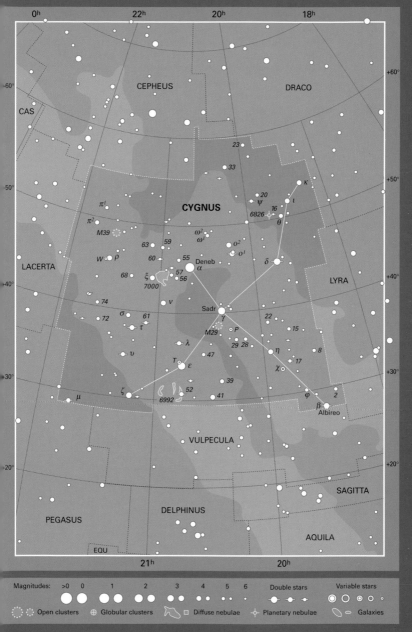

Magnitudes: >0   0   1   2   3   4   5   6   Double stars   Variable stars

Open clusters   Globular clusters   Diffuse nebulae   Planetary nebulae   Galaxies

χ (chi) Cyg, 19h 51m +32°.9, 350 l.y. away, is a red giant long-period variable of Mira type that varies every 400 days or so. At best it reaches mag. 3.3, brighter than any other star of its type except Mira itself, although R Hydrae can rival it. At its faintest χ Cyg drops to 14th mag. Its diameter is about 300 Suns.

ψ (psi) Cyg, 19h 56m +52°.4, 289 l.y. away, is a pair of white stars of mags. 5.0 and 7.5 divisible in small to moderate apertures.

61 Cyg, 21h 07m +38°.7, 11.4 l.y. away, is a showpiece pair of orange dwarf stars of mags. 5.2 and 6.1, orbiting each other in about 650 years, divisible in a small telescope or even binoculars. In addition to being among the closest stars to Earth, 61 Cygni was the first star to have its parallax measured, by the German astronomer Friedrich Wilhelm Bessel in 1838.

P Cyg, 20h 18m +38°.0, is an erratically variable blue supergiant that normally resides around 5th mag., although in the year 1600 it reached a peak of 3rd mag. Evidently the star is so large and luminous that it is close to instability. It has brightened gradually since the 18th century as it evolves into a red supergiant. Its distance is uncertain, but is probably several thousand light years.

W Cyg, 21h 36m +45°.4, 618 l.y. away, is a red giant that varies semi-regularly between 5th and 8th mags. with a rough period of 130 days.

M39 (NGC 7092), 21h 32m +48°.4, is a large, loose cluster of about 30 stars of 7th mag. and fainter, arranged in a triangle. It is visible in binoculars, and is even detectable with the naked eye under ideal conditions. It lies 950 l.y. away.

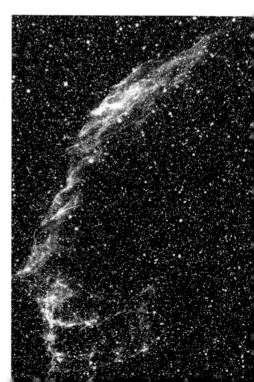

Drifting apart over thousands of years, the delicate traceries of the Veil Nebula are the tenuous remains of a star that shattered itself in a supernova explosion some 30,000 years ago. This is its brightest part, NGC 6992, which can be glimpsed in binoculars under clear skies. This arc of gas extends across more than two Moon diameters of sky, equivalent to a true length of nearly 50 light years. (Nigel Sharp, REU program/ AURA/NOAO/NSF)

The North America Nebula, NGC 7000, is a glowing cloud of gas shaped like the continent of North America, its particularly distinctive Gulf of Mexico formed by an obscuring cloud of dark dust. It lies near the star ξ (xi) Cygni, an orange supergiant of mag. 3.7 seen at left. (Philip Perkins)

NGC 6826, 19h 45m +50°.5, is an 8th-mag. planetary nebula about 3200 l.y. away, known as the Blinking Planetary because it appears to blink on and off. It is visible as a pale blue disk in 75-mm telescopes, but 150 mm is needed to show it to advantage. At its centre is a 10th-mag. star. Looking alternately at this star and away again produces the 'blinking' effect. NGC 6826 lies less than 1° from the wide double star 16 Cygni, which consists of two 6th-mag. creamy-white stars 70 l.y. away.

NGC 6992, 20h 56m +31°.7, is the brightest part of the Veil Nebula, the remnant of a supernova explosion about 30,000 years ago. Under ideal conditions NGC 6992 can be seen in binoculars as a faint arc. Another part of the nebula, NGC 6960, can be located in wide-angle telescopes with low power near the 4th-mag. star 52 Cyg, but the complete Veil Nebula, also known as the Cygnus Loop, is apparent only on long-exposure photographs, spanning nearly 3° of sky. Its estimated distance is about 2000 l.y.

NGC 7000, 20h 59m +44°.3, the North America Nebula, can be seen as a hook-shaped brightening in the Milky Way with the naked eye or binoculars in good skies. Despite its large size, 2° at its widest, it can be difficult to detect because of its low surface brightness. Long-exposure photographs such as the one above clearly show its shape, like the continent of North America. The nebula is about 1500 l.y. away, half the distance to the star Deneb. The nebula is believed to be lit up by an extremely hot 6th-mag. star that lies within it.

DELPHINUS   The Dolphin

A constellation which originated in Greek times, celebrating the long-standing relationship between humans and this intelligent sea creature. In legend, dolphins were the messengers of the sea-god Poseidon. They were credited with saving the life of Arion, a musician and poet, who was attacked on a ship. In this legend, the nearby constellation Lyra represents Arion's lyre. Delphinus has a distinctive shape, its four main stars forming a diamond known as Job's Coffin. Its two brightest stars are called Sualocin and Rotanev, which backwards read Nicolaus Venator, the Latinized form of the name Niccolò Cacciatore, who was assistant and successor to the Italian astronomer Giuseppe Piazzi at Palermo Observatory. Like the small neighbouring constellations Vulpecula and Sagitta, Delphinus lies in a rich area of the Milky Way and has become a favourite hunting-ground for novae.

α (alpha) Delphini, 20h 40m +15°.9, (Sualocin), mag. 3.8, is a blue-white main-sequence star 241 l.y. away.

β (beta) Del, 20h 38m +14°.6, (Rotanev), mag. 3.6, the constellation's brightest member, is a white star 97 l.y. away. It is a binary with an orbital period of 27 years, but the components are too close to divide in all but the largest amateur telescopes.

γ (gamma) Del, 20h 47m +16°.1, 102 l.y. away, is a showpiece double star consisting of golden and yellow-white stars of mags. 4.3 and 5.1, neatly separated in a small telescope. In the same telescopic field of view, but about 25 l.y. more distant, lies a faint binary, Σ 2725, consisting of stars of mags. 7.5 and 8.3.

The Naming of Variable Stars

In addition to the normal system of naming stars in a constellation (page 8), stars that vary in brightness have their own system of nomenclature. Stars that were already named when their variability was discovered, such as δ (delta) Cephei, β (beta) Persei or o (omicron) Ceti, retain their existing designation. Other variable stars are denoted by a system of one or two letters or, where that is insufficient, by the letter V and a number. The first nine variables in a constellation are given the letters R to Z. Then double letters are applied, from RR to RZ. Next the lettering runs from SS to SZ, and so on until ZZ is reached. Then the sequence goes from AA to AZ, BB to BZ, ending with QZ, at which point 334 variable stars have been named. (The letter J is omitted.) Further variables are designated V335, V336, and so on. Novae, too, are allocated variable-star designations. Hence the nova that erupted in Delphinus in 1967 became known as HR Delphini.

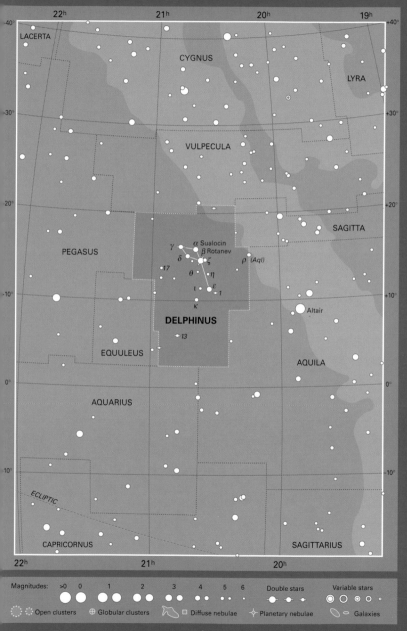

CYGNUS

LACERTA

LYRA

VULPECULA

SAGITTA

γ   α Sualocin
     β Rotanev
δ   ζ   ρ (Aql)
17   θ   η
PEGASUS   ι   ε 1
          κ

DELPHINUS

13

Altair

EQUULEUS

AQUILA

AQUARIUS

-10°

ECLIPTIC

CAPRICORNUS   SAGITTARIUS

Magnitudes: >0   0   1   2   3   4   5   6   Double stars   Variable stars

Open clusters   ⊕ Globular clusters   Diffuse nebulae   Planetary nebulae   Galaxies

DORADO   The Goldfish

This constellation, also known as the Swordfish, was introduced at the end of the 16th century by the Dutch navigators Pieter Dirkszoon Keyser and Frederick de Houtman. Its most notable feature is the Large Magellanic Cloud (LMC), the larger of the two satellite galaxies that accompany our Milky Way. In 1987 the first supernova visible to the naked eye since 1604, Supernova 1987A, erupted in the LMC, at 5h 35m −69°.3.

α (alpha) Doradus, 4h 34m −55°.0, mag. 3.3, is a blue-white star 176 l.y. away.

β (beta) Dor, 5h 34m −62°.5, a yellow-white supergiant 1040 l.y. away, is one of the brightest Cepheid variables, ranging from mag. 3.5 to 4.1 every 9 days 20 hours.

R Dor, 4h 37m −62°.1, 204 l.y. away, is a red giant that varies semi-regularly between mags. 4.8 and 6.6 with a period of around 11 months.

Large Magellanic Cloud (LMC), 5h 24m −69°, is a mini-galaxy 170,000 l.y. away, a satellite of the Milky Way, containing perhaps 10,000 million stars. To the naked eye it appears as a fuzzy elongated patch 6° in diameter, 12 times the apparent size of the Moon; its actual diameter is about 25,000 l.y. Binoculars and telescopes show individual stars, nebulae (notably the Tarantula Nebula, described below) and clusters.

NGC 2070, 5h 39m −69°.1, the Tarantula Nebula, is a looping cloud of hydrogen gas about 1000 l.y. in diameter in the Large Magellanic Cloud. To the naked eye the nebula appears as a fuzzy star, also known as 30 Doradus. The nebula's popular name, the Tarantula, comes from its spider-like shape. At the centre of the nebula is a cluster of supergiant stars, called R136, the light from which makes the nebula glow. The Tarantula is larger and brighter than any nebula in the Milky Way. If it were as close to us as the Orion Nebula, the Tarantula would fill the whole constellation of Orion, and would cast shadows on Earth.

The Large Magellanic Cloud, showing many loops of glowing gas in addition to star swarms. The large knot of bright pink gas at left of centre is the Tarantula Nebula, NGC 2070. (AURA/NOAO/NSF)

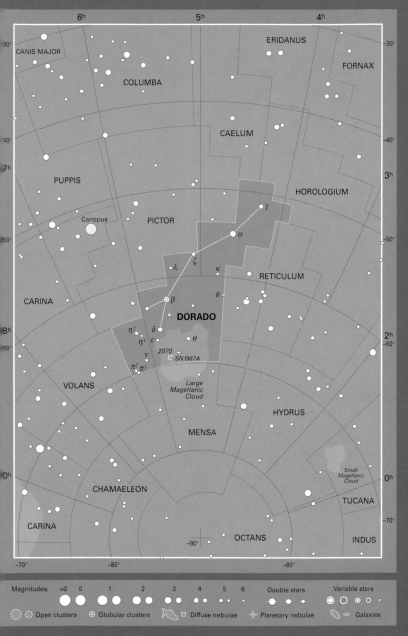

DRACO   The Dragon

Dragons feature in many ancient legends, so it is not surprising to find such a monster in the sky. This one is said to be Ladon, the dragon slain by Hercules as a prelude to stealing the golden apples from the garden of the Hesperides. In the sky, one foot of Hercules rests upon the dragon's head, while the creature's body lies coiled around the north celestial pole. Although Draco is one of the largest and most ancient constellations it is indistinct, with no stars brighter than 2nd magnitude. Draco contains the north pole of the ecliptic, i.e. one of the two points 90° from the plane of the Earth's orbit, at 18h 00m +66½°.

α (alpha) Draconis, 14h 04m +64°.4, (Thuban, 'serpent's head'), mag. 3.7, is a blue-white giant star 309 l.y. away. It was the pole star in about 2800 BC, but has lost that place to Polaris because of the effect of precession (see page 13).

β (beta) Dra, 17h 30m +52°.3, (Rastaban, 'serpent's head'), mag. 2.8, is a yellow giant 362 l.y. distant.

γ (gamma) Dra, 17h 57m +51°.5, (Etamin or Eltanin, 'the serpent'), mag. 2.2, is an orange giant 148 l.y. away, and the brightest star in the constellation. From observations of this star the English astronomer James Bradley discovered the effect known as the aberration of starlight in 1728.

μ (mu) Dra, 17h 05m +54°.5, (Alrakis), 88 l.y. away, is a close double star with matching cream-coloured components of mag. 5.6 and 5.7 which orbit every 670 years. The two stars are currently moving apart and becoming progressively easier to split in small apertures, although high magnification will be needed.

ν (nu) Dra, 17h 32m +55°.2, is a pair of identical white stars of mag. 4.9, easily visible in the smallest of telescopes and regarded as one of the finest binocular pairs. They lie 100 l.y. away.

ο (omicron) Dra, 18h 51m +59°.4, 322 l.y. away, is a mag. 4.6 orange giant star with a mag. 7.8 companion visible in small telescopes.

ψ (psi) Dra, 17h 42m +72°.1, 72 l.y. away, is a mag. 4.6 yellow-white star with a yellow mag. 5.8 companion visible in small telescopes or even binoculars.

16–17 Dra, 16h 36m +52°.9, is a wide pair of blue-white stars of mags. 5.1 and 5.5, 400 l.y. away, easily found in binoculars. Telescopes of 60 mm aperture with high magnification split the brighter star into a binary of mags. 5.4 and 6.5, making this a striking triple system.

39 Dra, 18h 24m +58°.6, 188 l.y away, is an impressive triple system, the two brightest members of which, at mags. 5.0 and 7.4, appear in binoculars as a wide blue-and-yellow pair. A telescope of 60 mm aperture with high magnification reveals that the brighter star has a closer mag. 8.0 companion.

40–41 Dra, 18h 00m +80°.0, is an easy pair of orange dwarf stars for small telescopes, mags. 5.7 and 6.1. They lie about 170 l.y. away.

NGC 6543, 17h 59m +66°.6, is a 9th-mag. planetary nebula 3500 l.y. away, one of the brightest of the class, showing in amateur telescopes as an irregular blue-green disk like an out-of-focus star. It is now known as the Cat's Eye Nebula from its appearance in Hubble Space Telescope photographs (see page 273).

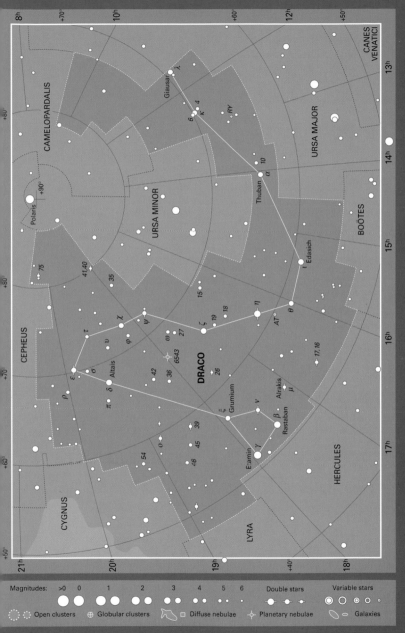

Magnitudes: >0 0 1 2 3 4 5 6   Double stars   Variable stars

Open clusters   Globular clusters   Diffuse nebulae   Planetary nebulae   Galaxies

EQUULEUS   The Little Horse

The second-smallest constellation in the sky, Equuleus seems to have originated with the Greek astronomer Ptolemy in the second century AD. Only the head of the horse is shown, next to the much larger horse Pegasus, and there are no legends that identify it.

α (alpha) Equulei, 21h 16m +5°.2, (Kitalpha, 'the section of the horse'), mag. 3.9, is a yellow giant 186 l.y. away.

γ (gamma) Equ, 21h 10m +10°.1, is a mag. 4.7 white star 115 l.y. away. A mag. 6.1 binocular companion, 6 Equ, is an unrelated background star.

1 Equ, 20h 59m +4°.3, also known as ε (epsilon) Equ, is a triple star 197 l.y. away. In small telescopes it appears as a white-and-yellow pair of mags. 5.4 and 7.4. The brighter component is also a close binary, of mags. 6.0 and 6.3; these stars orbit each other every 101 years. Currently they are closing together, and by 2015 will be too close to separate in amateur telescopes.

NGC 1300 in Eridanus is a barred spiral galaxy with tightly wound arms. (José Alfonso López Aguerri, M. Prieto, C. Muñoz-Tuñón and A. M. Varela/Instituto de Astrofísica de Canarias)

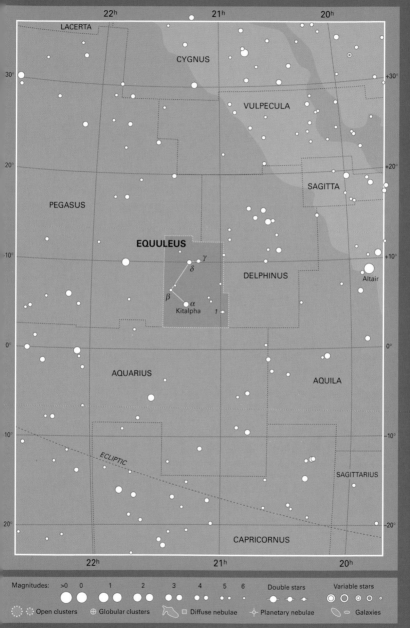

ERIDANUS   The River

An extensive constellation, the sixth-largest in the sky, but often over-looked because of its faintness. It meanders from Taurus in the north to Hydrus in the south. In mythology, Eridanus was the river into which Phaethon fell after trying to drive the chariot of his father, Helios the Sun god. But it also supposedly represents a real river. Early mythologists identified it with the Nile, but later Greek writers said it was the Po in Italy. Originally Eridanus included the stars of what is now Fornax and it stretched only as far as θ (theta) Eridani, which was then known as Achernar, from the Arabic meaning 'the river's end'; its present name, Acamar, comes from the same Arabic original as Achernar. In recent times Eridanus has been extended to nearly 60° south (below the horizon from Greece), and another star has been given the title Achernar. The constellation contains several interesting galaxies, all too distant and too faint to be picked up easily in amateur telescopes. One celebrated example is NGC 1300, located at 3h 19.7m, −19° 25′, a beautiful 10th-magnitude barred spiral pictured on page 144.

α (alpha) Eridani, 1h 38m −57°.2, (Achernar, 'the river's end'), mag. 0.5, is a blue-white main-sequence star 144 l.y. away.

β (beta) Eri, 5h 08m −5°.1, (Cursa, 'the footstool', referring to its position under the foot of Orion), mag. 2.8, is a blue-white star 89 l.y. away.

ε (epsilon) Eri, 3h 33m −9°.5, mag. 3.7, an orange main-sequence star 10.5 l.y. away, is among the most Sun-like of the nearby stars. It is orbited every 7 years by a planet of similar mass to Jupiter.

θ (theta) Eri, 2h 58m −40°.3, (Acamar), 161 l.y. away, is a striking pair of blue-white stars of mags. 3.2 and 4.3, divisible in small telescopes.

o² (omicron²) Eri, 4h 15m −7°.7, 16 l.y. away, also known as 40 Eridani, is a remarkable triple star. A small telescope shows that the mag. 4.4 orange primary, a star similar to the Sun, has a wide mag. 9.5 white dwarf companion, the most easily seen white dwarf in the sky. Small telescopes reveal that the white dwarf has an 11th-mag. companion which is a red dwarf, thereby completing a most interesting trio. The white dwarf and red dwarf orbit each other every 250 years, and remain easy to split until the latter half of the 21st century.

32 Eri, 3h 54m −3°.0, 290 l.y. away, is a beautiful double star for small telescopes, consisting of a yellow giant of mag. 4.8 and a blue-green mag. 6.1 companion.

39 Eri, 4h 14m −10°.3, 206 l.y. away, is an orange giant of mag. 4.9 with an 8th-mag. companion divisible in small telescopes.

p Eri, 1h 40m −56°.2, 27 l.y. away, is a beautiful wide duo of orange stars, mags. 5.8 and 5.9, with an orbital period of about 500 years.

NGC 1535, 4h 14m −12°.7, is a small 9th-mag. planetary nebula about 2000 l.y. away. Small telescopes show it, but apertures of 150 mm are needed to appreciate its blue-grey disk.

FORNAX   The Furnace

A barren constellation introduced in the 1750s by Nicolas Louis de Lacaille, originally under the name of Fornax Chemica, the Chemical Furnace. It contains a dwarf member of our Local Group of galaxies, called the Fornax Dwarf, approximately 500,000 l.y. from the Milky Way but too faint to see in amateur telescopes. Fornax also contains a compact cluster of galaxies about 75 million l.y. away, the brightest member of which is the 9th-mag. peculiar galaxy NGC 1316, also known as the radio source Fornax A, located at 3h 22.7m, −37° 12′.

α (alpha) Fornacis, 3h 12m −29°.0, 46 l.y. away, is a binary consisting of a yellow-white main-sequence star of mag. 3.9 with a deeper yellow mag. 6.5 companion of suspected variability. The pair orbit each other every 300 years or so and will remain divisible in small telescopes throughout the 21st century.

β (beta) For, 2h 49m −32°.4, mag. 4.5, is a yellow giant star 169 l.y. away.

NGC 1097, 2h 46m −30°.6, is a 9th-mag. barred spiral galaxy about 60 million l.y. away, visible in medium-sized telescopes.

NGC 1365, at 3h 33.6m, −36° 08′, a classic barred spiral galaxy of 10th mag., is a member of the Fornax galaxy cluster, 75 million l.y. away. (European Southern Observatory)

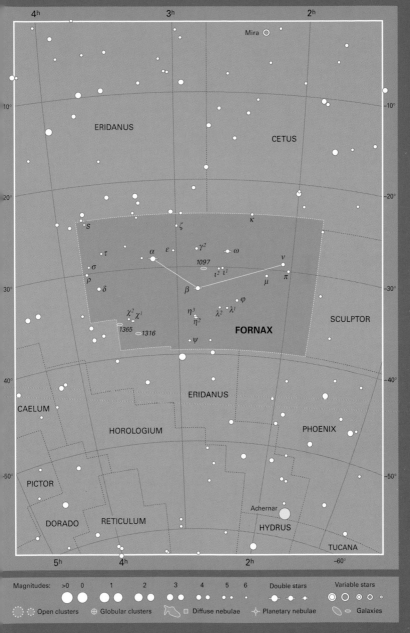

GEMINI   The Twins

A constellation that dates from ancient times, representing a pair of twins. We know them as Castor and Pollux, members of the Argonauts' crew and of mixed parentage: both were sons of Queen Leda of Sparta, but Castor's father was her husband King Tyndareus, while the father of Pollux was the god Zeus. The twins were the protectors of mariners, appearing in ships' rigging as the electrical phenomenon now known as St Elmo's fire. In the sky, the stars Castor and Pollux provide a useful yardstick for measuring angular distances – they are exactly 4°.5 apart. Gemini is a member of the zodiac, and the Sun passes through the constellation from late June to late July. Each year the Geminid meteors, one of the year's richest and brightest showers, radiate from a point near Castor, reaching a maximum around December 13–14, when up to 100 meteors per hour may be seen.

α (alpha) Geminorum, 7h 35m +31°.9, (Castor), 52 l.y. away, is an astounding multiple star, consisting of six separate components. To the naked eye it appears as a blue-white star of mag. 1.6. A 60-mm telescope with high magnification splits Castor into two components of mags. 1.9 and 3.0, which orbit each other every 470 years. Separation of these stars is increasing, and they will become progressively easier to split throughout the 21st century, after which they will start to close up again. Both stars are spectroscopic binaries. Small telescopes also show a wide red dwarf companion to Castor; this is itself an eclipsing binary of Algol type, varying between mags. 9.3 and 9.8 in 19.5 hours, completing the six-star system.

β (beta) Gem, 7h 45m +28°.0, (Pollux), mag. 1.2, is the brightest star in the constellation. Some astronomers have speculated that Pollux, being labelled β Geminorum, was once fainter than Castor and has since brightened, or Castor has faded. But the truth is that Johann Bayer, who allotted the Greek letters in 1603, did not distinguish carefully which star was the brighter, and so caused unnecessary confusion. Pollux is an orange giant 34 l.y. away.

γ (gamma) Gem, 6h 38m +16°.4, (Alhena), mag. 1.9, is a blue-white star 105 l.y. away.

δ (delta) Gem, 7h 20m +22°.0, (Wasat), 59 l.y. away, is a creamy-white star of mag. 3.5 with an orange dwarf companion of mag. 8.2. The brightness contrast makes the pair difficult in telescopes below about 75 mm aperture. Their estimated orbital period is over 1000 years.

ε (epsilon) Gem, 6h 44m +25°.1, mag. 3.1, is a yellow supergiant about 900 l.y. away. Binoculars, or a small telescope, reveal a wide companion of mag. 9.2.

ζ (zeta) Gem, 7h 04m +20°.6, 1200 l.y. away, is both a variable star and a binocular double. A yellow supergiant, it is a Cepheid variable, fluctuating between mags. 3.6 and 4.2 every 10.2 days. Binoculars or small telescopes reveal a wide mag. 7.6 companion, which is unrelated.

η (eta) Gem, 6h 15m +22°.5, 350 l.y. away, is another double–variable. A red giant star, it fluctuates in semi-regular manner between mags. 3.1 and 3.9 in about 233 days. It has a close 6th-mag. companion that orbits it every 500 years or so and requires a large telescope to distinguish it from the primary's glare.   ▶

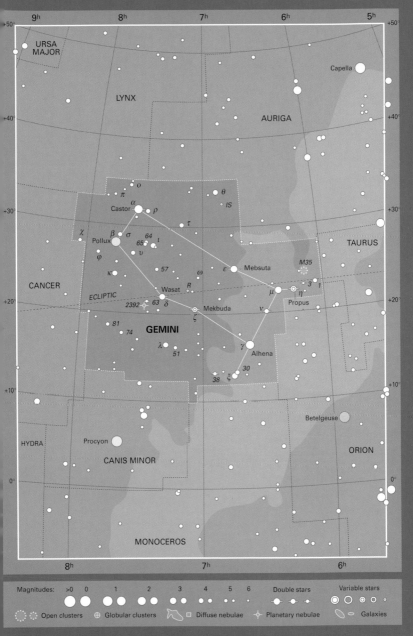

κ (kappa) Gem, 7h 44m +24°.4, 143 l.y. away, is a mag. 3.6 yellow giant, with an 8th-mag. companion made difficult in small telescopes because of the extreme brightness difference.

ν (nu) Gem, 6h 29m +20°.2, mag. 4.1, is a blue giant 500 l.y. away with a wide 8th-mag. companion visible in binoculars or small telescopes.

38 Gem, 6h 55m +13°.2, 91 l.y. away, is a double star for small telescopes, with white and yellow components of mags. 4.8 and 7.8.

M35 (NGC 2168), 6h 09m +24°.3, is a large, 5th-mag. open cluster visible to the naked eye or through binoculars, noticeably elongated in shape. It consists of about 200 stars covering a similar area of sky as the Moon, and is 2800 l.y. away. Even a small telescope at low magnification will show the members of this out-standing cluster to be arranged in curving chains. Nearby in the sky, but actually about 10,000 l.y. farther off, is NGC 2158, a very rich open cluster that appears as a small, faint patch of light requiring at least 100 mm aperture to distinguish.

NGC 2392, 7h 29m +20°.9, is an 8th-mag. planetary nebula known as the Eskimo or Clown Face Nebula because it looks somewhat like a face sur-rounded by a fringe when seen through a large telescope. A small telescope shows it as a blue-green ellipse about the same apparent size as the disk of Saturn, with a central star of 10th mag. NGC 2392 lies 3000 l.y. away.

GRUS   The Crane

One of the 12 constellations introduced at the end of the 16th century by the Dutch navigators Pieter Dirkszoon Keyser and Frederick de Houtman. It represents a water bird, the long-necked Crane, although it has also been depicted as a flamingo. The stars δ (delta) and μ (mu) Gruis are striking naked-eye doubles.

α (alpha) Gruis, 22h 08m −47°.0, (Alnair, 'the bright one'), mag. 1.7, is a blue-white star 101 l.y. away.

β (beta) Gru, 22h 43m −46°.9, is a red giant 170 l.y. away that varies between about mags. 2.0 and 2.3.

γ (gamma) Gru, 21h 54m −37°.4, mag. 3.0, is a blue giant 203 l.y. away.

δ^1 δ^2 (delta^1 delta^2) Gru, 22h 29m −43°.5, is a naked-eye pairing of two unrelated stars: δ^1, mag. 4.0, is a yellow giant 296 l.y. away; δ^2 is a mag. 4.1 red giant 325 l.y. away.

μ^1 μ^2 (mu^1 mu^2) Gru, 22h 16m −41°.3, is another naked-eye double of unrelated yellow giants that appear in the same line of sight by chance: μ^1 is of mag. 4.8, 262 l.y. away; μ^2 is of mag. 5.1, 240 l.y. away.

π^1 π^2 (pi^1 pi^2) Gru, 22h 23m −45°.9, is a binocular duo of unrelated stars. π^1 is a deep-red semi-regular variable that ranges between mags. 5.4 and 6.7 every 150 days or so; it lies 500 l.y. away. π^2 is a white giant of mag. 5.6, 132 l.y. away.

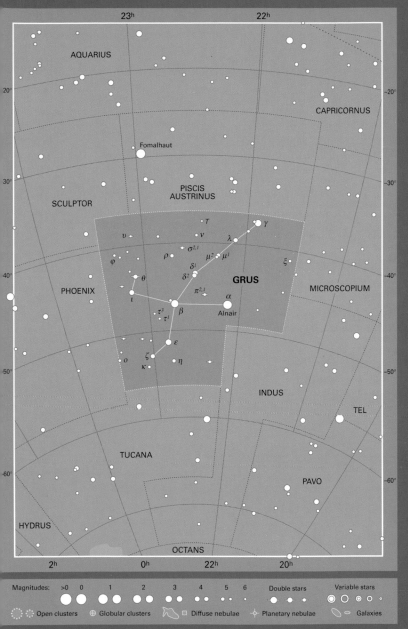

Magnitudes: >0 0 1 2 3 4 5 6   Double stars   Variable stars

Open clusters   ⊕ Globular clusters   Diffuse nebulae   Planetary nebulae   Galaxies

HERCULES

Hercules represents the Greek mythological hero famous for his twelve labours. Originally, though, the constellation was visualized as an anonymous kneeling man, with one foot on the head of the celestial dragon, Draco, which adjoins to the north. Some legends identify the constellation with the ancient Sumerian superman, Gilgamesh. Despite being the fifth-largest constellation, Hercules is not prominent. But it is stocked with an abundance of double stars for users of small telescopes, plus one of the brightest and richest globular clusters, M13, which is easily located on one side of the so-called Keystone, a quadrilateral of four stars that marks the pelvis of Hercules.

α (alpha) Herculis, 17h 15m +14°.4, (Rasalgethi, 'the kneeler's head'), about 400 l.y. distant, is a red giant star some 400 times the Sun's diameter, making it one of the largest stars known. Like most red giants it is erratically variable, in this case fluctuating from about mag. 3 to mag. 4. It is actually a double star, with a mag. 5.4 blue-green companion visible in small telescopes. The estimated orbital period of the pair is 3600 years.

β (beta) Her, 16h 30m +21°.5, (Kornephoros, 'club-bearer'), mag. 2.8, the brightest star in the constellation, is a yellow giant 148 l.y. away.

γ (gamma) Her, 16h 22m +19°.2, mag. 3.8, is a white giant star 195 l.y. away, with a wide, unrelated 10th-mag. companion visible in small telescopes.

δ (delta) Her, 17h 15m +24°.8, mag. 3.1, is a blue-white star 78 l.y. away. Small telescopes show a nearby mag. 8.2 star which is physically unrelated.

ζ (zeta) Her, 16h 41m +31°.6, 35 l.y. away, is a mag. 2.9 yellow-white star with a close mag. 5.7 orange companion that orbits it in 34.5 years. After being closest in 2001 the two stars open out, coming into the range of 220 mm apertures after 2010 and 150 mm when widest around 2025.

κ (kappa) Her, 16h 08m +17°.0, is a mag. 5.0 yellow giant 388 l.y. away with an unrelated mag. 6.3 orange giant companion, 470 l.y. away, easily seen in small telescopes.

ρ (rho) Her, 17h 24m +37°.1, 402 l.y. away, is a pair of blue-white giants of mags. 4.5 and 5.5, visible in small telescopes.

30 Her, 16h 29m +41°.9, also known as g Her, is a red giant that varies semi-regularly between mags. 4.3 and 6.3 every 3 months or so. It lies 361 l.y. away.

68 Her, 17h 17m +33°.1, also known as u Her, 865 l.y. away, is an eclipsing binary of Beta Lyrae type varying between mag. 4.7 and 5.4 every 2 days.

95 Her, 18h 02m +21°.6, 470 l.y. away, is a double star for small telescopes, consisting of two giant stars of mags. 4.9 and 5.2, appearing silver and gold.

100 Her, 18h 08m +26°.1, is an easy duo for small telescopes consisting of identical blue-white stars of mag. 5.8, distances 165 and 230 l.y., like a pair of celestial cat's eyes.

M13 (NGC 6205), 16h 42m +36°.5, is a 6th-mag. globular cluster of 300,000 stars, the brightest of its kind in northern skies. It can be seen with the naked eye and is unmistakable in binoculars, spanning half the apparent width of the ▶

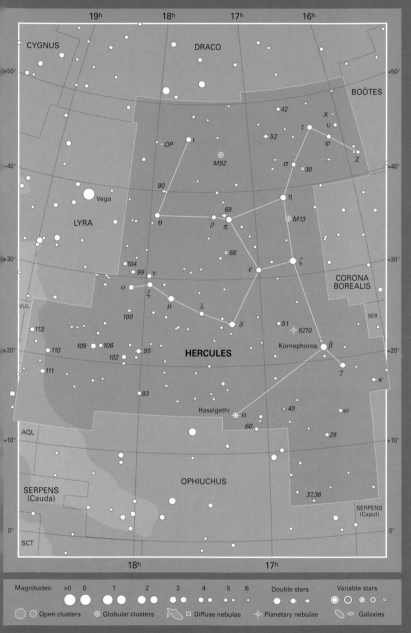

M13, *the great globular cluster in Hercules, is a swarm of 300,000 stars. (Simon Tulloch and Daniel Folha, Isaac Newton Group of Telescopes, La Palma)*

full Moon. The cluster lies 25,200 l.y. away and has a true diameter of over 100 l.y. A small telescope resolves individual stars throughout the cluster, giving a mottled, sparkling effect.

M92 (NGC 6341), 17h 17m +43°.1, is a globular cluster only slightly inferior to its more famous neighbour, M13, which overshadows it. M92 is easily seen in binoculars as a fuzzy star. It is smaller and more condensed than M13, and needs a larger telescope to resolve its stars. It lies 29,000 l.y. away and has an estimated age of around 14 billion years, making it the oldest globular known.

NGC 6210, 16h 45m +23°.8, is a 9th-mag. planetary nebula which a telescope of 75 mm or larger shows as a blue-green ellipse. It is about 4000 l.y. away.

HOROLOGIUM   The Pendulum Clock

One of the constellations representing mechanical instruments introduced in the 1750s by the Frenchman Nicolas Louis de Lacaille. As with so many of his constellations, Horologium is faint and obscure.

α (alpha) Horologii, 4h 14m −42°.3, mag. 3.9, is an orange giant star 117 l.y. away.

β (beta) Hor, 2h 59m −64°.1, mag. 5.0, is a giant white star 314 l.y. away.

R Hor, 2h 54m −49°.9, is a red giant variable of the Mira type, ranging between extremes of mag. 4.7 and 14.3 in 13 months or so. It lies about 1000 l.y. away.

TW Hor, 3h 13m −57°.3, is a deep-red pulsating variable of semi-regular type, ranging between mag. 5.5 and 6.0 with an approximate period of 5 months. It lies about 1300 l.y. away.

NGC 1261, 3h 12m −55°.2, is an 8th-mag. globular cluster 44,000 l.y. away.

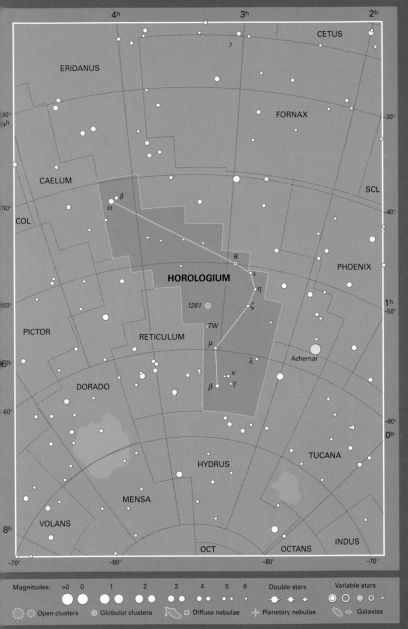

Magnitudes: >0 0 1 2 3 4 5 6   Double stars   Variable stars

Open clusters   ⊕ Globular clusters   Diffuse nebulae   Planetary nebulae   Galaxies

HYDRA   The Water Snake

The largest constellation in the sky, but by no means easy to identify because of its faintness. Apart from its brightest star, Alphard, which marks the heart of the Water Snake, Hydra's only readily recognizable feature is its head, an attractive group of six stars. Hydra winds its way from the head in the northern celestial hemisphere, on the borders of Cancer, to the tip of its tail, south of the celestial equator adjacent to Libra and Centaurus, a total length of over 100°. In mythology, Hydra is usually identified with the multi-headed monster slain by Hercules. Another legend links it with Corvus the Crow and Crater the Cup, which are found on its back, the bird having returned to the god Apollo with Hydra in its claws as an excuse for its aborted mission to fetch water in the cup.

α (alpha) Hydrae, 9h 28m −8°.7, (Alphard, 'the solitary one'), mag. 2.0, is an orange giant 177 l.y. away.

β (beta) Hya, 11h 53m −33°.9, mag. 4.3, is a blue-white star 365 l.y. away.

γ (gamma) Hya, 13h 19m −23°.2, mag. 3.0, is a yellow giant 132 l.y. away.

δ (delta) Hya, 8h 38m +5°.7, mag. 4.1, is a blue-white star 179 l.y. away.

ε (epsilon) Hya, 8h 47m +6°.4, 135 l.y. away, is a beautiful but difficult double star of contrasting colours, needing high power on a telescope of at least 75 mm aperture. The stars are yellow and blue, of mags. 3.4 and 6.7, and form a genuine binary with an orbital period of nearly 1000 years.   ▶

The face-on spiral galaxy M83 in Hydra, sometimes known as the Southern Pinwheel, is a swirl of stars and gas. See also page 160. (Bill Schoening/AURA/NOAO/NSF)

Magnitudes: >0   0   1   2   3   4   5   6   Double stars   Variable stars

Open clusters   ⊕ Globular clusters   Diffuse nebulae   Planetary nebulae   Galaxies

27 Hya, 9h 20m, −9°.6, mag. 4.8, is a white star 244 l.y. away with a wide mag. 7.0 companion, 202 l.y. away, visible in binoculars. Small telescopes split this companion into components of mags. 7 and 11.

54 Hya, 14h 46m −25°.4, 99 l.y. away, is an easy double for small telescopes, consisting of yellow and purple stars of mags. 5.3 and 7.4.

R Hya, 13h 30m −23°.3, is a red giant variable star similar to Mira in Cetus that fluctuates between 4th and 10th mags. every 390 days. At its best it can reach mag. 3.5, making it one of the brightest Mira stars. Its distance is about 2000 l.y.

U Hya, 10h 38m −13°.4, 528 l.y. away, is a deep-red variable star that fluctuates semi-regularly between mags. 4.2 and 6.6 every 115 days.

M48 (NGC 2548), 8h 14m −5°.8, is a large open cluster of about 80 stars, 2000 l.y. away, just visible to the naked eye under clear skies and a fine sight in binoculars. It is somewhat triangular in shape, wider than the apparent size of the full Moon, and is well shown by small telescopes under low power.

M68 (NGC 4590), 12h 39m −26°.7, is an 8th-mag. globular cluster visible as a fuzzy star in binoculars and just resolved with 100-mm apertures. It lies 31,000 l.y. away.

M83 (NGC 5236), 13h 37m −29°.9, is a large, face-on spiral galaxy of 8th mag., visible in a small telescope. It has a small, bright nucleus and signs of a central bar. Its spiral arms can be traced with an aperture of 150 mm. (See photograph on page 158.) M83 has been the site of more known supernovae than any other Messier object,. six in all.

NGC 3242, 10h 25m −18°.6, is a 9th-mag. planetary nebula of similar apparent size to the disk of Jupiter; it is popularly termed the 'ghost of Jupiter'. This often-overlooked object 2600 l.y. away is prominent enough to be picked up in small telescopes at low magnification as a blue-green disk, while larger instruments show a bright inner disk surrounded by a fainter halo.

HYDRUS   The Lesser Water Snake

The Dutch navigators Pieter Dirkszoon Keyser and Frederick de Houtman introduced this constellation at the end of the 16th century as a smaller southern counterpart to the great Water Snake, Hydra. Keyser and de Houtman's Hydrus, lying between the two Magellanic Clouds, almost bridges the gap between Eridanus and the south celestial pole. There is little to interest the casual observer.

α (alpha) Hydri, 1h 59m −61.6°, mag. 2.9, is a white main-sequence star 71 l.y. away.

β (beta) Hyi, 0h 26m −77°.3, mag. 2.8, the constellation's brightest member, is a yellow star 24 l.y. away.

γ (gamma) Hyi, 3h 47m −74°.2, mag. 3.2, is a red giant star 214 l.y. away.

π^1 π^2 (pi^1 pi^2) Hyi, 2h 14m −67°.8, is a binocular pair of red and orange giants, unrelated: π^1 is mag. 5.6 and lies 740 l.y. away; π^2, mag. 5.7, is 468 l.y. distant.

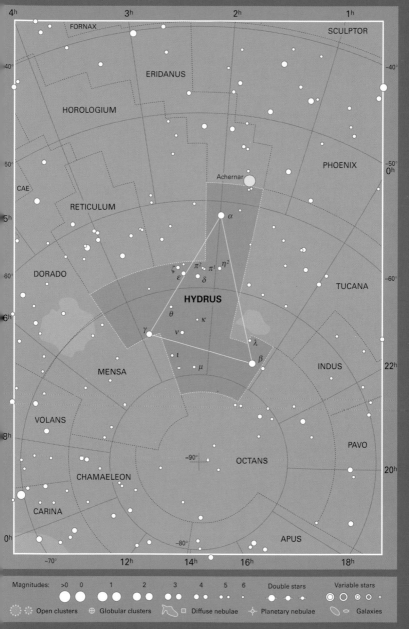

Magnitudes: >0 0 1 2 3 4 5 6 Double stars Variable stars

Open clusters ⊕ Globular clusters Diffuse nebulae Planetary nebulae Galaxies

INDUS   The Indian

A constellation representing an American native Indian, introduced at the end of the 16th century by the Dutch navigators Pieter Dirkszoon Keyser and Frederick de Houtman. None of its stars is brighter than 3rd magnitude.

α (alpha) Indi, 20h 38m −47°.3, mag. 3.1, is an orange giant star 101 l.y. away.

β (beta) Ind, 20h 55m −58°.5, mag. 3.7, is an orange giant 600 l.y. away.

δ (delta) Ind, 21h 58m −55°.0, mag. 4.4, is a white star 185 l.y. away.

ε (epsilon) Ind, 22h 03m −56°.8, mag. 4.7, is an orange dwarf somewhat smaller and cooler than the Sun. At a distance of 11.8 l.y., it is one of the Sun's closest neighbours.

θ (theta) Ind, 21h 20m −53°.4, 97 l.y. away, is a pair of white stars of mags. 4.5 and 7.0, divisible in a small telescope.

T Ind, 21h 20m −45°.0, is a deep-red variable star ranging semi-regularly between 5th and 7th magnitudes in 11 months or so. It lies about 1900 l.y. away.

Between Indus and Hydrus lies the Small Magellanic Cloud, an irregularly shaped splash of stars which is actually a satellite of our own Galaxy. For a description see page 246. (AURA/NOAO/NSF)

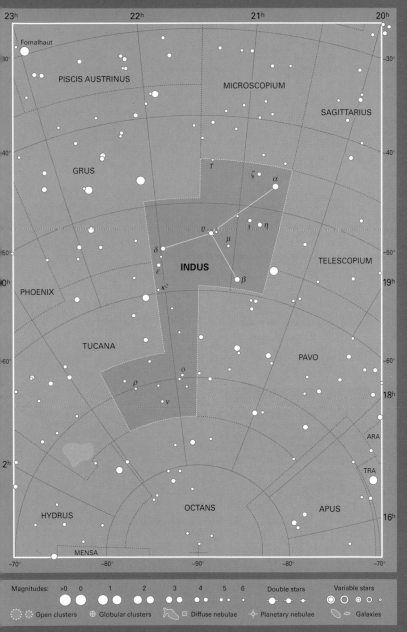

LACERTA   The Lizard

An inconspicuous constellation sandwiched between Cygnus and Andromeda, introduced in 1687 by the Polish astronomer Johannes Hevelius. An alternative constellation that once occupied this area was Sceptrum, the Sceptre and Hand of Justice, created in 1679 by the Frenchman Augustin Royer to commemorate King Louis XIV. In 1787 the German Johann Elert Bode called this region Frederick's Glory in honour of King Frederick the Great of Prussia. Both these alternatives have been discarded. The constellation's most celebrated object is BL Lacertae, location 22h 02.7m, +42° 17′, originally thought to be a peculiar 14th-magnitude variable star. It is the prototype of a group of objects, the BL Lac objects or Lacertids, believed to be giant elliptical galaxies with variable centres, lying far off in the Universe and evidently related to quasars. Three bright novae appeared within the boundaries of Lacerta during the 20th century.

α (alpha) Lacertae, 22h 31m +50°.3, mag. 3.8, is a blue-white main-sequence star 102 l.y. away.

β (beta) Lac, 22h 24m +52°.2, mag. 4.4, is a yellow giant 170 l.y. away.

NGC 7243, 22h 15m +49°.9, is a scattered open cluster for small telescopes, consisting of a few dozen stars of 8th mag. and fainter, 2500 l.y. away.

Johannes Hevelius (1611–1687)

Johannes Hevelius of Danzig was one of the finest observers of his day. His masterwork was a catalogue of 1564 stars, published posthumously in 1690. The catalogue was accompanied by a set of sky maps, beautifully engraved by Hevelius himself, on which he introduced seven constellations still in use today: Canes Venatici, Lacerta, Leo Minor, Lynx, Scutum, Sextans and Vulpecula. Another important cartographic product of Hevelius was his map of the Moon, published in 1647 in his book *Selenographia*. It was the first major Moon map, and introduced the first system of lunar nomenclature. He named lunar formations after features on Earth: for instance, the crater Copernicus he called Etna, while Tycho was Mount Sinai and Mare Imbrium the Mediterranean. Only a few of the names given by Hevelius remain, such as the lunar Alps and Apennines; his names have mostly been discarded in favour of the system introduced by Giovanni Battista Riccioli (1598–1671), who named the craters after famous philosophers and astronomers. Fittingly enough, the craters commemorating Hevelius and Riccioli are found close together on the Moon.

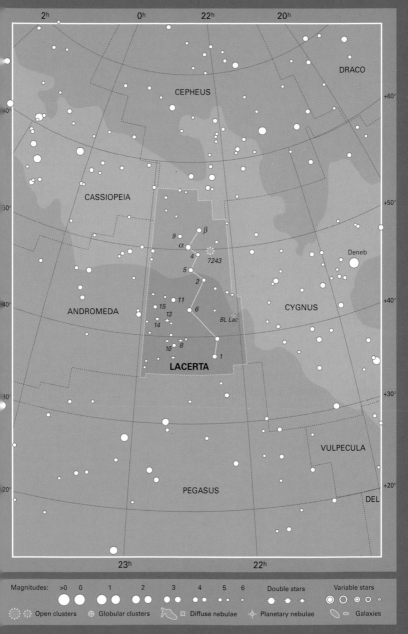

2h 0h 22h 20h

DRACO

CEPHEUS

+60°

CASSIOPEIA

+50°

β

9
α
4   7243
5
2

11   CYGNUS   +40°
15
13
14   6   BL Lac

Deneb

10   8
1

ANDROMEDA   +40°

LACERTA

+30°

VULPECULA

+30°

PEGASUS   +20°

DEL

20°

23h   22h

Magnitudes: >0 0 1 2 3 4 5 6   Double stars   Variable stars

Open clusters   Globular clusters   Diffuse nebulae   Planetary nebulae   Galaxies

LEO   The Lion

One of the few constellations that looks like the figure it is supposed to represent – in this case, a crouching lion. The Lion's head is outlined by the so-called Sickle of six stars from ε (epsilon) to α (alpha) Leonis; the Lion's body stretches out behind, its tail marked by β (beta) Leonis. Mythologically, this is the Lion reputedly slain by the hero Hercules as the first of his 12 labours. The Sun passes through the constellation from mid-August to mid-September. Every November, the Leonid meteors radiate from a point near γ (gamma) Leonis. Usually the numbers seen are low, peaking at about 10 per hour on November 17–18, but activity increases dramatically at 33-year intervals when the parent comet, Tempel–Tuttle, returns to perihelion. Leo contains the third-nearest star to the Sun, CN Leonis (also called Wolf 359), a red dwarf 7.8 l.y. away. It is of magnitude 13.5, although it is a flare star and can occasionally brighten by up to a magnitude; it is located at 10h 56.5m, +7° 01′, near the border with Sextans. Leo contains numerous distant galaxies, the brightest of which are mentioned on page 168, plus three faint dwarf members of our Local Group, beyond the reach of amateur telescopes.

α (alpha) Leonis, 10h 08m +12°.0, (Regulus, 'the little king'), mag. 1.4, is a blue-white main-sequence star 77 l.y. away. It has a wide companion of mag. 7.7, visible in binoculars or small telescopes.

β (beta) Leo, 11h 49m +14°.6, (Denebola, 'the lion's tail'), mag. 2.1, is a blue-white star 36 l.y. away.

γ (gamma) Leo, 10h 20m +20°.0, (Algieba, 'the forehead'), 126 l.y. away, consists of a pair of golden-yellow giant stars of mags. 2.3 and 3.6, orbiting in about 600 years. They form an exceptionally handsome double in small telescopes, one of the finest in the sky. In binoculars, an unrelated mag. 4.8 yellowish foreground star, 40 Leonis, is visible nearby.

δ (delta) Leo, 11h 14m +20°.5, (Zosma), mag. 2.6, is a blue-white star 58 l.y. away. ▶

Trails of Leonid meteors sweep through the Big Dipper in this 43-second exposure, made from Kitt Peak, Arizona, during the great Leonid storm of 1966. (Dave McLean, AURA/NOAO/NSF)

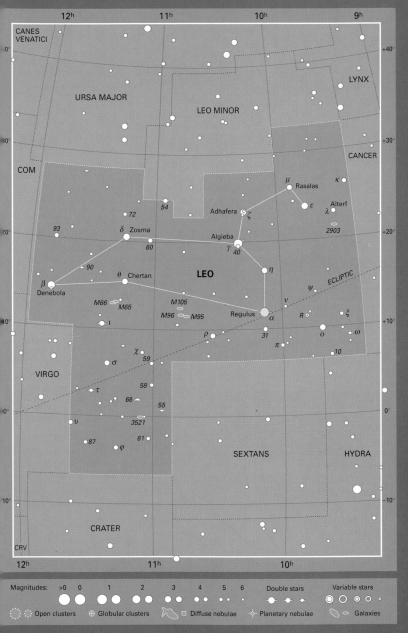

Magnitudes: >0 0 1 2 3 4 5 6 Double stars Variable stars

Open clusters ⊕ Globular clusters Diffuse nebulae Planetary nebulae Galaxies

ε (epsilon) Leo, 9h 46m +23°.8, mag. 3.0, is a yellow giant 251 l.y. away.

ζ (zeta) Leo, 10h 17m +23°.4, (Adhafera), mag. 3.4, is a giant white star 260 l.y. away. To the north of it, binoculars show 35 Leonis, an unrelated foreground star of mag. 6.0. To the south a wider third star, 39 Leonis, mag. 5.8 and also in the foreground, can be seen in binoculars, making this an optical triple.

ι (iota) Leo, 11h 24m +10°.5, 79 l.y. away, is a close and difficult double star. It appears to the naked eye as a yellow-white star of mag. 4.0, but actually consists of components of mags. 4.1 and 6.7 orbiting every 183 years. Currently the stars are moving apart; apertures of 100 mm should be sufficient to separate them until 2010 after which they become progressively easier, coming within range of all but the very smallest apertures when at their widest from 2053 to 2063.

τ (tau) Leo, 11h 28m +2°.9, 621 l.y. away, is a yellow giant of mag. 5.0 with an 8th-mag. companion visible in binoculars and small telescopes.

54 Leo, 10h 56m +24°.7, 289 l.y. away, is a double for small telescopes, consisting of blue-white components of mags. 4.5 and 6.3.

R Leo, 9h 48m +11°.4, is a red giant variable of Mira type, lying about 330 l.y. away. It appears strongly red when at maximum and is roughly 450 times larger than our Sun. R Leonis normally varies between 6th and 10th mags. with an average period of 310 days, but on occasion can become as bright as mag. 4.4.

M65, M66 (NGC 3623, NGC 3627), 11h 19m +13°.0, a pair of spiral galaxies about 20 million l.y. away. At 9th mag. they can be detected in large binoculars under clear conditions, but at least 100 mm aperture at low power is required for their elongated shape and condensed centres to be clearly seen.

M95, M96 (NGC 3351, NGC 3368), 10h 44m +11°.7, 10h 47m +11°.8, a pair of spiral galaxies of 10th and 9th mag. respectively, about 25 million l.y. away, visible as circular nebulosities in small telescopes. Larger apertures show that M95 actually has a central bar. About 1° away lies the smaller M105 (NGC 3379), 10h 48m +12°.6, a 9th-mag. elliptical galaxy at a similar distance.

LEO MINOR   The Lesser Lion

Leo Minor lies between the larger and brighter constellations of Leo to the south and Ursa Major to the north. Johannes Hevelius, the Polish astronomer, introduced Leo Minor in 1687. There is little of interest in it. Surprisingly there is no star labelled α (alpha). This is the result of an error by the English astronomer Francis Baily, who assigned Greek letters to the stars of Leo Minor in 1845 but failed to letter the brightest star through an oversight.

β (beta) Leonis Minoris, 10h 28m +36°.7, mag. 4.2, is a yellow giant star 146 l.y. away.

46 LMi, 10h 53m +34°.2, mag. 3.8, the brightest star in the constellation, is an orange giant 98 l.y. away.

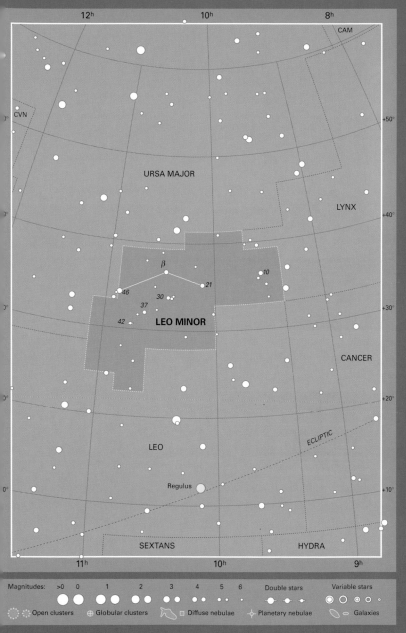

LEPUS   The Hare

Lepus is a constellation known since ancient Greek times. It represents a hare, cunningly located at the feet of its hunter, Orion, and pursued endlessly across the sky by Canis Major, the hunter's dog. The Hare is also associated in many legends with the Moon. For instance, the familiar figure of the man in the Moon is sometimes interpreted as a hare or rabbit, so perhaps Lepus is another incarnation of the lunar hare. Lepus is overshadowed by Orion's brilliance, but is not without interest for amateur observers.

α (alpha) Leporis, 5h 33m −17°.8, (Arneb, 'hare'), mag. 2.6, is a white supergiant star, about 1300 l.y. away.

β (beta) Lep, 5h 28m −20°.8, (Nihal), mag. 2.8, is a yellow giant star 159 l.y. away.

γ (gamma) Lep, 5h 44m −22°.5, 29 l.y. away, is an attractive binocular duo consisting of a yellow star of mag. 3.6 with an orange companion of mag. 6.2.

δ (delta) Lep, 5h 51m −20°.9, mag. 3.8, is a yellow giant 112 l.y. away.

ε (epsilon) Lep, 5h 05m −22°.4, mag. 3.2, is an orange giant 227 l.y. away.

κ (kappa) Lep, 5h 13m −12°.9, 560 l.y. away, is a mag. 4.4 blue-white star with a close mag. 7.4 companion, difficult to see in the smallest telescopes because of the magnitude contrast.

R Lep, 5h 00m −14°.8, 820 l.y. away, is an intensely red star known as Hind's Crimson Star after the English observer John Russell Hind, who described it in 1845 as 'like a drop of blood on a black field'. R Leporis is a Mira-type variable that ranges from mag. 5.5 at its brightest to as faint as 12th mag. in a period of around 430 days.

RX Lep, 5h 11m −11°.8, 447 l.y. away, is a red giant that varies semi-regularly between mags. 5.0 and 7.4 every 2 months or so.

M79 (NGC 1904), 5h 24m −24°.5, is a small but rich globular cluster 44,000 l.y. away, visible as a fuzzy 8th-mag. star in small telescopes. Nearby in the same low-power field is the multiple star Herschel 3752, consisting of a mag. 5.4 primary with two companions, a close one of mag. 6.6 and a wide one of mag. 9.1, all visible in small telescopes.

NGC 2017, 5h 39m −17°.8, is a small but remarkable star cluster, also known as the multiple star Herschel 3780. Modest amateur telescopes reveal a group of five well-spaced stars ranging from 6th to 10th mag. In addition, the brightest star has a mag. 7.9 companion that requires a telescope of at least 200 mm aperture to split, while an aperture of at least 100 mm shows that one of the 9th-mag. stars is a close double. There is also a 12th-mag. component which should be visible with 100 mm or upwards, so this is actually a group of at least eight stars. However, it is not a true cluster because the stars lie at various distances and are moving in different directions.

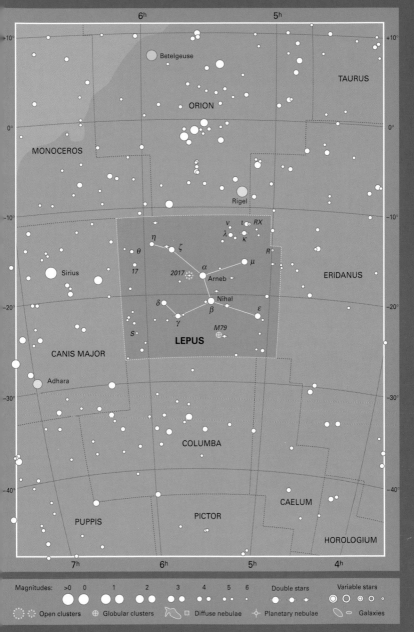

Magnitudes: >0 0 1 2 3 4 5 6   Double stars   Variable stars

Open clusters   Globular clusters   Diffuse nebulae   Planetary nebulae   Galaxies

LIBRA   The Scales

A small, faint constellation of the zodiac, through which the Sun passes during November. The ancient Greeks knew it as the Claws of the Scorpion, an extension of neighbouring Scorpius, and this identification lives on in the names of the stars α (alpha), β (beta) and γ (gamma) Librae (see below). But the Romans made it into a separate constellation in the time of Julius Caesar in the first century BC. Since then the Scales have come to be regarded as the symbol of justice, held aloft by the goddess of justice, Astraeia, in the shape of the neighbouring figure of Virgo. Libra once contained the September equinox, the point at which the Sun passes south of the celestial equator each year. Because of precession, this point moved into neighbouring Virgo around 730 BC, but the September equinox is still sometimes referred to as the First Point of Libra. Although faint, Libra contains several stars of interest.

α (alpha) Librae, 14h 50m −16°.0, (Zubenelgenubi, 'the southern claw'), 77 l.y. away, is a wide binocular double consisting of a blue-white star of mag. 2.7 with a white companion of mag. 5.2.

β (beta) Lib, 15h 17m −9°.4, (Zubeneschamali, 'the northern claw'), mag. 2.6, the brightest in the constellation, is celebrated as one of the few bright stars to show a distinct greenish tinge. It lies 160 l.y. away.

γ (gamma) Lib, 15h 36m −14°.8, (Zubenelakrab, 'the scorpion's claw'), mag. 3.9, is an orange giant 152 l.y. away.

δ (delta) Lib, 15h 01m −8°.5, 304 l.y. away, is an eclipsing variable of the Algol type. It varies between mags. 4.9 and 5.9 in 2 days 8 hours.

ι (iota) Lib, 15h 12m −19°.8, is a multiple star 377 l.y. away. Its main blue-white component, of mag. 4.5, has a wide mag. 9.4 companion that is difficult to see in the smallest telescopes because of the brightness difference. An aperture of 75 mm or above with high magnification will split this fainter companion into two stars of 10th and 11th mags. The brightest component of ι Lib is itself a binary with a 23-year period, but too close for amateur telescopes. Binoculars show a mag. 6.1 star nearby called 25 Librae, which is a foreground object, 219 l.y. away.

μ (mu) Lib, 14h 49m −14°.1, 235 l.y. away, is a close double star consisting of components of mags. 5.7 and 6.8, divisible in a telescope of 75 mm aperture.

48 Lib, 15h 58m −14°.3, mag. 4.9, is a shell star similar to γ (gamma) Cassiopeiae and Pleione in Taurus. It is a blue supergiant 513 l.y. away with an abnormally high speed of rotation that causes it to throw off rings of gas from its equator, varying irregularly by a few tenths of a magnitude as it does so. It also bears the variable-star designation FX Lib.

NGC 5897, 15h 17m −21°.0, is a large but loosely scattered 9th-mag. globular cluster 40,000 l.y. away, unspectacular in small instruments.

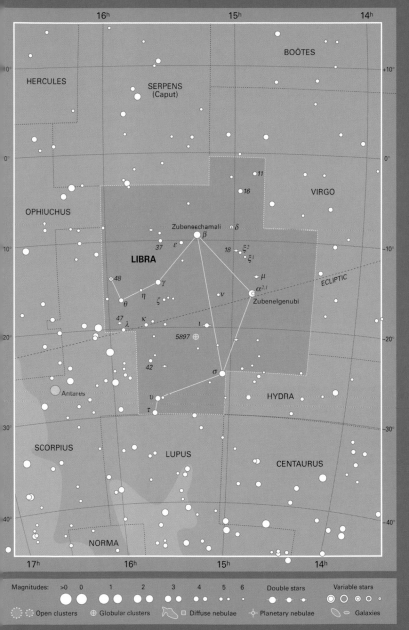

LUPUS   The Wolf

Lupus is stocked with numerous interesting objects, although it is often overlooked in favour of its more spectacular neighbours Scorpius and Centaurus. The constellation was regarded by the Greeks and Romans as an unspecified wild animal, impaled on a pole by Centaurus the Centaur. Its identification as a wolf seems to have become common in Renaissance times. Lupus lies in the Milky Way and is rich in double stars.

α (alpha) Lupi, 14h 42m −47°.4, mag. 2.3, is a blue giant star 548 l.y. away.

β (beta) Lup, 14h 59m −43°.1, mag. 2.7, is a blue giant 524 l.y. away.

γ (gamma) Lup, 15h 35m −41°.2, mag. 2.8, is a blue-white star 570 l.y. away. It is a close binary with a 190-year orbital period, splittable only in apertures of 200 mm and above.

ε (epsilon) Lup, 15h 23m −44°.7, 504 l.y. away, is a blue-white star of mag. 3.4 with a wide mag. 8.8 companion visible in a small telescope. The primary is itself a close double, splittable only in large apertures.

η (eta) Lup, 16h 00m −38°.4, 493 l.y. away, is a double star consisting of a mag. 3.4 blue-white primary with a mag. 7.9 companion, not easy to see in a small telescope because of the magnitude contrast.

κ (kappa) Lup, 15h 12m −48°.7, 188 l.y. away, is an easy double star for small telescopes, consisting of blue-white components of mags. 3.9 and 5.7.

μ (mu) Lup, 15h 19m −47°.9, 291 l.y. away, is a multiple star. Small telescopes reveal a mag. 4.3 blue-white primary with a wide mag. 6.9 companion. But in telescopes of at least 100 mm, with high magnification, the primary itself is seen to be double, consisting of two near-identical stars of mags. 5.0 and 5.1.

ξ (xi) Lup, 15h 57m −34°.0, 200 l.y. away, is a neat pair of mag. 5.1 and 5.6 blue-white stars, well seen in a small telescope.

π (pi) Lup, 15h 05m −47°.1, 500 l.y. away, appears to the naked eye as mag. 3.9, but telescopes above 75 mm aperture show that it consists of two close blue-white stars of mags. 4.6 and 4.7.

GG Lup, 15h 19m −40°.8, 514 l.y., is an eclipsing binary of Algol type, ranging between mags. 5.6 and 6.1 with a period of 1.85 days.

NGC 5822, 15h 05m −54°.3, is a large, loose open cluster of about 150 faint stars, 2400 l.y. away, visible in binoculars or small telescopes.

NGC 5986, 15h 46m −37°.8, is an 8th-mag. globular cluster, 33,000 l.y. distant, visible as a rounded patch in a small telescope.

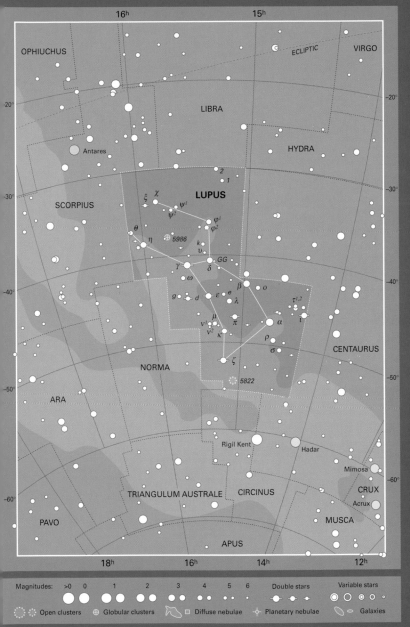

LYNX   The Lynx

A decidedly obscure constellation, despite its considerable size (larger than Gemini, for instance). It was introduced in 1687 by the Polish astronomer Johannes Hevelius to fill the gap between Ursa Major and Auriga. He named it Lynx because, it is said, only the lynx-eyed would be able to see its stars – a reference to the fact that his own eyesight was exceptionally keen. Despite its faintness, owners of small telescopes will find many exquisite double stars within it.

α (alpha) Lyncis, 9h 21m +34°.4, mag. 3.1, is a red giant 222 l.y. away.

5 Lyn, 6h 27m +58°.4, 680 l.y. away, is an orange giant star of mag. 5.2 with a wide, unrelated mag. 7.9 companion, visible in a small telescope.

12 Lyn, 6h 46m +59°.4, 229 l.y. away, is a fascinating triple star. A small telescope will show a mag. 5.0 blue-white star with a fainter mag. 7.2 companion. Telescopes of 75 mm aperture and over reveal that the brighter component is itself a binary, of mags. 5.5 and 6.1, which orbit each other every 700 years.

15 Lyn, 6h 57m +58°.4, 170 l.y. away, is a close double star for telescopes of 150 mm aperture and above. The components are of mags. 4.7 and 5.8, the brighter star appearing a deep yellow colour.

19 Lyn, 7h 23m +55°.3, is an attractive triple star for small telescopes, with blue-white components of mags. 5.8 and 6.9. The very wide third star is of mag. 7.6. The trio lies about 500 l.y. away.

38 Lyn, 9h 19m +36°.8, 122 l.y. away, is a pair of mag. 3.9 and 6.3 blue-white stars, difficult in the smallest telescopes because of their closeness.

41 Lyn, 9h 29m +45°.6, 288 l.y. away, is actually over the border in Ursa Major (hence this designation has now fallen out of use, but we retain it here for identification purposes). It is a yellow giant of mag. 5.4, and small telescopes reveal a wide companion of mag. 8.0. A 10th-mag. star nearby forms a triangle, making this an apparent triple.

NGC 2419, 7h 38m +38°.9, is an unusually remote globular cluster, of the type known as an intergalactic tramp. It is of 11th mag. and lies about 330,000 l.y. from the centre of our Galaxy, farther than either of the Magellanic Clouds.

Flamsteed Numbers

Apart from α Lyncis, all the main stars in Lynx are referred to not by Greek letters but by so-called Flamsteed numbers. These numbers originate from a catalogue of 2935 stars, *Historia Coelestis Britannica*, compiled by the first Astronomer Royal of England, John Flamsteed (1646–1719). The catalogue was published posthumously in 1725. Flamsteed listed the stars in each constellation in order of right ascension. The numbers that are now known as Flamsteed numbers were not actually allocated by Flamsteed himself, but were added to the stars in his catalogue by later astronomers.

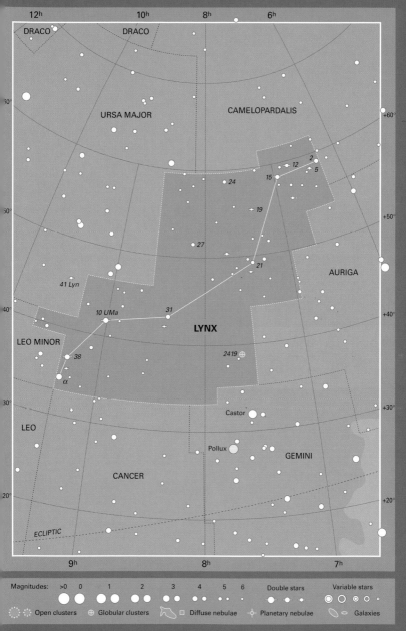

LYRA   The Lyre

A constellation dating from ancient times, representing the stringed instrument invented by Hermes and subsequently given by his half-brother Apollo to the great musician Orpheus. This constellation has also been visualized as an eagle or vulture. Although small, Lyra is bright and prominent. It contains the fifth-brightest star in the sky, Vega, which forms one corner of the Summer Triangle (Deneb in Cygnus and Altair in Aquila mark the other two corners). Our Sun's motion around the Galaxy is carrying us in the general direction of Vega at a velocity of 20 km/s relative to nearby stars. Because of precession, Vega will be the pole star between AD 13,000 and 14,000, although coming no closer than 5°.7 to the pole. The Lyrid meteor shower emanates from this constellation each year, reaching a peak of about 10 per hour on April 21–22.

α (alpha) Lyrae, 18h 37m +38°.8, (Vega, 'the swooping eagle'), mag. 0.03, is a brilliant blue-white main-sequence star 25 l.y. away. It is the fifth-brightest star in the sky, and is surrounded by a disk of dust from which planets may be forming.

β (beta) Lyr, 18h 50m +33°.4, (Sheliak, 'the harp'), 882 l.y. away, is a remarkable multiple star. Small telescopes easily resolve it as a double star of cream and blue components. The fainter, blue star is of mag. 7.2, while the brighter star is an eclipsing binary that varies between mags. 3.3 and 4.4 in 12.9 days. β Lyrae is the prototype of a class of eclipsing variables in which the stars are so close together that gravity distorts them into egg-shapes, and hot gas spirals off them into space.

γ (gamma) Lyr, 18h 59m +32°.7, (Sulafat, 'the tortoise'), mag. 3.2, is a blue-white giant 635 l.y. away.

δ¹ δ² (delta¹ delta²) Lyr, 18h 54m +37°.0, is a wide naked-eye or binocular double consisting of two unrelated stars: δ¹, blue-white, mag. 5.6, 1080 l.y. away; and δ², a red giant 899 l.y. away, which varies erratically from mag. 4.2 to 4.3. ►

The Double Double, ε (epsilon) Lyrae, as seen through a telescope. (Wil Tirion)

178

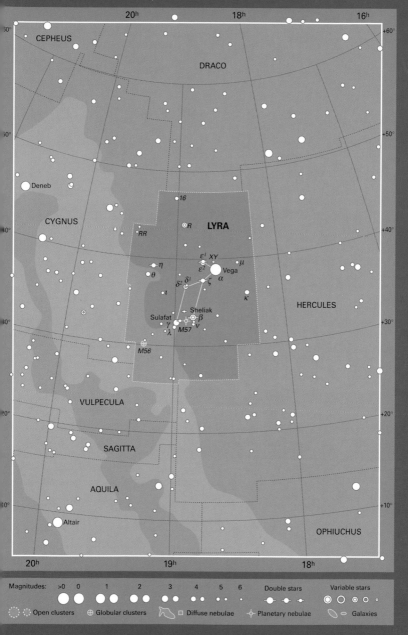

Magnitudes: >0 0 1 2 3 4 5 6   Double stars   Variable stars

Open clusters   ⊕ Globular clusters   Diffuse nebulae   ✦ Planetary nebulae   Galaxies

ϵ^1 ϵ^2 (epsilon[1] epsilon[2]) Lyr, 18h 44m +39°.7, 161 l.y. away, is a celebrated quadruple star known popularly as the Double Double. Binoculars or even keen eyesight separate it into two stars, ϵ^1 and ϵ^2, of mags. 4.7 and 4.6 respectively. But a telescope of 60 to 75 mm aperture and high magnification reveals that each star is itself double, the two pairs being oriented almost at right angles to each other (see diagram on page 178). The ϵ^1 pair has mags. of 5.0 and 6.1 and an orbital period of over 1000 years; the ϵ^2 pair, slightly closer together, has mags. of 5.2 and 5.5, and a period of about 600 years. Quadruple stars are rare, and this is the finest of them.

ζ (zeta) Lyr, 18h 45m +37°.6, 152 l.y. away, is a double star, easily split in small telescopes or binoculars into components of mags. 4.4 and 5.7.

η (eta) Lyr, 19h 14m +39°.1, 1040 l.y. away, is a blue-white star of mag. 4.4 with a wide mag. 9.1 companion visible in a small telescope.

R Lyr, 18h 55m +43°.9, 350 l.y. away, is a red giant that varies semi-regularly between mags. 3.9 and 5.0 every 6 or 7 weeks.

RR Lyr, 19h 25m +42°.8, 745 l.y. away, is the prototype of an important class of variable stars used as 'standard candles' for indicating distances in space. RR Lyrae variables are often found in globular clusters and are thus known as cluster-type variables. Related to Cepheid variables, they are giant stars that pulsate in size, varying by about one magnitude usually in less than a day. RR Lyrae itself varies from mag. 7.1 to 8.1 in 13.6 hours.

M57 (NGC 6720), 18h 54m +33°.0, the Ring Nebula, is a famous 9th-mag. planetary nebula 2000 l.y. away, conveniently placed between β (beta) and γ (gamma) Lyrae. On photographs taken through large telescopes it looks like a celestial smoke ring. A small telescope shows it as a noticeably elliptical misty disk, but a larger aperture is needed to see the central hole. It is one of the brightest planetary nebulae and appears larger in the sky than the planet Jupiter. (For a photograph see page 273.)

MENSA   The Table Mountain

The faintest of all the constellations, with no star brighter than magnitude 5.0. Mensa was introduced by the Frenchman Nicolas Louis de Lacaille to commemorate Table Mountain at the Cape of Good Hope from where he surveyed the southern skies in 1751–52. Part of the Large Magellanic Cloud strays from neighbouring Dorado over the border into Mensa, possibly reminding Lacaille of the cloud that frequently caps the real Table Mountain. Unfortunately, there is little else of interest here.

α (alpha) Mensae, 6h 10m −74°.8, mag. 5.1, is a yellow star similar to the Sun, 33 l.y. away.

β (beta) Men, 5h 03m −71°.3, mag. 5.3, is a yellow giant 642 l.y. away.

γ (gamma) Men, 5h 32m −76°.3, mag. 5.2, is an orange giant 101 l.y. away.

η (eta) Men, 4h 55m −74°.9, mag. 5.5, is an orange giant 712 l.y. away.

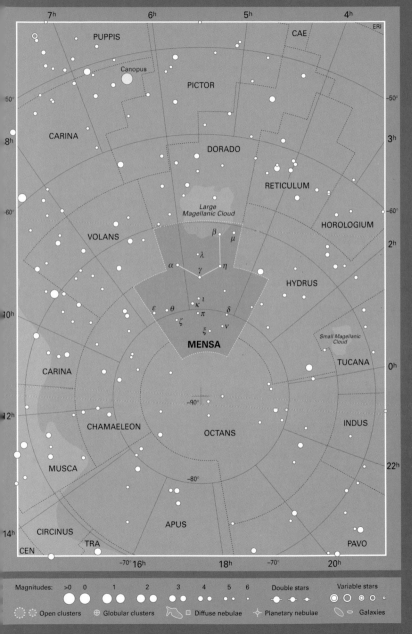

MICROSCOPIUM   The Microscope

Another of the southern hemisphere constellations representing scientific instruments that were introduced in the 1750s by the Frenchman Nicolas Louis de Lacaille. As with so many of his constellations, Microscopium is little more than a filler, encompassing a few faint stars between better-known figures.

α (alpha) Microscopii, 20h 50m −33°.8, mag. 4.9, is a yellow giant star 381 l.y. away. It has a 10th-mag. companion, visible in small telescopes.

γ (gamma) Mic, 21h 01m −32°.3, mag. 4.7, is a yellow giant 224 l.y. away.

ε (epsilon) Mic, 21h 18m −32°.2, mag. 4.7, is a blue-white main-sequence star 165 l.y. away.

The Rosette Nebula, NGC 2237, in Monoceros, perhaps the most strikingly beautiful nebula in the heavens, surrounds the star cluster NGC 2244. See page 184. (Nigel Sharp/AURA/NOAO/NSF)

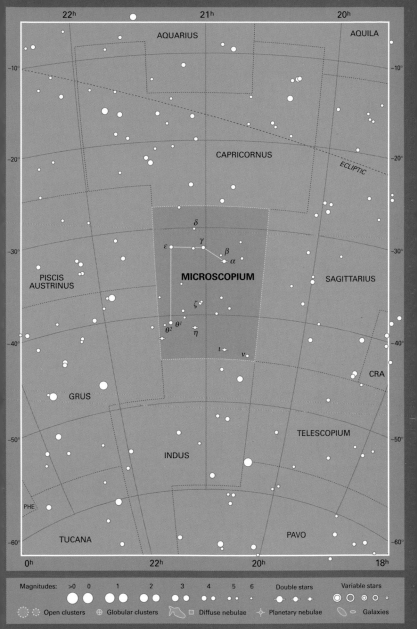

MONOCEROS   The Unicorn

A faint but fascinating constellation between Orion and Canis Minor, introduced by the Dutch theologian and astronomer Petrus Plancius in 1613, apparently because of references to a unicorn in the Old Testament of the Bible. Its location in the Milky Way ensures that it is well stocked with nebulae and clusters. Among the constellation's most celebrated features is Plaskett's Star, a mag. 6.1 spectroscopic binary named after the Canadian astronomer John S. Plaskett, who found in 1922 that it is the most massive pair of stars known; according to current data, it consists of two blue supergiants of masses 43 and 51 times that of the Sun, orbiting each other every 14.4 days. Plaskett's Star lies at 6h 37.4m, +6° 08′, near the open cluster NGC 2244, of which it may be an outlying member.

α (alpha) Monocerotis, 7h 41m −9°.6, mag. 3.9, is an orange giant star 144 l.y. away.

β (beta) Mon, 6h 29m −7°.0, 690 l.y. away, is rated as perhaps the finest triple star in the heavens. The smallest of telescopes should separate the three components, of mags. 4.6, 5.0 and 5.4, on a steady night. They form a curving arc of blue-white stars, the faintest two being the closest together.

δ (delta) Mon, 7h 12m −0°.5, mag. 4.2, is a blue-white star 375 l.y. away. It has a wide, unrelated naked-eye companion, 21 Mon, of mag. 5.5.

8 Mon, 6h 24m +4°.6, also known as ε (epsilon) Mon, is an easy double star for small telescopes, consisting of unrelated yellow and blue-white components of mags. 4.4 and 6.7, distances 128 and 79 l.y., in an attractive low-power field.

S Mon, 6h 41m +9°.9, also known as 15 Mon, is an intensely luminous blue-white star of mag. 4.7, situated 2600 l.y. away in the star cluster NGC 2264 (see page 186). It is a double star, with a close companion of mag. 7.6 visible in a small telescope. S Mon is slightly variable, fluctuating erratically by about 0.1 mag.

M50 (NGC 2323), 7h 03m −8°.3, is an open cluster of about 80 stars, half the apparent size of the full Moon, visible in binoculars and small telescopes. Apertures of 100 mm or so resolve it into a ragged patch of stars of 8th mag. and fainter, with a reddish star near its southern edge. M50 is 3300 l.y. away.

NGC 2232, 6h 27m −4°.7, is a scattered cluster of 20 stars for binoculars, containing the mag. 5.1 blue-white star 10 Mon. The cluster, which covers the same area of sky as the full Moon, lies 1200 l.y. away.

NGC 2237, NGC 2244, 6h 32m +4°.9, is a complex combination of a faint diffuse nebula, known as the Rosette Nebula, and a cluster of stars, all about 5000 l.y. away. Long-exposure photographs show the nebula as a pink loop, twice the apparent diameter of the full Moon. Visual observations with large amateur telescopes reveal only the brightest parts of the nebula, each of which is given a separate NGC number. The associated cluster, NGC 2244, consists of stars that have been born from the Rosette Nebula's gas; it is just visible to the naked eye and is an easy binocular object. The six most prominent stars of the cluster form a rectangular shape, although the brightest of them, 12 Mon, mag. 5.9, is not a true member but an unrelated foreground star. The cluster is ▶

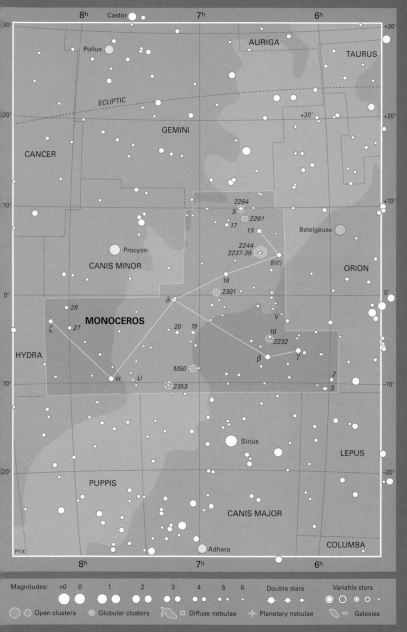

Magnitudes: >0 0 1 2 3 4 5 6 Double stars Variable stars

Open clusters Globular clusters Diffuse nebulae Planetary nebulae Galaxies

likely to be the only part of this celebrated object visible in a small telescope, but the pale outline of the nebula can just be made out in binoculars under clear, dark skies. (See photograph on page 182.)

NGC 2261, 6h 39m +8°.7, Hubble's Variable Nebula, is a small, faint, fan-shaped nebula containing the remarkable variable star R Mon. Its erratic brightness fluctuations, from mag. 9.5 down to about 12th mag., may be caused by the pangs of its birth from the surrounding nebula. This star and its nebula are open to study only by larger amateur telescopes; the star's distance is uncertain but it may be associated with the nearby NGC 2264 complex (see below), which would place it at about 2600 l.y.

NGC 2264, 6h 41m +9°.9, is another combination of star cluster and nebula. The cluster, visible in binoculars, has about 40 members, including the 5th-mag. S Mon (see page 184). The nebula, known as the Cone Nebula because of its tapered shape, shows up well only on long-exposure photographs and is beyond the reach of amateur telescopes. The distance of NGC 2264 is 2600 l.y.

NGC 2301, 6h 52m +0°.5, is a binocular cluster of 80 or so stars of 8th mag. and fainter, the brightest of them arranged in a vertical chain. It lies 2500 l.y. away.

NGC 2353, 7h 15m −10°.3, is an open cluster for small telescopes, consisting of about 30 stars of 9th mag. and fainter seemingly arranged in a spiral pattern. It lies about 3900 l.y. away.

MUSCA   The Fly

A small southern constellation lying at the foot of the Southern Cross. It is one of the 12 constellations introduced at the end of the 16th century by the Dutch navigators Pieter Dirkszoon Keyser and Frederick de Hout-man, originally under the name of Apis, the Bee. There is little of note in Musca other than part of the dark Coalsack Nebula, which spills into it from neighbouring Crux.

α (alpha) Muscae, 12h 37m −69°.1, mag. 2.7, is a blue-white star 306 l.y. away.

β (beta) Mus, 12h 46m −68°.1, 311 l.y. away, is a close pair of stars of mags. 3.6 and 4.0, requiring 100 mm aperture and high magnification to split. The orbital period of the pair is 400 years or so.

δ (delta) Mus, 13h 02m −71°.5, mag. 3.6, is an orange giant 91 l.y. away.

θ (theta) Mus, 13h 08m −65°.3, is a double star of mags. 5.6 and 7.6 for small telescopes. The brighter star is a blue supergiant, while its companion is a Wolf–Rayet star, a rare type of very hot star; it is the second-brightest such star in the sky, γ (gamma) Velorum being the brightest of all.

NGC 4833, 13h 00m −70°.9, is a fairly large, 7th-mag. globular cluster 18,000 l.y. away, visible in binoculars and small telescopes and resolvable into stars with an aperture of 100 mm.

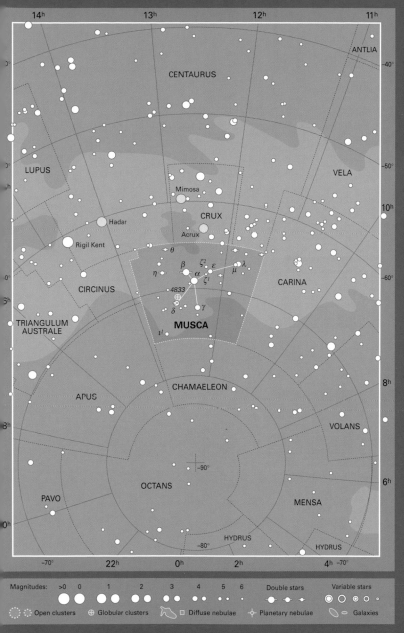

NORMA   The Set Square

A superfluous constellation invented in the 1750s by Nicolas Louis de Lacaille, originally under the name Norma et Regula, the set square and ruler. Its constituent stars were previously part of Ara, Lupus and Scorpius. Since Lacaille's time the boundaries of Norma have been altered, so that the stars which were α (alpha) and β (beta) Normae have been reabsorbed into Scorpius, where they are now anonymous. Norma lies in a rich region of the Milky Way.

$γ^2$ (gamma^2) Normae, 16h 20m −50°.2, mag. 4.0, is a yellow giant 128 l.y. away, the brightest star in the constellation. Next to it lies the far more distant yellow-white supergiant $γ^1$ (gamma^1) Normae, mag. 5.0, around 1500 l.y. away.

δ (delta) Nor, 16h 06m −45°.2, mag. 4.7, is a white star 123 l.y. away.

ε (epsilon) Nor, 16h 27m −47°.6, 400 l.y. away, is a double star with components of mags. 4.5 and 6.7 visible in small telescopes.

$ι^1$ (iota^1) Nor, 16h 04m −57°.8, 140 l.y. away, appears in small telescopes as a double star of mags. 4.6 and 8.1. In addition, the brighter star is itself a very close binary, divisible only in very large telescopes, with a 27-year orbital period.

NGC 6087, 16h 19m −57°.9, is a loose, large binocular cluster of about 40 stars, 3000 l.y. away, with chains of stars extending from it like a spider's legs. At the centre lies its brightest star, the Cepheid variable S Nor, which ranges from mag. 6.1 to 6.8 in 9.8 days.

At 15h 51.7m −51° 31′ in Norma lies the satisfyingly symmetrical planetary nebula known variously as Shapley 1 (Sp 1), PK 329+02.1 or RCW 100. Of 13th mag., it requires larger apertures to be seen. The central star is 14th mag. (Anglo–Australian Telescope Board)

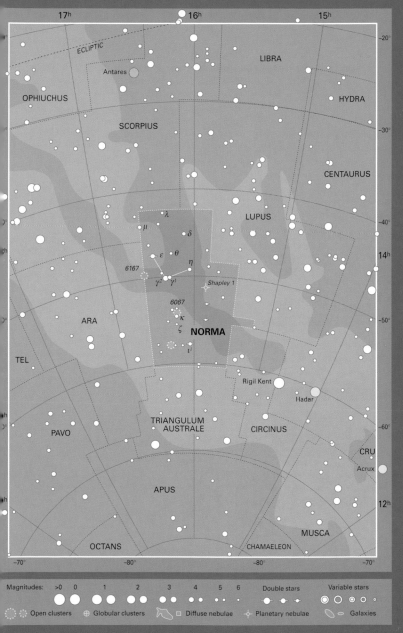

Magnitudes: >0 0 1 2 3 4 5 6   Double stars   Variable stars

Open clusters   Globular clusters   Diffuse nebulae   Planetary nebulae   Galaxies

OCTANS   The Octant

The constellation that contains the south pole of the sky. Despite this privileged position, Octans is faint and unremarkable. The south celestial pole forms a near-equilateral triangle with 5th-mag. τ (tau) and χ (chi) Octantis. There is no southern equivalent of Polaris, the north pole star; the nearest naked-eye star to the southern pole is 5th-mag. σ (sigma) Octantis. Currently this star lies about 1° from the pole, but the distance is increasing due to precession (see chart below); it was closest to the pole, under ¾°, around 1860. Octans commemorates the instrument known as the octant, a forerunner of the sextant, invented by the Englishman John Hadley and used by him for measuring star positions. The constellation itself was introduced in the 1750s by Nicolas Louis de Lacaille, and its dullness is a memorial to his lack of imagination.

α (alpha) Octantis, 21h 05m −77°.0, mag. 5.1, is a spectroscopic binary consisting of white and yellow giants 148 l.y. away.

β (beta) Oct, 22h 46m −81°.3, mag. 4.1, is a white star 140 l.y. away.

δ (delta) Oct, 14h 27m −83°.7, mag. 4.3, is an orange giant 279 l.y. away.

ε (epsilon) Oct, 22h 20m −80°.4, also known as BO Oct, is a red giant that varies semi-regularly between mags. 4.6 and 5.3 with an approximate period of 8 weeks. It lies 268 l.y. away.

λ (lambda) Oct, 21h 51m −82°.7, 435 l.y. away, is a double star with yellow and white components of mags. 5.5 and 7.2, individually visible in small telescopes.

ν (nu) Oct, 21h 41m −77°.4, mag. 3.7, the brightest star in the constellation, is an orange giant 69 l.y. away.

σ (sigma) Oct, 21h 09m −89°.0, mag. 5.4, the nearest naked-eye star to the south celestial pole, is a white giant 270 l.y. away.

The movement of the south celestial pole over 800 years as a result of precession. (Wil Tirion)

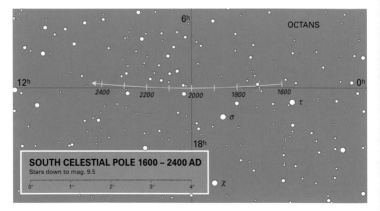

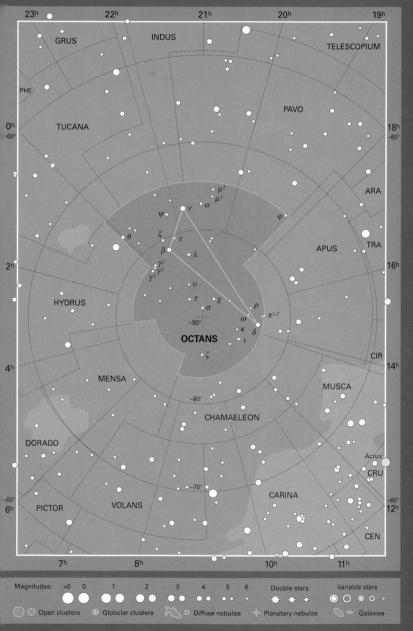

OPHIUCHUS The Serpent Holder

An ancient constellation, representing a man encoiled by a serpent (the constellation Serpens). Ophiuchus is usually identified as Aesculapius, a mythical healer who was a forerunner of Hippocrates; his reputed powers included the ability to raise the dead. The serpent he holds is a symbol of this power, since snakes are seemingly reborn every year when they shed their skin. Perhaps the most celebrated star in Ophiuchus is the mag. 9.5 red dwarf Barnard's Star, 5.9 l.y. away, the second-closest star to the Sun. It lies at 17h 57.8m, +4° 42′, and is named after the American astronomer E. E. Barnard who found in 1916 that it has the greatest proper motion of any star, covering the apparent diameter of the Moon every 180 years. The southernmost regions of Ophiuchus extend into rich starfields of the Milky Way, in the direction of the centre of the Galaxy; consequently the constellation is replete with star clusters. Ophiuchus was the site of the last supernova seen to erupt in our Galaxy, popularly called Kepler's Star, which appeared in 1604 at 17h 30.6m, −21° 29′, reaching mag. −3.

α (alpha) Ophiuchi, 17h 35m +12°.6, (Rasalhague, 'head of the serpent collector'), mag. 2.1, is a white main-sequence star 47 l.y. away.

β (beta) Oph, 17h 43m +4°.6, (Cebalrai, 'the shepherd's dog'), mag. 2.8, is an orange giant 82 l.y. away.

γ (gamma) Oph, 17h 48m +2°.7, mag. 3.7, is a blue-white main-sequence star 95 l.y. away.

δ (delta) Oph, 16h 14m −3°.7, (Yed Prior, 'the preceding star of the hand'), mag. 2.7, is a red giant 170 l.y. away.

ε (epsilon) Oph, 16h 18m −4°.7, (Yed Posterior, 'the following star of the hand'), mag. 3.2, is a yellow giant 108 l.y. away.

ζ (zeta) Oph, 16h 37m −10°.6, mag. 2.5, is a blue main-sequence star 458 l.y. away.

η (eta) Oph, 17h 10m −15°.7, (Sabik), mag. 2.4, is a blue-white main-sequence star 84 l.y. away.

ρ (rho) Oph, 16h 26m −23°.4, 400 l.y. away, is a striking multiple star for small instruments. The brightest component is mag. 5.0, with a close companion of mag. 5.7 visible in a small telescope under high magnification; either side of this pair are wide binocular companions of mags. 6.7 and 7.3.

τ (tau) Oph, 18h 03m −8°.2, 170 l.y. away, is a close pair of cream-coloured stars, mags. 5.2 and 5.9, orbiting every 280 years and gradually closing; they currently require at least 75 mm aperture and will need 100 mm by the year 2025.

36 Oph, 17h 15m −26°.6, 20 l.y. away, is a pair of mag. 5.1 orange dwarf stars split by small apertures. Their calculated orbital period is about 500 years.

70 Oph, 18h 05m +2°.5, 16 l.y. away, is a celebrated double star, consisting of yellow and orange components of mags. 4.2 and 6.0 which orbit each other in a period of 88 years. They are easily divisible in the smallest apertures and will continue to be so throughout the first half of the 21st century, being at their widest around 2025.

◄

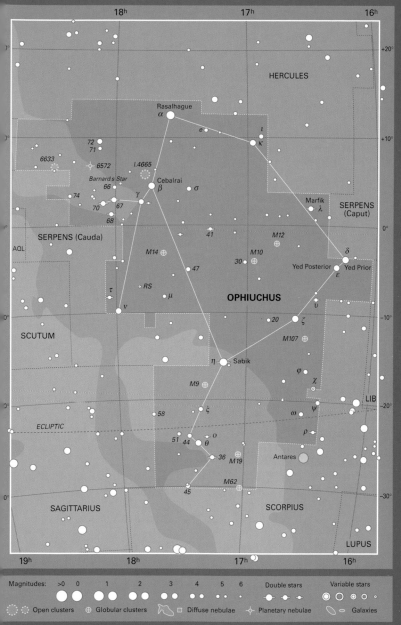

RS Oph, 17h 50m −6°.7, is a recurrent nova seen to have erupted five times, a record it shares with T Pyxidis and the fainter U Scorpii. Normally around 12th mag., RS Oph flared up to naked-eye brightness in 1898, 1933, 1958, 1967 and 1985.

M10 (NGC 6254), 16h 57m −4°.1, is a 7th-mag. globular cluster visible in binoculars or a small telescope. It is 14,000 l.y. away, somewhat closer than its neighbour M12. Individual stars can be resolved with telescopes of 75 mm aperture.

M12 (NGC 6218), 16h 47m −1°.9, is a 7th-mag. globular cluster 18,000 l.y. away, visible in binoculars or a small telescope. In small apertures it appears slightly larger and less easy to resolve than its neighbour M10, and its stars are more loosely scattered. There are several other globular clusters in Ophiuchus worthy of attention, but M10 and M12 are the finest.

NGC 6572, 18h 12m +6°.8, is a 9th-mag. planetary nebula 2000 l.y. away, visible in at least 75 mm aperture as a tiny blue-green ellipse.

NGC 6633, 18h 28m +6°.6, is a scattered binocular cluster of about 30 stars covering an area similar to that of the full Moon. It is 950 l.y. away.

IC 4665, 17h 46m +5°.7, is a loose and irregular open cluster of two dozen or so stars of 7th mag. and fainter, 1100 l.y. away, larger than the apparent size of the Moon and best seen in binoculars.

ORION   The Hunter

Without doubt the brightest and grandest constellation of all, crammed with objects of interest for all sizes of instrument. Orion's impressiveness stems in large measure from the fact that it contains an area of star formation in a nearby arm of our Galaxy, centred on the famous Orion Nebula. The Orion Nebula, M42, marks the Hunter's sword, hanging from his belt. The belt itself is formed by a line of three bright stars. Orion is depicted as brandishing a club and shield at the snorting Taurus, the Bull. Legend tells that the boastful Orion was stung to death by a scorpion, and was placed in the sky so that he sets as his slayer, represented by the constellation Scorpius, rises. Each year the Orionid meteors radiate from a point near the border with Gemini. As many as 25 Orionid meteors per hour may be seen around October 22.

α (alpha) Orionis, 5h 55m +7°.4, (Betelgeuse, corrupted from an Arabic name referring to a hand), 427 l.y. away, is a red supergiant star about 500 times the size of the Sun, so large that it is unstable. It fluctuates erratically in size, changing in brightness as it does so from mag. 0.0 to 1.3, making it the most obviously variable of all first-magnitude stars; the average brightness is around mag. 0.5.

β (beta) Ori, 5h 15m −8°.2, (Rigel, 'foot'), at mag. 0.2 the brightest star in Orion, is a blue-white supergiant 773 l.y. away; note its colour contrast with Betelgeuse. Rigel has a mag. 6.8 companion, difficult to see in small telescopes, particularly in poor seeing, because of glare from the primary. ▶

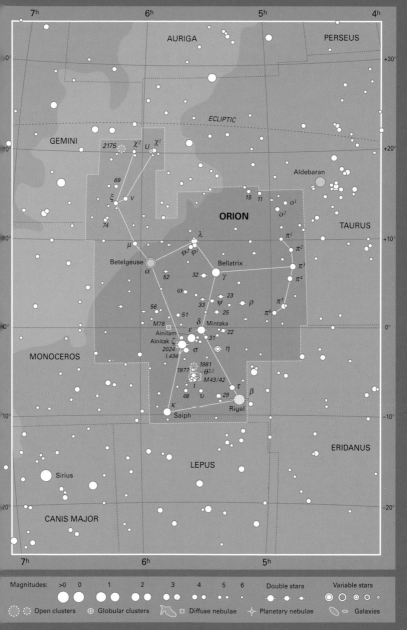

γ (gamma) Ori, 5h 25m +6°.3, (Bellatrix, 'the female warrior'), mag. 1.6, is a blue giant star 243 l.y. away.

δ (delta) Ori, 5h 32m −0°.3, (Mintaka, 'belt'), about 2000 l.y. away, is a complex multiple star. It is a blue giant which appears of mag. 2.2 to the naked eye. Binoculars or small telescopes reveal a wide companion of mag. 6.9. The brighter star is also an eclipsing binary that varies by about 0.1 mag. every 5.7 days.

ε (epsilon) Ori, 5h 36m −1°.2, (Alnilam, 'the string of pearls'), mag. 1.7, is a blue supergiant about 1300 l.y. away.

ζ (zeta) Ori, 5h 41m −1°.9, (Alnitak, 'the girdle'), 820 l.y. away, is a blue supergiant that appears to the naked eye as mag. 1.7. But telescopes of 75 mm aperture and above reveal a close companion of mag. 3.9 that is estimated to orbit it every 1500 years. There is also a much wider 10th-mag. star.

η (eta) Ori, 5h 24m −2°.4, 900 l.y. away, is a complex multiple–variable. A telescope of at least 100 mm aperture at high magnification is needed to show that it consists of two close stars, of mags. 3.8 and 4.8. The brighter star is also an eclipsing binary, varying by 0.3 mag. every 8 days.

θ^1 (theta¹) Ori, 5h 35m −5°.4, about 1500 l.y. away, is a multiple star at the heart of the Orion Nebula, from which it has recently formed and which it now illuminates. This star is popularly known as the Trapezium, because a small telescope shows four stars here; but 100 mm aperture also reveals two others, of 11th mag. The four main stars of the Trapezium are of mags. 5.1, 6.7, 6.7 and 8.0. Nearby lies θ^2 (theta²) Orionis, a binocular double of mags. 5.1 and 6.4.

ι (iota) Ori, 5h 35m −5°.9, 1300 l.y. away, is a double star on the southern edge of the Orion Nebula, divisible in a small telescope. Its components are of mags. 2.8 and 6.9. Also visible in the same field of view is a wider double of blue-white stars, Σ 747, of mags. 4.8 and 5.7.

κ (kappa) Ori, 5h 48m −9°.7, (Saiph, 'sword'), mag. 2.1, is a blue supergiant 722 l.y. away.

λ (lambda) Ori, 5h 35m +9°.9, is a blue giant of mag. 3.5 with a mag. 5.6 companion visible in small telescopes under high magnification. The stars lie about 1060 l.y. away.

σ (sigma) Ori, 5h 39m −2°.6, 1150 l.y. away, is perhaps the most impressive of all Orion's stellar treasures. To the naked eye it appears as a blue-white star of mag. 3.8, but small telescopes reveal much more. On one side of the star are blue-white companions of mags. 6.8 and 6.6, the wider of which can be glimpsed in binoculars; it is an eclipsing binary with a range of about 0.1 mag. On the opposite side is a closer 9th-mag. companion, which is more difficult to see because of glare from the primary. The effect is like a planet with moons. To complete the picture, in the same telescopic field of view is a faint triple star called Σ 761, consisting of a narrow triangle of 8th- and 9th-mag. stars. An extraordinarily rich and unexpected sight, to be returned to again and again.

U Ori, 5h 56m +20°.2, is a huge red giant variable of Mira type, many hundreds of times the diameter of the Sun, whose brightness ranges between 5th and 13th mags. in about 1 year.

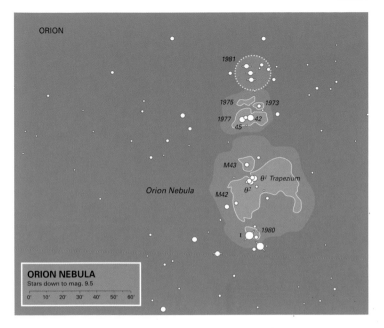

Detail chart of the Orion Nebula region (Wil Tirion)

M42, M43 (NGC 1976, NGC 1982), 5h 35m −5°.4, the Orion Nebula, is one of the greatest deep-sky wonders – a gigantic cloud of gas and dust, 1500 l.y. away and 20 l.y. in diameter, from which a star cluster is being born. Behind the visible part of the nebula, illuminated by the stars of the Trapezium (see θ Ori), radio and infrared astronomers have detected an even larger dark cloud in which more stars are forming. M42 covers an area greater than 1° × 1°, and is indisputably the finest diffuse nebula in the sky, clearly visible to the naked eye as a hazy cloud. Binoculars and small telescopes reveal some of the more prominent wreaths and swirls of gas, which become more complex and breathtaking with increasing aperture. Although colour photographs depict the nebula as red and blue (see page 264), to the eye it appears distinctly greenish because of the different colour sensitivities of photographic film and the human eye. A dark lane of dust separates M42 from M43, a smaller and rounder patch to the north that is really part of the same huge gas cloud; M43 is centred on a 7th-mag. star.

M78 (NGC 2068), 5h 47m +0°.0, is a small, elongated reflection nebula looking like a short-tailed comet, with a 10th-mag. double star at its head.

NGC 1977, 5h 36m −4°.9, 1500 l.y. away, is an elongated nebulosity just north of the Orion Nebula, centred on the mag. 4.6 blue giant star 42 Orionis, also known as c Orionis. This object would be more celebrated were it not so over-shadowed by M42.

NGC 1981, 5h 35m −4°.4, is a scattered cluster of about 20 stars of 6th mag. and fainter, 1300 l.y. away, to the north of the nebulosity NGC 1977. Included in this cluster is the double star Σ 750, a neat pair of 7th- and 8th-mag. stars.

NGC 2024, 5h 41m −2°.4, is a mushroom-shaped cloud of gas about ½° wide surrounding the star ζ (zeta) Orionis. Running south from ζ Ori is a strip of nebulosity, IC 434, into which is indented the celebrated Horsehead Nebula, a dark cloud of obscuring dust shaped like a horse's head (see page 266). Although long-exposure photographs show NGC 2024 and the Horsehead Nebula well, they are notoriously difficult to detect with amateur telescopes.

PAVO   The Peacock

A constellation introduced at the end of the 16th century by the Dutch navigators Pieter Dirkszoon Keyser and Frederick de Houtman. It is one of several celestial birds in the region, which include Apus, Tucana, Grus and Phoenix. In Greek mythology the peacock was sacred to Hera, goddess of the heavens, from whose breast the Milky Way sprang. Hera set a creature with a hundred eyes called Argos to watch over a white heifer, into which she guessed her husband Zeus had turned one of his illicit lovers, the nymph Io. At the request of Zeus, Hermes slew the watchful Argos and released the heifer. Hera placed the hundred eyes of Argos on the peacock's tail.

α (alpha) Pavonis, 20h 26m −56°.7, (Peacock), mag. 1.9, is a blue-white star 183 l.y. away.

β (beta) Pav, 20h 45m −66°.2, mag. 3.4, is a white star 138 l.y. away.

δ (delta) Pav, 20h 09m −66°.2, mag. 3.6, is a yellow star 20 l.y. away.

η (eta) Pav, 17h 46m −64°.7, mag. 3.6, is an orange giant 371 l.y. away.

κ (kappa) Pav, 18h 57m −67°.2, 544 l.y. away, is one of the brightest Cepheid variables. It is a yellow-white supergiant, varying between mags. 3.9 and 4.8 in 9.1 days.

ξ (xi) Pav, 18h 23m −61°.5, 420 l.y. away, is a red giant of mag. 4.4 with a close companion of mag. 8.6, lost in the primary's glare in small telescopes.

SX Pav, 21h 29m −69°.5, 396 l.y. away, is a red giant that varies semi-regularly between mags. 5.3 and 6.0 every 7 weeks or so.

NGC 6744, 19h 10m −63°.9, is a 9th-mag. spiral galaxy for small telescopes, with a short central bar and widespread arms. Photographs show it to be one of the largest barred spirals, but in modest apertures only the brightest central portion will be seen. It is about 25 million l.y. away.

NGC 6752, 19h 11m −60°.0, is a large 6th-mag. globular cluster, visible in binoculars and resolvable in 75-mm telescopes, covering half the apparent diameter of the Moon. A double star of 8th and 9th mags. at one edge is a foreground object. The cluster lies 15,000 l.y. away.

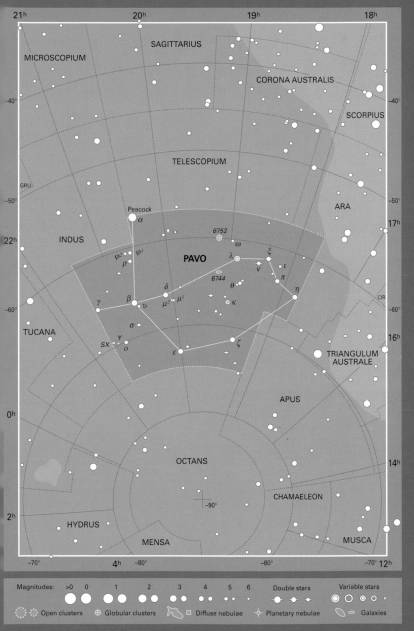

21ʰ 20ʰ 19ʰ 18ʰ

MICROSCOPIUM

SAGITTARIUS

CORONA AUSTRALIS

-40°

SCORPIUS

TELESCOPIUM

GRU

ARA

-50°

17ʰ

Peacock
α
6752
ω
ξ
λ
t
ν
π
φ² φ¹
ρ
PAVO
6744
θ
η
δ
κ
γ
β
ν
μ² μ¹
σ
SX
γ
ο
ε
ζ

INDUS

22ʰ

TUCANA

-60°

CIR

16ʰ

TRIANGULUM
AUSTRALE

0ʰ

APUS

OCTANS
-90°

CHAMAELEON

14ʰ

2ʰ

HYDRUS

MENSA

MUSCA

-70° 12ʰ

4ʰ -80° -80° -70°

Magnitudes: >0 0 1 2 3 4 5 6 Double stars Variable stars

Open clusters ⊕ Globular clusters □ Diffuse nebulae ✧ Planetary nebulae Galaxies

PEGASUS

The winged horse of Greek mythology, born from the blood of Medusa after she was slain by Perseus, who lies nearby in the sky. The most famous feature of Pegasus is the Great Square, outlined by four stars. One of these stars, once known as δ (delta) Pegasi, is nowadays assigned to Andromeda. The Great Square of Pegasus is over 15° wide and 13° high, yet contains surprisingly few naked-eye stars for such a large area. The brightest are: υ (upsilon), mag. 4.4; τ (tau), mag. 4.6; ψ (psi), mag. 4.6; 56, mag. 4.8; φ (phi), mag. 5.1; 71, mag. 5.3; and 75, mag. 5.5.

α (alpha) Pegasi, 23h 05m +15°.2, (Markab, 'shoulder'), mag. 2.5, is a blue-white giant star 140 l.y. away.

β (beta) Peg, 23h 04m +28°.1, (Scheat, 'shin'), 199 l.y. away, is a red giant that varies from mag. 2.3 to 2.7 with no definite period.

γ (gamma) Peg, 0h 13m +15°.2, (Algenib, 'the side'), mag. 2.8, is a blue-white star 333 l.y. away. It is a pulsating variable of the β Cephei type, but its fluctuations every 3 hours 40 minutes are only 0.1 mag., too slight to be discernible with the naked eye.

ε (epsilon) Peg, 21h 44m +9°.9, (Enif, 'nose'), 670 l.y. away, is an orange super-giant of mag. 2.4. A small telescope, or even good binoculars, reveals a wide bluish mag. 8.4 companion star. Larger telescopes also show an 11th-mag. companion closer to ε Peg, making this an apparent triple system. ε Peg is catalogued as variable because of two unexplained fluctuations: one night in November 1847 it was seen about a magnitude fainter than normal, while on a night in September 1972 it was seen at about magnitude 0.7 for four minutes before fading back to normal.

ζ (zeta) Peg, 22h 41m +10°.8, (Homam), mag. 3.4, is a blue-white star 209 l.y. away.

η (eta) Peg, 22h 43m +30°.2, (Matar), mag. 2.9, is a yellow giant 215 l.y. away.

π (pi) Peg, 22h 10m +33°.2, is a very wide binocular duo of white and yellow stars, both giants. They are of mags. 4.3 and 5.6, and lie 252 and 283 l.y. away respectively.

51 Peg, 22h 57m +20°.8, mag. 5.5, is a yellow main-sequence star 50 l.y. away, similar to the Sun. In 1995 it became the first star other than the Sun known to have a planet; the planet's estimated mass is about half that of Jupiter.

M15 (NGC 7078), 21h 30m +12°.2, is an outstanding 6th-mag. globular cluster, 33,000 l.y. distant, at the limit of naked-eye visibility but easily seen in binoculars; a 6th-mag. star nearby is a sure guide to its location. A small telescope shows it as a glorious misty sight in an attractive field. Apertures of 150 mm or so resolve its outer regions into a mottled ground of sparkling stars, and larger telescopes show stars all the way to the bright and condensed core.

NGC 7331, 22h 37m +34°.4, is a 10th-mag. spiral galaxy seen nearly edge-on, visible under good conditions in apertures of 100 mm or so as an elongated smudge. It lies about 50 million l.y. away.

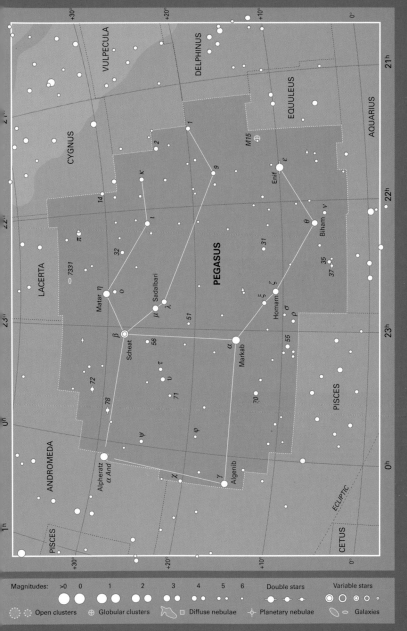

PEGASUS

VULPECULA
DELPHINUS
CYGNUS
EQUULEUS
AQUARIUS
LACERTA
ANDROMEDA
PISCES
CETUS

M15
Enif ε
θ Biham ν
35
37
31
Matar η
o
μ Sadalbari λ
51
ξ ζ Homam
ρ σ
β Scheat
56
τ υ
α Markab
55
71
70
φ
ψ
72
78
χ γ Algenib
Alpheratz α And

7331
π
32
14
κ
ι
2
γ
9

ECLIPTIC

Magnitudes: >0 0 1 2 3 4 5 6   Double stars   Variable stars

Open clusters   Globular clusters   Diffuse nebulae   Planetary nebulae   Galaxies

PERSEUS

Perseus was the hero of Greek mythology who rescued the chained maiden Andromeda from the clutches of the sea monster Cetus. Prior to that, Perseus had slain Medusa the Gorgon, whose head he is pictured holding in one hand. The Gorgon's head is marked by the winking star Algol, sometimes imagined as Medusa's evil eye. Perseus lies in a rich part of the Milky Way and is well worth sweeping with binoculars; note, in particular, the Double Cluster. In 1901 Nova Persei flared up to magnitude 0.2 at 3h 31.2m, +43° 54', throwing off a shell of gas that is now visible in large telescopes. NGC 1499, the California Nebula, so named because its shape resembles California, spans the apparent width of five full Moons north of 4th-magnitude ξ (xi) Persei, an exceedingly hot blue giant or supergiant which illuminates it. Despite its considerable size the California Nebula is elusive visually, but shows up well on long-exposure photographs. At 3h 19.8m, +41° 31', lies the radio source Perseus A, associated with the 12th-magnitude peculiar galaxy NGC 1275, which is at the centre of the Perseus cluster of galaxies, 250 million l.y. away. Near γ (gamma) Persei lies the radiant of the Perseid meteors, the most glorious meteor shower of the year: around August 12–13 as many as 75 bright meteors can be seen flashing from Perseus each hour.

α (alpha) Persei, 3h 24m +49°.9, (Mirphak, 'elbow', or Algenib, 'the side'), mag. 1.8, is a yellow-white supergiant 592 l.y. away. Binoculars reveal a brilliant scattering of stars covering 3° in this region forming a loose cluster, Melotte 20. α Per itself seems to form the head of a snaking chain of stars.

β (beta) Per, 3h 08m +41°.0, (Algol, 'the demon'), 93 l.y. away, is one of the most celebrated variable stars in the sky. It is the prototype of the eclipsing binary class of variables, in which two close stars periodically eclipse each other as they orbit their common centre of gravity. Algol's eclipses occur every 2.87 days, when the star's apparent brightness sinks from mag. 2.1 to 3.4 before returning to normal about 10 hours later.

γ (gamma) Per, 3h 05m +53°.5, mag. 2.9, is a yellow giant 256 l.y. away. It is an eclipsing binary with the unusually long period of 14.6 years, dimming by 0.3 mag. for 10 days. This star's variability was first detected in 1990.

δ (delta) Per, 3h 43m +47°.8, mag. 3.0, is a blue giant 528 l.y. away.

ε (epsilon) Per, 3h 58m +40°.0, 538 l.y. away, is a blue-white star of mag. 2.9 with an unrelated mag. 7.6 companion, difficult to see through the smallest telescopes because of the magnitude contrast.

ζ (zeta) Per, 3h 54m +31°.9, 980 l.y. away, is a blue supergiant of mag. 2.9 with a mag. 9.5 companion visible in a small telescope.

η (eta) Per, 2h 51m +55°.9, about 1300 l.y. away, is an orange supergiant of mag. 3.8 with a mag. 8.5 blue companion that forms an attractive double for small telescopes. The field of view contains a sprinkling of background stars.

ρ (rho) Per, 3h 05m +38°.8, 325 l.y. away, is a red giant that varies between mags. 3.3 and 4.0 in semi-regular fashion every 7 weeks or so.

▶

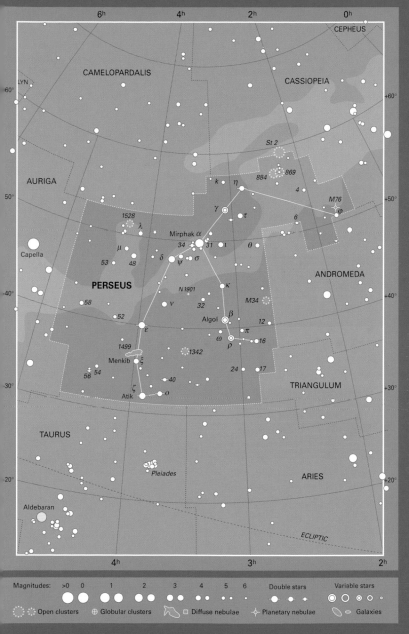

CEPHEUS

CAMELOPARDALIS

CASSIOPEIA

LYN

St 2

884   869

4

AURIGA

κ   η

M76

φ

1528

λ

γ

τ

6

Mirphak α

34 31 ι

θ

Capella

μ

53   48

δ

ψ   σ

PERSEUS

ANDROMEDA

58

ν

N 1901

32

κ

M34

52

ε

Algol β

12

1499

ω

π

16

Menkib ξ

1342

ρ

56 54

ζ

40

24

17

Atik

o

TRIANGULUM

TAURUS

Pleiades

ARIES

Aldebaran

ECLIPTIC

Magnitudes: >0   0   1   2   3   4   5   6   Double stars   Variable stars

Open clusters   ⊕ Globular clusters   Diffuse nebulae   Planetary nebulae   Galaxies

M34 (NGC 1039), 2h 42m +42°.8, is a bright star cluster at the limit of naked-eye visibility. It is far less rich and condensed than the Double Cluster, containing about 60 stars splashed over an area larger than the apparent size of the full Moon. Binoculars resolve it into stars, and it is well seen in a small telescope, many of the stars seeming to form pairs. M34 lies about 1500 l.y. away.

M76 (NGC 650–1), 1h 42m +51°.6, the Little Dumbbell, is a planetary nebula, the faintest object on Messier's list with a magnitude of about 10 and hence difficult to see, although it can be picked up in 100 mm apertures on a dark night. It is relatively large for a planetary nebula, similar in size to the Ring Nebula in Lyra, but smaller than its namesake the Dumbbell Nebula in Vulpecula, which its elongated shape resembles. Each end of the Little Dumbbell has a separate NGC number. It lies about 3500 l.y. away.

NGC 869, NGC 884, 2h 19m +57°.2, 2h 22m +57°.1, the famous Double Cluster in Perseus, also known as h and χ (chi) Persei. They are two open star clusters visible to the naked eye and superb in binoculars, each covering an area equal to the full Moon. NGC 869 is the brighter and richer of the pair, containing an estimated 200 stars, while NGC 884 appears more scattered. They both lie about 7200 l.y. away and are relatively young, only a few million years old. Small telescopes have an advantage when observing these objects, for at low powers both clusters fit into the same field of view, which is not always the case with larger and more powerful telescopes. Most of the stars in the clusters are blue-white, but there are several red giants to be spotted in and around NGC 884. In binoculars, note a curving chain of stars leading northwards towards a large starry patch known as the cluster Stock 2 (abbreviated St 2). This whole area is a breathtaking sight in all apertures. (For a photograph, see page 270.)

PHOENIX   The Phoenix

An inconspicuous constellation near the southern end of Eridanus, representing the mythical bird that was regularly reborn from its own ashes. It was introduced at the end of the 16th century by the Dutch navigators Pieter Dirkszoon Keyser and Frederick de Houtman in an area that had been known by the Arabs as the Boat, moored on the bank of the river Eridanus.

α (alpha) Phoenicis, 0h 26m –42°.3, (Ankaa), mag. 2.4, is an orange giant star 77 l.y. away.

β (beta) Phe, 1h 06m –46°.7, 185 l.y. away, appears to the naked eye as a yellow star of mag. 3.3. In fact, it is a close double with well-matched components of mags. 4.0 and 4.2, currently too close for resolution in small telescopes.

γ (gamma) Phe, 1h 28m –43°.6, mag. 3.4, is an orange giant 234 l.y. away.

ζ (zeta) Phe, 1h 08m –55°.2, 280 l.y. away, is a complex variable and multiple star. The main star is a blue-white eclipsing binary of Algol type that fluctuates between mags. 3.9 and 4.4 every 1.67 days. It has an 8th-mag. companion visible in a small telescope. There is also a much closer 7th-mag. companion visible only in large apertures.

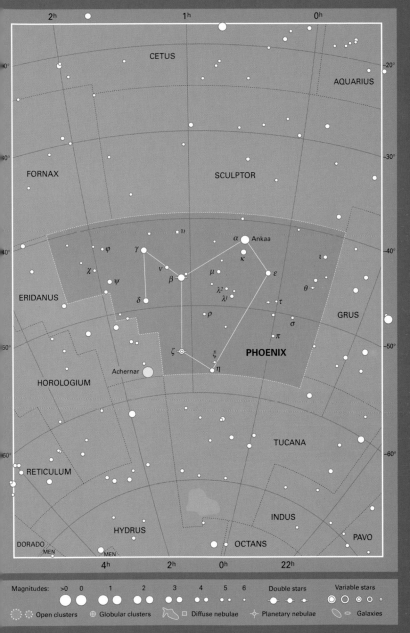

PICTOR   The Painter's Easel

A faint constellation overshadowed by the neighbouring brilliant star Canopus in Carina on one side and the Large Magellanic Cloud in Dorado on the other. The constellation was invented in the 1750s by Nicolas Louis de Lacaille, who originally called it Equuleus Pictoris, which has since been shortened. At 5h 11.7m, –45° 01′ lies the mag. 8.9 red dwarf known as Kapteyn's Star, 12.8 l.y. away, named after the Dutch astronomer who discovered in 1897 that it has the second-largest proper motion of any known star (the record is held by Barnard's Star in Ophiuchus). Kapteyn's Star moves 1° in 415 years; see illustration below.

α (alpha) Pictoris, 6h 48m –61°.9, mag. 3.2, is a white star 99 l.y. away.

β (beta) Pic, 5h 47m –51°.1, mag. 3.9, is a blue-white main-sequence star 63 l.y. away. This star became famous in 1984 when astronomers photographed a disk of dust and gas around it, thought to be a planetary system in the process of formation.

γ (gamma) Pic, 5h 50m –56°.2, mag. 4.5, is an orange giant 174 l.y. away.

δ (delta) Pic, 6h 10m –55°.0, is a blue-white star about 1700 l.y. away. It is an eclipsing binary of the β (beta) Lyrae type, varying from mag. 4.7 to 4.9 every 1.67 days.

ι (iota) Pic, 4h 51m –53°.5, 120 l.y. away, is an easy double of mags. 5.6 and 6.4 for small telescopes.

Chart showing the proper motion of Kapteyn's Star over a period of 200 years. (Wil Tirion)

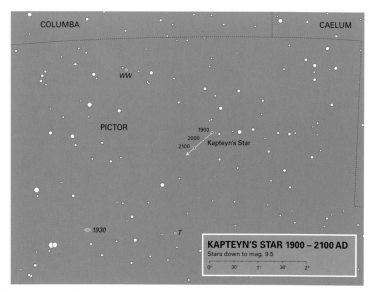

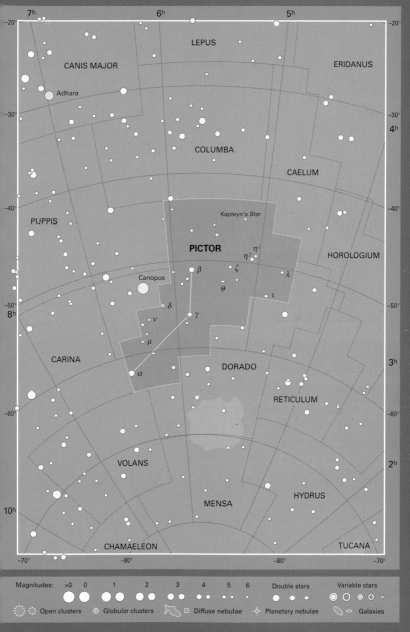

PISCES   The Fishes

An ancient constellation representing two fishes tied by their tails, the knot being marked by the star α (alpha) Piscium. One legend identifies the constellation with Aphrodite and her son Eros, who swam away from the attack of the monster Typhon in the guise of fishes. The Sun is in Pisces from mid-March to late April, so the constellation contains the March (or *vernal*) equinox – the point at which the Sun crosses the celestial equator into the northern celestial hemisphere each year. This point originally lay in neighbouring Aries, but precession has now carried it into Pisces; eventually, in just under 600 years' time, it will pass on into Aquarius. The constellation's most distinctive feature is a ring of seven stars called the Circlet; clockwise from north they are θ (theta), 7, γ (gamma), κ (kappa), λ (lambda), TX (or 19) and ι (iota) Piscium.

α (alpha) Piscium, 2h 02m +2°.8, (Alrescha, 'the cord'), 139 l.y. away, appears to the naked eye as a star of mag. 3.8, but in fact is a challenging double with an orbital period of over 900 years. Its components, of mags. 4.2 and 5.2, are gradually closing but will remain within range of 100-mm aperture telescopes for at least the first half of the 21st century. The brighter star is a spectroscopic binary, and the fainter one may be also. Their colour is bluish-white, although some observers see the brighter star as greenish.

β (beta) Psc, 23h 04m +3°.6, mag. 4.5, is a blue-white star 493 l.y. away.

γ (gamma) Psc, 23h 17m +3°.3, mag. 3.7, is a yellow giant 131 l.y. away. ▶

M74 in Pisces is a spiral galaxy with loosely wound arms. See page 210. (Todd Boroson/AURA/NOAO/NSF)

Magnitudes: >0 0 1 2 3 4 5 6   Double stars   Variable stars

Open clusters   Globular clusters   Diffuse nebulae   Planetary nebulae   Galaxies

ζ (zeta) Psc, 1h 14m +7°.6, 148 l.y. away, is a wide double of mags. 5.2 and 6.4, divisible in the smallest telescopes.

η (eta) Psc, 1h 31m +15°.3, mag. 3.6, is the brightest star in the constellation. It is a yellow giant, 294 l.y. away.

κ (kappa) Psc, 23h 27m +1°.3, mag. 4.9, is a blue-white star 162 l.y. away. It forms a wide binocular double with 9 Piscium, mag. 6.3, actually a background star.

ρ (rho) Psc, 1h 26m +19°.2, mag. 5.3, is a white star 85 l.y. away that forms an easy binocular duo with the unrelated orange giant 94 Piscium, mag. 5.5, 307 l.y. away.

ψ^1 (psi[1]) Psc, 1h 06m +21°.5, is a wide pair of blue-white stars of mags. 5.3 and 5.6 about 230 l.y. away, visible in small telescopes or even good binoculars.

TV Psc, 0h 28m +17°.9, 490 l.y. away, is a red giant that varies semi-regularly between mags. 4.7 and 5.4 every 7 weeks.

TX Psc, 23h 46m +3°.5, also known as 19 Psc, 760 l.y. away, is a deep-red irregular variable star, visible with the naked eye or binoculars. It fluctuates between about mags. 4.8 and 5.2.

M74 (NGC 628), 1h 37m +15°.8, is a face-on spiral galaxy of 9th mag., 30 million l.y. away. In dark conditions it can be glimpsed through a small telescope as a pale disk with a starlike nucleus and faint foreground stars dotted across it, but it needs an aperture of at least 150 mm to be well seen. (See photograph on page 208.)

PISCIS AUSTRINUS   The Southern Fish

Sometimes called Piscis Australis, this constellation has been known since ancient times and is often represented as a fish drinking the flow of water from the urn of neighbouring Aquarius. This fish was said to be the parent of the two zodiacal fishes, represented by Pisces.

α (alpha) Piscis Austrini, 22h 58m −29°.6, (Fomalhaut, 'the fish's mouth'), mag. 1.2, is a blue-white main-sequence star 25 l.y. away. It is surrounded by a disk of cool dust from which a planetary system may be forming.

β (beta) PsA, 22h 32m −32°.3, 148 l.y. away, is a wide double star consisting of a mag. 4.3 blue-white primary and a mag. 7.7 companion visible in small telescopes.

γ (gamma) PsA, 22h 53m −32°.9, 222 l.y. away, is a double star of mags. 4.5 and 8.0, made difficult to split in small telescopes by the magnitude contrast.

η (eta) PsA, 22h 01m −28°.5, 1000 l.y. away, is a close pair of blue-white stars of mags. 5.8 and 6.8, divisible with an aperture of 100 mm and high power.

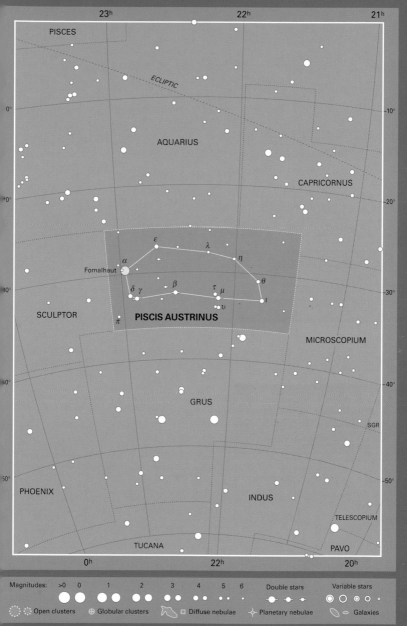

PUPPIS   The Stern

This is the largest of the three sections into which the ancient constellation of Argo Navis, the ship of the Argonauts, was dismembered in 1763 by Nicolas Louis de Lacaille; the other sections are Carina and Vela. When Argo Navis was split up its stars retained their existing Greek letters, and it so happens that the labelling of the stars assigned to Puppis begins with ζ (zeta). Puppis lies in the Milky Way and contains rich starfields for sweeping with binoculars.

ζ (zeta) Puppis, 8h 04m −40°.0, (Naos, 'ship'), mag. 2.2, is a brilliant blue supergiant about 1400 l.y. away, the hottest (and hence bluest) of all naked-eye stars, with a surface temperature of about 40,000°C.

ξ (xi) Pup, 7h 49m −24°.9, mag. 3.3, is a yellow supergiant about 1350 l.y. away. Binoculars reveal a wide, unrelated mag. 5.3 yellow companion, 321 l.y. away.

π (pi) Pup, 7h 17m −37°.1, mag. 2.7, is an orange supergiant 1100 l.y. away.

ρ (rho) Pup, 8h 08m −24°.3, mag. 2.8, is a yellow-white star 63 l.y. away. It is a variable of the δ (delta) Scuti type, fluctuating by 0.2 mag. with a period of 3 hours 23 minutes.

k Pup, 7h 39m −26°.8, 454 l.y. away, is a striking double star with blue-white components of mags. 4.5 and 4.6 easily divisible in small telescopes.

L¹ L² Pup, 7h 13m −45°.2, is an optical double consisting of two unrelated and contrasting stars. L¹ is mag. 4.9, a blue-white star 182 l.y. away. L² is a red giant semi-regular variable, 198 l.y. away, that fluctuates between 3rd and 6th mags. every 140 days or so.

V Pup, 7h 58m −49°.2, is an eclipsing binary of the β (beta) Lyrae type, about 1200 l.y. away. It varies from mag. 4.4 to 4.9 with a period of 35 hours.

M46 (NGC 2437), 7h 42m −14°.8, is a 6th-mag. open cluster of about 100 faint stars of remarkably uniform brightness, most of them around 10th mag. With its neighbour M47 it is visible to the naked eye as a brighter knot in the Milky Way. In binoculars it appears as a smudgy patch two-thirds the size of the full Moon, while a small telescope shows it as a sprinkling of stardust. M46 lies 5200 l.y. away. On its northern edge lies the 10th-mag. planetary nebula NGC 2438. This is not a member of the cluster, but is a foreground object about 3000 l.y. from us.

M47 (NGC 2422), 7h 37m −14°.5, is a scattered naked-eye cluster of similar apparent size to the full Moon. It contains three dozen or so stars, the brightest being of mag. 5.7. M47 lies about 1500 l.y. away, less than one-third the distance of its richer neighbour M46.

M93 (NGC 2447), 7h 45m −23°.9, is a 6th-mag. binocular cluster, 3600 l.y. away, consisting of 80 stars of 8th mag. and fainter, arranged in a wedge-shape.

NGC 2451, 7h 45m −38°.0, is a large and bright open cluster of 40 stars 1050 l.y. away, well seen in binoculars, centred on the mag. 3.6 orange giant c Puppis.

NGC 2477, 7h 52m −38°.5, is a large 6th-mag. open cluster consisting of a swarm of faint stars, looking in binoculars like a loose globular, seemingly with arms. This excellent object would undoubtedly have featured on Messier's list had he lived farther south. It lies 4000 l.y. away.

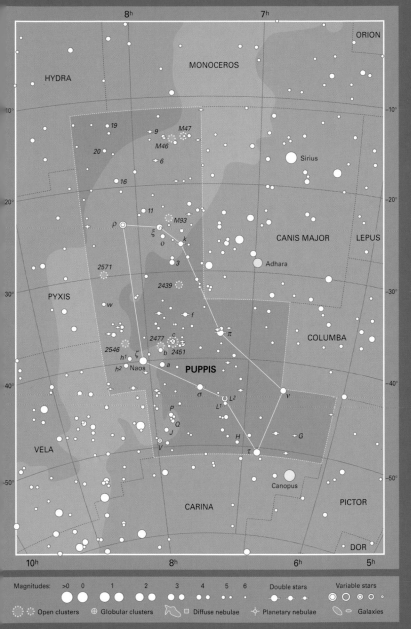

PYXIS   The Compass

A small constellation invented by Nicolas Louis de Lacaille in the 1750s, representing a magnetic compass. It lies near Puppis, the stern of the Argonauts' ship, and was formed from stars that Ptolemy had catalogued as being part of the mast of Argo; but that ship, of course, would not have had a magnetic compass, so Pyxis cannot really be considered a genuine part of the old Argo. This area has also been known as Malus, the mast of Argo. Pyxis contains no objects of particular interest to users of small telescopes, despite the fact that it lies in the Milky Way.

α (alpha) Pyxidis, 8h 44m −33°.2, mag. 3.7, is a blue giant star 845 l.y. away.

β (beta) Pyx, 8h 40m −35°.3, mag. 4.0, is a yellow giant 388 l.y. away.

γ (gamma) Pyx, 8h 51m −27°.7, mag. 4.0, is an orange giant 209 l.y. away.

T Pyx, 9h 05m −32°.4, is a recurrent nova that has undergone five recorded eruptions, in 1890, 1902, 1920, 1944 and 1966. Normally it is of mag. 14, but brightens to 6th or 7th magnitude. Further outbursts may be expected.

Nicolas Louis de Lacaille (1713–1762)

Lacaille, a French astronomer, was the first person to map the southern skies comprehensively, as a result of which he became known as the Father of Southern Astronomy. He directed an expedition of the French Academy of Sciences to the Cape of Good Hope in 1750–54, where he systematically surveyed the southern celestial hemisphere, listing over 10,000 stars. The accurate positions of 2000 of these, along with a star map, were published posthumously in 1763 under the title *Coelum Australe Stelliferum*. Lacaille is usually best remembered for the 14 new constellations he introduced, representing instruments used in science and the fine arts: Antlia, Caelum, Circinus, Fornax, Horologium, Mensa, Microscopium, Norma, Octans, Pictor, Pyxis, Reticulum, Sculptor and Telescopium. Lacaille also dismantled one constellation, Robur Carolinum, Charles's Oak; this was formed in 1678 by Edmond Halley from some of the stars of Argo Navis to commemorate the oak tree in which his patron, King Charles II, hid after his defeat by Oliver Cromwell at the Battle of Worcester. It is said that this inspired piece of flattery earned Halley his master's degree from Oxford by the king's express command. Lacaille, less impressed, uprooted the oak and returned its stars to Argo Navis.

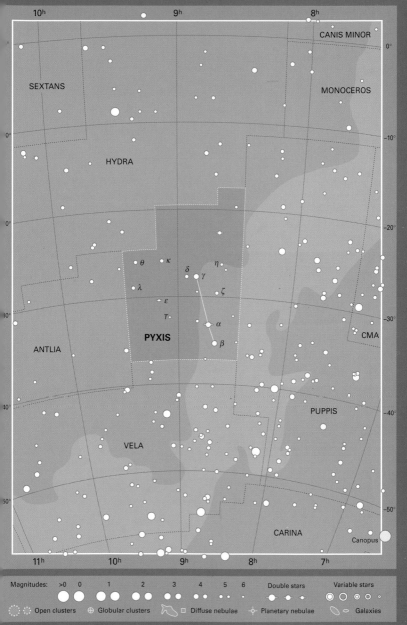

RETICULUM   The Net

A constellation introduced in the 1750s by Lacaille to commemorate a grid-like device known as a reticle which he used for measuring star positions during his surveys of the southern sky. It lies near the Large Magellanic Cloud, but is not prominent.

α (alpha) Reticuli, 4h 14m −62°.5, mag. 3.3, is a yellow giant star 163 l.y. away.

β (beta) Ret, 3h 44m −64°.8, mag. 3.8, is an orange star 100 l.y. away.

ζ¹ ζ² (zeta¹ zeta²) Ret, 3h 18m −62°.5, 39 l.y. away, is a wide naked-eye or binocular double of near-identical yellow main-sequence stars similar to the Sun, of mags. 5.5 and 5.2 respectively.

Across the border of Reticulum with neighbouring Dorado lies this attractive 9th-mag. spiral galaxy, NGC 1566. It is a type known as a Seyfert galaxy, with a bright, variable centre. (SAAO)

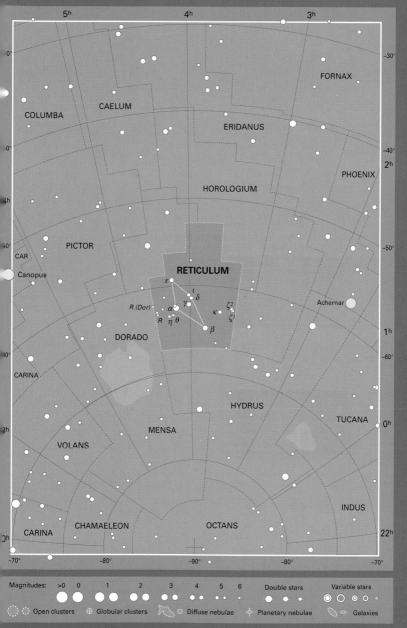

SAGITTA   The Arrow

Despite its diminutive size – it is the third-smallest constellation – this arrow-shaped group is distinctive enough to have been recognized by the ancient Greeks. In the sky, the Arrow seems to be flying between Cygnus the Swan and Aquila the Eagle; in one legend, the Arrow was shot by Hercules. Like its neighbour Vulpecula, Sagitta lies in a rich part of the Milky Way.

α (alpha) Sagittae, 19h 40m +18°.0, mag. 4.4, is a yellow giant 473 l.y. away.

β (beta) Sge, 19h 41m +17°.5, mag. 4.4, is a yellow giant 467 l.y. away.

γ (gamma) Sge, 19h 59m +19°.5, mag. 3.5, the brightest star in the constellation, is an orange giant 274 l.y. away.

δ (delta) Sge, 19h 47m +18°.5, mag. 3.7, is a red giant 448 l.y. away.

ζ (zeta) Sge, 19h 49m +19°.1, 326 l.y. away, is a blue-white star of mag. 5.0 with a 9th-mag. companion, divisible in a small telescope.

S Sge, 19h 56m +16°.6, is a yellow supergiant Cepheid variable, ranging between mag. 5.2 and 6.0 in 8.4 days.

VZ Sge, 20h 00m +17°.5, 746 l.y. away, is a pulsating red giant that varies irregularly between mags. 5.3 and 5.6.

WZ Sge, 20h 08m +17°.7, is a recurrent nova that flared up from 15th mag. to 7th or 8th mag. in 1913, 1946 and 1978; its location is worth checking in case of another outburst.

M71 (NGC 6838), 19h 54m +18°.6, is an 8th-mag. globular cluster 13,000 l.y. away, visible as a small, somewhat elongated misty patch in binoculars or a small telescope and resolvable with 100 mm aperture. Its stars are scattered and it lacks a central condensation, so it looks more like a dense open cluster than a typical globular.

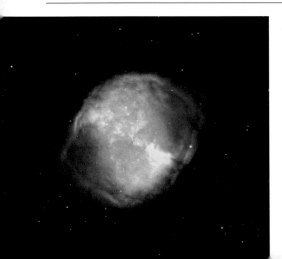

Just north of Sagitta's border, in neighbouring Vulpecula, lies the Dumbbell Nebula, M27, a shell of gas thrown off by a dying star. It is visible in binoculars and small telescopes. (See page 260.) (From The IAC Morphological Catalog of Northern Galactic Planetary Nebulae, by A. Manchado, M.A. Guerrero, L. Stanghellini & M. Serra-Ricart/IAC)

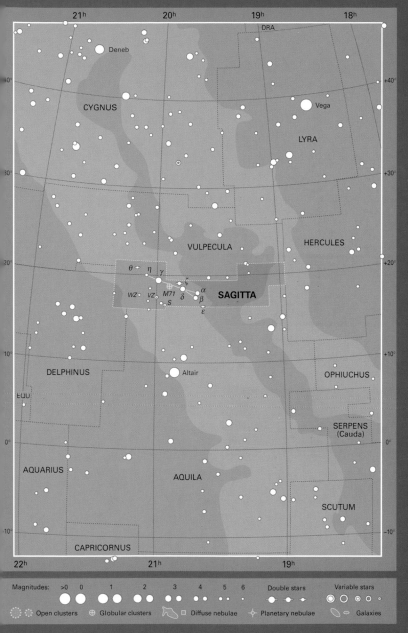

SAGITTARIUS   The Archer

An ancient constellation depicting a centaur, half man, half beast, with a raised bow and arrow. It is an older constellation than the other celestial centaur, Centaurus, and is different in character. Whereas Centaurus is identified as a scholarly, beneficent creature, Sagittarius is depicted with a threatening look, aiming his arrow at the heart of Scorpius, the Scorpion. The bow is marked by the stars λ (lambda), δ (delta) and ε (epsilon) Sagittarii, while γ (gamma) is the tip of the arrow. The main stars of Sagittarius are often visualized as outlining the shape of a teapot, while λ (lambda), φ (phi), σ (sigma), τ (tau) and ζ (zeta) Sagittarii form a ladle shape known as the Milk Dipper – a suitable implement to dip into this rich region of the Milky Way.

The centre of our Galaxy lies in Sagittarius, so the Milky Way starfields are particularly rich here, as well as in neighbouring Scutum and Scorpius. The actual centre of the Galaxy is marked by a radio and infra-red source known as Sagittarius A, at 17h 46.1m, −28° 51′. The main attraction of Sagittarius is its clusters and nebulae. Messier catalogued a total of 15 objects in Sagittarius, more than in any other constellation; only a selection of them can be mentioned here. The Sun passes through the constellation from mid-December to mid-January, and thus lies in Sagittarius at the December solstice, its farthest point south of the equator.

α (alpha) Sagittarii, 19h 24m −40°.6, (Rukbat, 'knee', or Alrami, 'the archer'), mag. 4.0, is one of several instances in which the star labelled α in a constellation is not the brightest. It is a blue-white main-sequence star 170 l.y. away.

$β^1$ $β^2$ (beta^1 beta^2) Sgr, 19h 23m −44°.5, (Arkab, from the Arabic for 'Achilles tendon'), is a pair of unrelated naked-eye stars. $β^1$ Sgr, a blue-white star of mag. 4.0, 378 l.y. away, has a mag. 7.2 companion visible in a small telescope. $β^2$ Sgr is a white star of mag. 4.3, 139 l.y. away. All three stars appear in the same line of sight by chance.

γ (gamma) Sgr, 18h 06m −30°.4, (Alnasl, 'the point', i.e. of the arrow), mag. 3.0, is an orange giant 96 l.y. away.

δ (delta) Sgr, 18h 21m −29°.8, (Kaus Media, 'middle of the bow'), mag. 2.7, is an orange giant 306 l.y. away.

ε (epsilon) Sgr, 18h 24m −34°.4, (Kaus Australis, 'southern part of the bow'), mag. 1.8, is the brightest star in Sagittarius. It is a blue-white giant 145 l.y. away.

λ (lambda) Sgr, 18h 28m −25°.4, (Kaus Borealis, 'northern part of the bow'), mag. 2.8, is an orange giant 77 l.y. away.

σ (sigma) Sgr, 18h 55m −26°.3, (Nunki), mag. 2.1, is a blue-white star 224 l.y. away.

W Sgr, 18h 05m −29°.6, is a yellow supergiant Cepheid variable ranging between mags. 4.3 and 5.1 with a period of 7.6 days. It lies about 2100 l.y. away.

X Sgr, 17h 48m −27°.8, is a yellow-white giant Cepheid variable ranging between mags. 4.2 and 4.9 in 7 days. It lies about 1100 l.y. away.

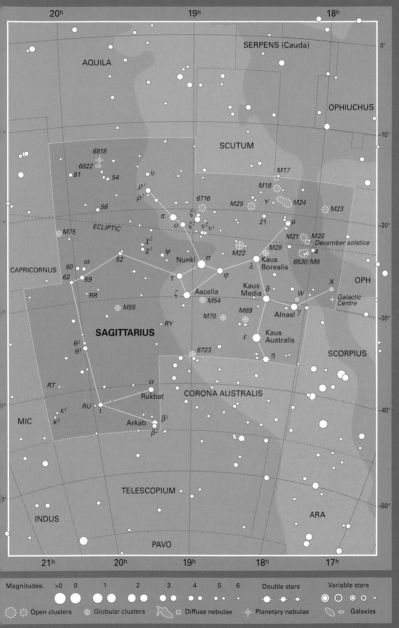

Magnitudes: >0 0 1 2 3 4 5 6   Double stars   Variable stars

Open clusters   Globular clusters   Diffuse nebulae   Planetary nebulae   Galaxies

Among the Milky Way starfields of Sagittarius are two outstanding nebulae: M20, the Trifid Nebula (top), and M8, the Lagoon Nebula. The Trifid consists of a pink emission nebula and a blue reflection nebula. (Nik Szymanek)

Y Sgr, 18h 21m −18°.9, is a yellow-white giant Cepheid that varies between mags. 5.3 and 6.2 in 5.8 days. It lies about 1300 l.y. away.

RR Sgr, 19h 56m −29°.2, is a red giant variable of Mira type, with a range from mag. 5.4 to 14.0 and a period of about 11 months. Its distance is too great to measure accurately.

RY Sgr, 19h 17m −33°.5, is the southern equivalent of R Coronae Borealis – a star like a reverse nova that normally shines at around 6th mag., but which can suddenly and unpredictably drop to 14th mag.

M8 (NGC 6523), 18h 04m −24°.4, the Lagoon Nebula, is a famous gaseous nebula, visible to the naked eye, elongated in shape and encompassing the star cluster NGC 6530. M8 is a fine object for binoculars or telescopes, covering the area of three full Moons, with a dark rift down its centre. In the eastern half of the nebula is NGC 6530, a cluster of about 25 stars of 7th mag. and fainter, formed recently from the surrounding gas. The other (western) side of the nebula is dominated by two main stars, the brighter of which is the blue super-giant 9 Sgr, mag. 5.9. Long-exposure photographs show the nebula as an intense red, but visually it appears milky-white. M8 is about 5000 l.y. away.

M17 (NGC 6618), 18h 21m −16°.2, the Omega, Horseshoe or Swan Nebula, is another gaseous nebula. Binoculars show it as a wedge-shaped object of similar apparent width to the full Moon, while in larger instruments it appears arch-shaped, variously likened to a Greek capital omega (Ω), a horseshoe or a swan, which accounts for its range of popular names. M17 lies 5000 l.y. away. About 1° south of it is M18 (NGC 6613), 4000 l.y. distant, a small, loose cluster of 20 stars of 9th mag. and fainter, unimpressive in binoculars.

M20 (NGC 6514), 18h 03m −23°.0, the Trifid Nebula, is a cloud of glowing gas far less impressive visually than photographically. Moderate-sized telescopes show it as only a diffuse patch of light centred on the double star HN 40, of 8th and 9th mags., which was evidently born from it and now illuminates it. The Trifid Nebula gets its name from three dark lanes of dust that trisect it, well shown on photographs but elusive in small apertures. It lies 5000 l.y. away, the same as M8.

M21 (NGC 6531), 18h 05m −22°.5, is a spider-like open cluster in the same low-power field of view as M20, containing about 70 stars of 7th mag. and fainter, 4000 l.y. away.

M22 (NGC 6656), 18h 36m −23°.9, is a large, rich 5th-mag. globular cluster, one of the finest in the entire heavens and ranked third only to ω (omega) Centauri and 47 Tucanae. Visible to the naked eye, M22 is an excellent binocular object and a fine sight in small telescopes, which reveal its noticeably elliptical outline. A telescope of 75 mm aperture will begin to resolve its outer regions, and larger apertures show its brightest stars to be reddish. Its nucleus is not as condensed as that of many other globulars. M22 lies 10,000 l.y. away.

M23 (NGC 6494), 17h 57m −19°.0, is a widely spread open cluster of fairly uniform appearance, just at the limit of resolution in binoculars. It is elongated in shape, consisting of a field of 10th- and 11th-mag. stars, some arranged in arcs. It lies 2100 l.y. away.

M24, 18h 18m −18°.5, is a rich and extensive Milky Way starfield south of M17 and M18, grainy and shimmering in binoculars. Some observers restrict the name M24 to a small cluster of faint stars in its northern half known also as NGC 6603, but this is not what Messier meant. The whole Milky Way star cloud in this region measures about 2° × 1° and is one of the most prominent parts of the Milky Way to the naked eye.

M25 (IC 4725), 18h 32m −19°.2, 2300 l.y. away, is a scattered cluster of about 30 stars, well seen in binoculars. The most prominent members form two bars across the cluster's centre. Its brightest star is U Sgr, a yellow supergiant Cepheid which varies from mag. 6.3 to 7.1 with a period of 6 days 18 hours.

M55 (NGC 6809), 19h 40m −31°.0, is a 6th-mag. globular cluster, nebulous-looking in binoculars with little central condensation. Small telescopes resolve individual stars and show a dark notch on one side. M55 lies 19,000 l.y. away.

SCORPIUS   The Scorpion

A resplendent constellation, lying in a rich area of the Milky Way and packed with exciting objects for users of small telescopes. In mythology, Scorpius was the scorpion whose sting killed Orion. In the sky Orion still flees from the scorpion, for Orion sets below the horizon as Scorpius rises. Originally, in ancient Greek times and earlier, Scorpius was a much larger constellation, but in the first century BC the stars that comprised its claws were made into the separate constellation of Libra by the Romans. Scorpius clearly resembles the creature after which it is named, with a distinctive curve of stars forming its stinging tail. Its heart is marked by the star Antares, a name that can be translated either as 'rival of Mars' or 'like Mars', a reference to its strong red colour. North of β (beta) Scorpii, at 16h 19.9m, −15° 38′, lies the strongest X-ray source in the sky, Scorpius X-1. This has been identified with a 13th-magnitude spectroscopic binary star 2300 l.y. away. The Sun passes briefly through Scorpius during the last week of November. A scattering of bright stars centred about 500 l.y. from us extends from Scorpius via Lupus into Centaurus and Crux; this is known as the Scorpius–Centaurus association.

α (alpha) Scorpii, 16h 29m −26°.4, (Antares), 604 l.y. away, is a red supergiant 400 times the diameter of the Sun. It is a semi-regular variable, fluctuating between mags. 0.9 and 1.2 with an approximate period of 5 years. Antares has a mag. 5.4 blue companion that requires at least 75 mm aperture and the steadiest atmospheric conditions to be visible against the primary's glare. The orbital period of the companion around Antares is estimated to be nearly 900 years.

β (beta) Sco, 16h 05m −19°.8, (Graffias, 'claws', or Acrab, 'scorpion') is a striking double star divisible in the smallest telescopes, consisting of unrelated blue-white main-sequence stars of mags. 2.6 and 4.9, distances 530 and about 1100 l.y.

δ (delta) Sco, 16h 00m −22°.6, (Dschubba, 'forehead'), mag. 2.3, is a blue-white star 402 l.y. away.

ε (epsilon) Sco, 16h 50m −34°.3, mag. 2.3, is an orange giant 65 l.y. away.

ζ^1 ζ^2 (zeta^1 zeta^2) Sco, 16h 54m −42°.4, is a naked-eye double of unrelated stars, ζ^2 being a mag. 3.6 orange giant 151 l.y. away, and ζ^1 a blue supergiant that varies erratically between mags. 4.7 and 4.9; ζ^1 is probably an outlying member of the open cluster NGC 6231 (see page 227).

θ (theta) Sco, 17h 37m −43°.0, mag. 1.9, is a white giant 272 l.y. away with a mag. 5.3 companion visible in small telescopes.

λ (lambda) Sco, 17h 34m −37°.1, (Shaula, 'sting'), mag. 1.6, is a blue-white star 703 l.y. away.

μ^1 μ^2 (mu^1 mu^2) Sco, 16h 52m −38°.0, is a naked-eye double of unrelated stars. μ^1, 822 l.y. away, is an eclipsing binary that varies from mag. 2.9 to 3.2 in 34 hours 43 minutes; μ^2 is a blue-white star of mag. 3.6, 517 l.y. away. ▶

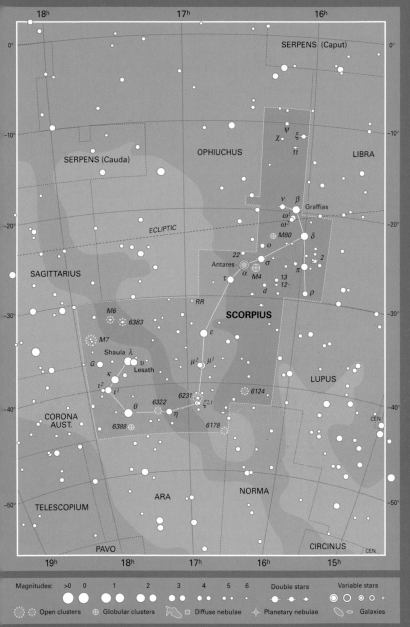

M6 and M7 are two glorious open clusters. Right: M6 is popularly known as the Butterfly Cluster because of its shape, although there is also a strong resemblance to the outline of a bird. Most of its stars are blue, with the exception of its brightest member, the orange giant BM Scorpii, seen at the left of the cluster. (Nigel Sharp, Mark Hanna/AURA/NOAO/NSF)

Left: M7 is a large naked-eye cluster, seen against a backdrop of shimmering Milky Way starfields, unlike M6 to the north which has a much darker background. M7 is only just over half the distance of M6, so its individual members appear brighter than those of M6, and are scattered over a much wider area. (Nigel Sharp, REU program/ AURA/NOAO/NSF)

ν (nu) Sco, 16h 12m –19°.5, 437 l.y. away, is a quadruple star similar to the famous Double Double in Lyra. A small telescope, or even powerful binoculars, shows ν Sco as a wide double, with blue-white components of mags. 4.0 and 6.3. Telescopes of 75 mm and above reveal at high magnification that the fainter star is itself a close double of mags. 6.7 and 7.7. The brighter star is an even closer double of mags. 4.3 and 5.4, requiring an aperture of 100 mm to split.

ξ (xi) Sco, 16h 04m –11°.4, about 95 l.y. away, is a celebrated multiple star. A small telescope shows it as a white star of mag. 4.2 with a mag. 7.3 orange companion; also visible in the same field is a fainter and wider pair, Σ 1999, composed of mag. 7.4 and 8.0 stars that are gravitationally connected to ξ Sco. Therefore, at first sight, ξ Sco looks like another Double Double. But the brighter star is itself a close pair, consisting of yellow-white stars of mags. 4.8 and 5.1 orbiting each other with a period of 46 years. They were closest in 1996 and are currently moving apart; after the year 2006 they should be divisible with 150 mm aperture, and 100 mm should separate them by 2015.

ω^1 ω^2 (omega¹ omega²) Sco, 16h 07m −20°.7, is a pair of unrelated stars distinguishable with the naked eye: ω^1, a blue-white main-sequence star of mag. 4.0, 424 l.y. away, and ω^2, a yellow giant of mag. 4.3, 265 l.y. away.

RR Sco, 16h 57m −30°.6, is a red giant variable of the Mira type, ranging between 5th and 12th mag. every 9 months or so. It lies about 1150 l.y. away.

M4 (NGC 6121), 16h 24m −26°.5, is a 6th-mag. globular cluster appearing almost as large as the full Moon. It looks like a woolly ball in binoculars, but is not as easy to spot as its magnitude would suggest because its light is spread over a large area. In 100-mm telescopes individual stars are resolved and there is a noticeable bar of stars running north–south across its centre. M4 is more loosely scattered than many globulars, and does not have a strong central condensation. It is one of the closest globulars to us, 7000 l.y. away.

M6 (NGC 6405), 17h 40m −32°.2, is an impressive 4th-mag. open cluster of about 80 stars arranged in radiating chains, popularly called the Butterfly Cluster. Binoculars and small telescopes resolve the main stars which form the butterfly shape. The brightest of them, BM Sco, on one of the 'wings', is an orange giant semi-regular variable which ranges between 5th and 7th mags. every 27 months or so. M6 lies 1600 l.y. away.

M7 (NGC 6475), 17h 54m −34°.8, the most southerly of the Messier objects, is a huge, scattered 3rd-mag. open cluster of 80 or so stars individually of 6th mag. and fainter, visible to the naked eye as a brighter knot in the Milky Way. The apparent diameter is over twice that of the full Moon, and it is easily resolved in binoculars. With M6 at the edge of the same binocular field, this is an exceptionally rich sight. The central group of stars is arranged in an X-shape with scattered outliers that form triangular surroundings like a Christmas tree, all against the backdrop of a dense star cloud. An outstanding cluster and a classic for small apertures. M7 lies 950 l.y. away, and is not related to M6.

M80 (NGC 6093), 16h 17m −23°.0, is a small, 7th-mag. globular cluster visible in binoculars or a small telescope, appearing like the fuzzy head of a comet. It lies 27,000 l.y. away.

NGC 6231, 16h 54m −41°.8, is a naked-eye open cluster of over 100 stars in a rich area of the Milky Way that is well worth sweeping with binoculars. The brightest stars of the group are of 6th mag. and give the impression of a mini-Pleiades in binoculars and small telescopes. The 5th-mag. blue supergiant ζ^1 (zeta¹) Sco (see page 224) is probably an outlying member of this cluster, which lies 6500 l.y. away. NGC 6231 is connected to a larger, scattered cluster of fainter stars visible in binoculars, known both as Trumpler 24 and Harvard 12, which lies 1° to the north. The chain of stars linking the two clusters delineates one of the spiral arms of our Galaxy.

SCULPTOR   The Sculptor

One of the faint and half-forgotten constellations introduced in the 1750s by the French astronomer Nicolas Louis de Lacaille to fill in the southern skies. It represents a sculptor's studio. Sculptor contains the south pole of our Galaxy, 90° from the plane of the Milky Way. In this direction we can look out into deep space, unobscured by stars or dust, and see many faint galaxies. Among these is a very faint member of our Local Group, the Sculptor Dwarf, detectable only on long-exposure photographs taken through large telescopes.

α (alpha) Sculptoris, 0h 59m −29°.4, mag. 4.3, is a blue-white giant 672 l.y. away.

β (beta) Scl, 23h 33m −37°.8, mag. 4.4, is a blue-white star 178 l.y. away.

γ (gamma) Scl, 23h 19m −32°.5, mag. 4.4, is an orange giant 179 l.y. away.

δ (delta) Scl, 23h 49m −28°.1, mag. 4.6, is a blue-white main-sequence star 143 l.y. away.

ε (epsilon) Scl, 1h 46m −25°.1, 89 l.y. away, is a binary of mags. 5.3 and 8.6, visible in small telescopes. The estimated orbital period is over 1000 years.

κ¹ (kappa¹) Scl, 0h 09m −28°.0, 224 l.y. away, is a tight pair of white stars, of mags. 6.1 and 6.2, at the limit of resolution of a 75-mm telescope.

R Scl, 1h 27m −32°.5, is a deep-red semi-regular variable star that ranges from about mag. 5.8 to 7.7 with a period of about a year. It lies about 1500 l.y. away.

S Scl, 0h 15m −32°.0, is a red giant variable of Mira type, ranging from mag. 5.5 to 13.6 in around a year. It lies about 1500 l.y. away.

NGC 55, 0h 15m −39°.2, is an 8th-mag. spiral galaxy seen nearly edge-on, so it appears elongated. One half is more prominent that the other. It is similar in size and shape to NGC 253 (see below), although not quite as bright. Its distance is estimated at 6 million l.y.

NGC 253, 0h 48m −25°.3, is a 7th-mag. spiral galaxy seen nearly edge-on, hence appearing cigar-shaped. Nearly ½° long, it can be picked up in binoculars but at least 100 mm aperture is required to distinguish the mottling caused by dust clouds in its spiral arms. Its estimated distance is 9 million l.y. (see photo below).

NGC 253, a spiral galaxy seen nearly edge-on and hence appearing elliptical. It has no central bulge, but its arms are a seething mass of stars and dust. (Todd Boroson/ AURA/NOAO/NSF)

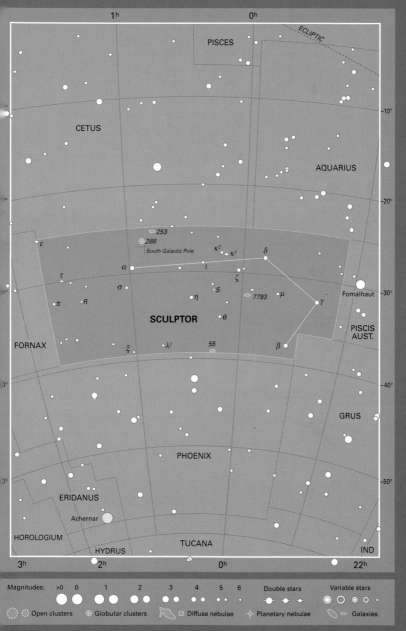

SCUTUM   The Shield

A faint constellation between Aquila and Serpens, introduced in 1684 by the Polish astronomer Johannes Hevelius under the title Scutum Sobiescianum, Sobieski's Shield, in honour of his patron, King John III Sobieski. Rich Milky Way star fields are the main attraction here, notably the Scutum star cloud in the northern half of the constellation, about 6° across and reputedly the brightest part of the Milky Way outside Sagittarius. The prominent open cluster M11 lies near a notch of dark nebulosity at the northern edge of the Scutum star cloud.

α (alpha) Scuti, 18h 35m −8°.2, mag. 3.8, is an orange giant 174 l.y. away.

δ (delta) Sct, 18h 42m −9°.1, 187 l.y. distant, is the prototype of a rare class of variable stars that pulsate in size every few hours, producing small-amplitude brightness changes. δ Scuti itself is a white giant that varies from mag. 4.6 to 4.8 with a period of 4 hours 39 minutes.

R Sct, 18h 48m −5°.7, is a pulsating orange supergiant about 1400 l.y. away that varies between mags. 4.2 and 8.6 every 5 months or so.

M11 (NGC 6705), 18h 51m −6°.3, the Wild Duck Cluster, is a showpiece open cluster of about 200 stars, half the apparent width of the full Moon. At 6th mag. it is at the limit of naked-eye visibility, but binoculars show it as a misty patch. In a telescope with a magnification of around ×100 it breaks up into a sparkling field of faint stardust. The cluster gets its popular name from the fact that its brightest members form a distinct fan-shape, resembling a flight of ducks, an effect which is noticeable visually through a telescope but which becomes lost on long-exposure photographs. An 8th-mag. star, slightly brighter than the rest, lies at the fan's apex, with a double star nearby. M11 is 6500 l.y. away.

M26 (NGC 6694), 18h 45m −9°.4, is an open cluster for small telescopes, of similar size to M11 but containing only about two dozen stars and hence much fainter. It lies about 5000 l.y. away.

Among the rich Milky Way starfields of Scutum lies M11, the Wild Duck cluster. In small instruments it appears to be shaped like a fan or arch, because of a lower density of bright stars towards one side, the right in this image. (Nigel Sharp, REU program/AURA/ NOAO/NSF)

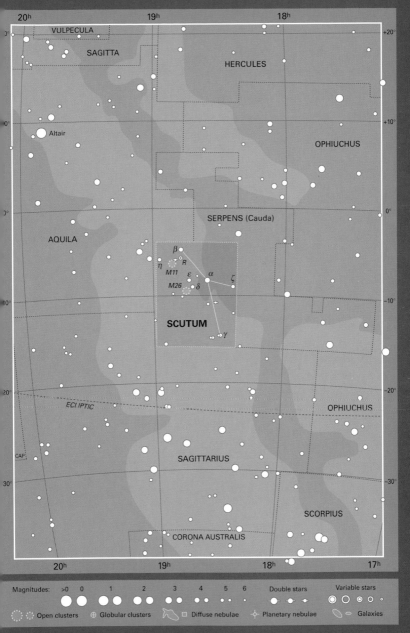

SERPENS   The Serpent

An ancient constellation, representing a snake wound around the body of Ophiuchus. Serpens is actually split into two halves, one either side of Ophiuchus: Serpens Caput, the head, which is the larger and more prominent half; and Serpens Cauda, the tail. It is the only constellation to be split in two, but both halves count as one constellation.

α (alpha) Serpentis, 15h 44m +6°.4, (Unukalhai, 'the serpent's neck'), mag. 2.6, is an orange giant star 73 l.y. away.

β (beta) Ser, 15h 46m +15°.4, 153 l.y. distant, is a blue-white main-sequence star in the serpent's head of mag. 3.7, with a 10th-mag. companion visible in small telescopes. An unrelated mag. 6.7 background star, 29 Ser, is visible just north of it in binoculars.

γ (gamma) Ser, 15h 56m +15°.7, mag. 3.8, is a white main-sequence star 36 l.y. away.

δ (delta) Ser, 15h 35m +10°.5, 210 l.y. away, is white star of mag. 4.2 with a close mag. 5.2 companion visible in a small telescope at high magnification.

η (eta) Ser, 18h 21m −2°.9, mag. 3.2, is an orange giant star 62 l.y. away.

θ (theta) Ser, 18h 56m +4°.2, (Alya), 132 l.y. distant, is an elegant pair of white stars of mags. 4.6 and 5.0, easily split in the smallest telescopes.

ν (nu) Ser, 17h 21m −12°.8, 193 l.y. away, is a blue-white star of mag. 4.3 with a wide mag. 8.3 companion visible in binoculars or small telescopes. ▶

M16 and the surrounding Eagle Nebula in Serpens, a spectacular combination of star cluster and gas cloud. (Bill Schoening/AURA/NOAO/NSF)

| Magnitudes: | >0 | 0 | 1 | 2 | 3 | 4 | 5 | 6 | Double stars | Variable stars |

Open clusters · Globular clusters · Diffuse nebulae · Planetary nebulae · Galaxies

τ^1 (tau[1]) Ser, 15h 26m +15°.4, 920 l.y. away, a red giant of mag. 5.2, is the brightest member of a loose scattering of eight stars of 6th mag. near β Ser, all visible in binoculars.

R Ser, 15h 51m +15°.1, is a red giant variable of Mira type, ranging between mags. 5.2 and 14.4 in approximately a year. It lies about 900 l.y. away.

M5 (NGC 5904), 15h 19m +2°.1, is a 6th-mag. globular cluster 26,000 l.y. away, visible in binoculars or small telescopes. It is rated as one of the finest globulars in the northern sky, second only to the famous M13 in Hercules. A telescope of 100 mm aperture reveals its brilliant, condensed centre and mottled outer regions with curving chains of stars. Close to M5 (but actually in the foreground) lies 5 Ser, a mag. 5.0 yellow-white star with a mag. 10 companion.

M16 (NGC 6611), 18h 19m −13°.8, is a hazy-looking open star cluster 8500 l.y. distant, of similar apparent size to the full Moon, embedded in the larger Eagle Nebula. The cluster is a grouping of about 60 stars of 8th mag. and fainter at the limit of resolution in binoculars. Small telescopes show that most of its members congregate in a V-shape in the northern half. The surrounding Eagle Nebula adds a touch of haziness to the cluster when seen in binoculars. The nebula is too faint to be seen well in amateur telescopes, but shows up beautifully on long-exposure photographs (see page 232).

IC 4756, 18h 39m +5°.4, is a scattered open cluster of 8th-mag. stars and fainter, about two Moon diameters wide, visible in binoculars. It lies 1300 l.y. away.

SEXTANS   The Sextant

A faint and insignificant constellation south of Leo, introduced in 1687 by the Polish astronomer Johannes Hevelius. It commemorates the instrument he used for measuring star positions. Hevelius continued to make naked-eye sightings of star positions with his sextant long after telescopes were available.

α (alpha) Sextantis, 10h 08m −0°.4, mag. 4.5, is a blue-white giant star 287 l.y. away.

β (beta) Sex, 10h 30m −0°.6, mag. 5.1, is a blue-white main-sequence star 345 l.y. away.

γ (gamma) Sex, 9h 53m −8°.1, mag. 5.1, is a blue-white main-sequence star 262 l.y. away.

δ (delta) Sex, 10h 30m −2°.7, mag. 5.2, is a blue-white main-sequence star 300 l.y. away.

17, 18 Sex, 10h 10m −8°.4, is a neat pairing of unrelated stars, mags. 5.9 and 5.6, distances 527 and 473 l.y., easily divisible in binoculars.

NGC 3115, 10h 05m −7°.7, is a 9th-mag. elliptical galaxy known as the Spindle Galaxy, 14 million l.y. away. Moderate-sized amateur telescopes show its elongated outline and brighter centre.

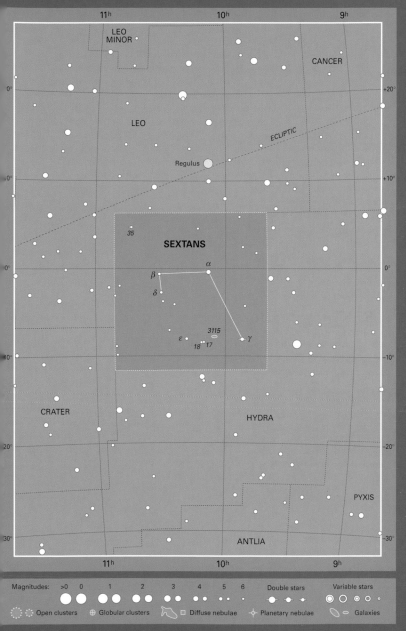

TAURUS   The Bull

One of the most ancient constellations, recognized since the dawn of civilization. In Greek mythology, Taurus represents the animal disguise adopted by Zeus to carry off Princess Europa to Crete. Only the front half of the bull is depicted in the sky, its face being formed by the V-shaped cluster of stars known as the Hyades. Its glinting red eye is marked by the star Aldebaran, and its long horns are tipped by the stars β (beta) and ζ (zeta) Tauri. In addition to the Hyades, Taurus contains the celebrated star cluster of the Pleiades, or Seven Sisters. In Taurus occurred the famous supernova that was seen from Earth in AD 1054 and gave rise to the Crab Nebula, M1. At 4h 22.0m, +19° 32′, lies the faint Hind's Variable Nebula, NGC 1554–5, discovered in the 19th century by the English astronomer John Russell Hind; within this nebula lies the 10th-mag. star T Tauri, 576 l.y. away, prototype of a class of irregular variables believed to be stars in the process of formation. Each year, the Taurid meteors radiate from south of the Pleiades, reaching a maximum of about 10 per hour around November 4. The Sun passes through the constellation from mid-May to late June, and is in Taurus at the June solstice; precession carried the position of the June solstice into Taurus from Gemini at the end of 1989.

α (alpha) Tauri, 4h 36m +16°.5, (Aldebaran, 'the follower', i.e. of the Pleiades), is an orange giant irregular variable that fluctuates between about mags. 0.75 and 0.95. Although it appears to be part of the Hyades cluster, it is in fact an unrelated foreground star, 65 l.y. away.

β (beta) Tau, 5h 26m +28°.6, (Alnath or Elnath, 'the butting one'), mag. 1.7, is a blue-white giant 131 l.y. away.

ζ (zeta) Tau, 5h 38m +21°.1, 417 l.y. away, is a blue giant that is slightly variable, ranging erratically between about mags. 2.9 and 3.2.

θ^1 θ^2 (theta^1 theta^2) Tau, 4h 29m +15°.9, is a naked-eye or binocular double in the Hyades cluster, consisting of yellow and white giants of mags. 3.8 and 3.4 respectively, distances 158 and 149 l.y. θ^2 is the brightest member of the Hyades.

κ^1 κ^2 (kappa^1 kappa^2) Tau, 4h 25m +22°.3, are a pair of white stars of mags. 4.2 and 5.3 that form a naked-eye or binocular duo, 153 and 144 l.y. away respectively. Both are outlying members of the Hyades.

λ (lambda) Tau, 4h 01m +12°.5, 370 l.y. away, is an eclipsing binary of the Algol type, varying between mags. 3.4 and 3.9 with a period of 4 days.

σ^1 σ^2 (sigma^1 sigma^2) Tau, 4h 39m +15°.8, is a wide binocular double of blue-white stars in the Hyades, mags. 5.1 and 4.7 and distances 152 and 159 l.y. respectively.

φ (phi) Tau, 4h 20m +27°.4, 342 l.y. away, is an optical double divisible in small telescopes, consisting of a mag. 5.0 orange giant and a white star of mag. 8.4.

χ (chi) Tau, 4h 23m +25°.6, 268 l.y. away, is a double star for small telescopes, with blue and gold components of mags. 5.4 and 7.6.

►

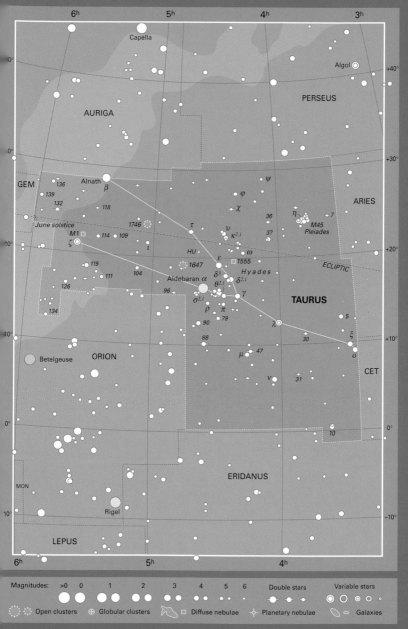

| Magnitudes: | >0 | 0 | 1 | 2 | 3 | 4 | 5 | 6 | Double stars | Variable stars |

Open clusters · Globular clusters · Diffuse nebulae · Planetary nebulae · Galaxies

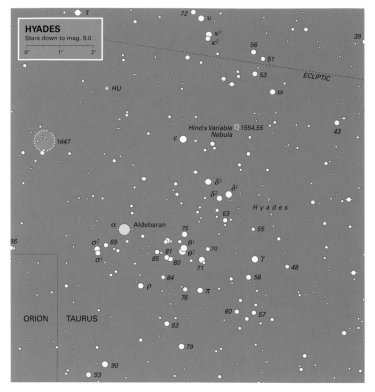

Detail chart of the Hyades. (Wil Tirion)

The Hyades, 4h 27m +16°, is a large and bright open cluster of about 200 stars covering over 5° of sky. The brightest members form a distinctive V-shape, easily visible to the naked eye. In mythology the Hyades were the daughters of Atlas and Aethra, and half-sisters of the Pleiades. Because of its size, the cluster is best studied with binoculars rather than a telescope. The bright star Aldebaran is not a member of the Hyades, but is superimposed on it by chance; the brightest true member is actually θ² (theta²) Tauri (see page 236). The centre of the cluster lies 150 l.y. away; its distance is important, for it marks the first step in our distance scale of the Galaxy.

M1 (NGC 1952), 5h 35m +22°.0, is the celebrated Crab Nebula, the remains of a star that exploded as a supernova. It can be glimpsed through binoculars on clear, dark nights. Despite its fame, the Crab Nebula is a disappointing object for small telescopes, appearing as an elliptical 8th-mag. wisp of nebulosity. At the centre of the nebula, beyond the reach of amateur telescopes, is a 16th-mag. object, the remains of the star that exploded. This faint object is now known to be a pulsar. The Crab Nebula and pulsar lie about 6500 l.y. away.

M45, 3h 47m +24°, the Pleiades, is the brightest and most famous star cluster in the sky; it is popularly termed the Seven Sisters, after a group of mythological nymphs, the daughters of Atlas and Pleione. Approximately seven stars are visible to the naked eye, covering three full Moon widths of sky; binoculars bring dozens more into view. About 100 stars belong to the cluster, which is centred 378 l.y. away. Unlike the stars of the Hyades, which are older and more evolved, the Pleiades formed within the last 50 million years and include many young blue giants. The brightest member is η (eta) Tauri (Alcyone), mag. 2.9. Other prominent members are 16 Tau (Celaeno), mag. 5.5; 17 Tau (Electra), mag. 3.7; 19 Tau (Taygeta), mag. 4.3; 20 Tau (Maia), mag. 3.9; 21 Tau (Asterope), mag. 5.8; 23 Tau (Merope), mag. 4.1; 27 Tau (Atlas), mag. 3.6; and BU Tau (Pleione), a shell star that throws off rings of gas at irregular intervals, causing it to fluctuate unpredictably between mags. 4.8 and 5.5. The whole of the Pleiades cluster is embedded in a faint nebulosity. This nebula is noticeable on long-exposure photographs (see page 240), and under very clear conditions its brightest part, around Merope, may be glimpsed in binoculars or small telescopes. This nebulosity was long thought to be the remains of the cloud from which the stars formed, but now it seems more likely that it is an entirely separate cloud into which the stars have since drifted by chance.

Detail chart of the Pleiades. (Wil Tirion)

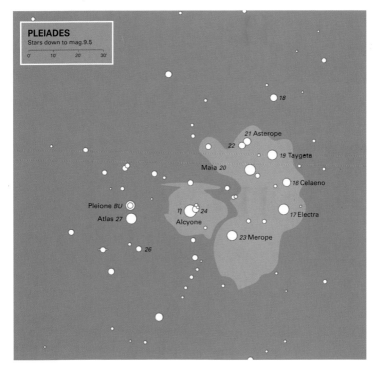

M45, the Pleiades in Taurus, is the most glorious star cluster in the entire sky. Light from the hot, young stars is reflected from surrounding dust, producing a blue nebulosity that is brightest near Merope, bottom centre. (Philip Perkins)

TELESCOPIUM The Telescope

A constellation invented in the 1750s by the Frenchman Nicolas Louis de Lacaille to honour the most important of astronomical instruments. As with so many of Lacaille's constellations, it is faint and contrived, containing little to interest owners of small telescopes.

α (alpha) Telescopii, 18h 27m −46°.0, mag. 3.5, is a blue-white star 249 l.y. away.

$δ^1$ $δ^2$ (delta^1 delta^2) Tel, 18h 32m −45°.9, is a pair of blue-white stars of mags. 4.9 and 5.1, visible separately in binoculars. They are unrelated to each other, being 800 and 1120 l.y. from us respectively.

ε (epsilon) Tel, 18h 11m −46°.0, mag. 4.5, is a yellow giant 409 l.y. away.

ζ (zeta) Tel, 18h 29m −49°.1, mag. 4.1, is a yellow giant 127 l.y. away.

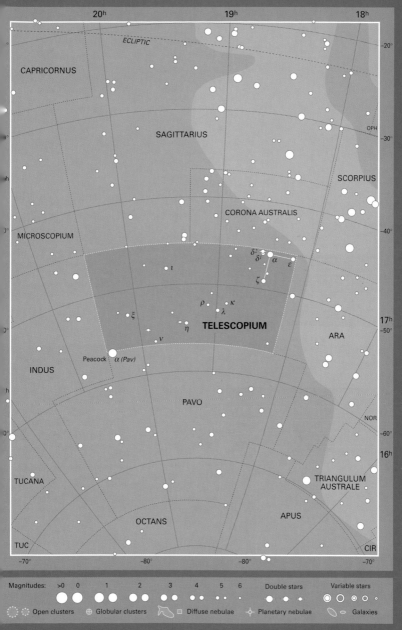

Magnitudes: >0 0 1 2 3 4 5 6   Double stars   Variable stars

Open clusters   Globular clusters   Diffuse nebulae   Planetary nebulae   Galaxies

TRIANGULUM   The Triangle

A small but distinctive constellation lying between Andromeda and Aries, consisting of three main stars that form a thin delta shape; the Greeks referred to it as Deltoton. Its most important feature is the spiral galaxy M33, the third-largest member of our Local Group of galaxies, after the Andromeda Galaxy and our own Milky Way.

α (alpha) Trianguli, 1h 53m +29°.6, mag. 3.4, is a yellow-white star 64 l.y. away.

β (beta) Tri, 2h 10m +35°.0, mag. 3.0, the constellation's brightest member, is a white star 124 l.y. away.

γ (gamma) Tri, 2h 17m +33°.8, mag. 4.0, is a blue-white star 118 l.y. away.

6 Tri, 2h 12m +30°.3, 305 l.y. away, is a mag. 5.2 yellow giant with a close mag. 6.6 companion visible in a small telescope.

R Tri, 2h 37m +34°.3, is a red giant variable of Mira type, ranging from mag. 5.4 to 12.6 every 9 months or so. It lies about 1300 l.y. away.

M33 (NGC 598), 1h 34m +30°.7, is a spiral galaxy 2.7 million l.y. away in our Local Group. Presented almost face-on, it covers a larger area of sky than the full Moon. Despite its size and proximity it is not prominent visually because its light is spread over such a large area. M33 is best picked up on a dark night in binoculars or a small telescope with low power to enhance the contrast. Unlike most galaxies, it does not have a noticeably stellar nucleus. Quite large amateur telescopes are needed to trace the spiral arms.

The spiral galaxy M33 in Triangulum is one of the closest galaxies to us. (Bill Schoening/AURA/NOAO/NSF)

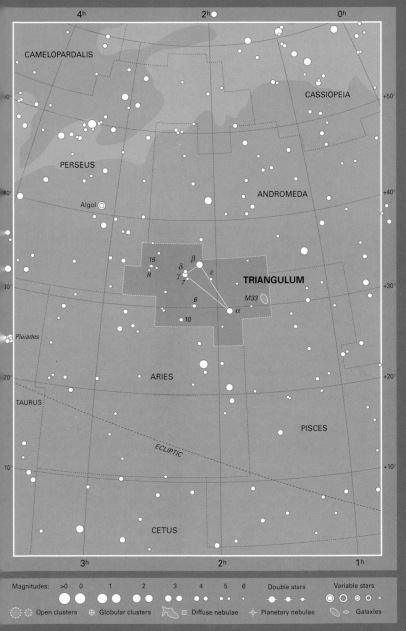

CAMELOPARDALIS

CASSIOPEIA

+50°

PERSEUS

ANDROMEDA

+40°

Algol

15
R
δ
γ 7
β
ε
TRIANGULUM
6
M33
α
10

+30°

Pleiades

ARIES

+20°

TAURUS

PISCES

ECLIPTIC

+10°

CETUS

Magnitudes:	>0	0	1	2	3	4	5	6	Double stars	Variable stars

Open clusters ⊕ Globular clusters Diffuse nebulae ✦ Planetary nebulae Galaxies

TRIANGULUM AUSTRALE
The Southern Triangle

A small but readily distinguishable constellation near α (alpha) Centauri, introduced at the end of the 16th century by the Dutch navigators Pieter Dirkszoon Keyser and Frederick de Houtman. The French astronomer Nicolas Louis de Lacaille visualized it as a surveyor's level. Its three main stars are brighter than those of its northern equivalent, Triangulum.

α (alpha) Trianguli Australis, 16h 49m −69°.0, (Atria), mag. 1.9, is an orange giant star 415 l.y. away.

β (beta) TrA, 15h 55m −63°.4, mag. 2.8, is a white star 40 l.y. away.

γ (gamma) TrA, 15h 19m −68°.7, mag. 2.9, is a blue-white star 183 l.y. away.

NGC 6025, 16h 04m −60.5°, is a binocular cluster, elongated in shape, containing about 60 stars of 7th mag. and fainter, 2500 l.y. away.

Johann Bayer (1572–1625)

Johann Bayer was a German lawyer in Augsburg and an amateur astronomer who, in 1603, published the first star atlas that covered the entire sky, *Uranometria*. He based his maps of the northern heavens mainly on the observations made by the great Danish astronomer Tycho Brahe, while his information on the southern skies came from the work of the Dutch navigator Pieter Dirkszoon Keyser. In addition to the 48 constellations known since ancient times, Bayer showed the 12 new constellations around the southern celestial pole invented by Keyser and his countryman Frederick de Houtman: Apus, Chamaeleon, Dorado, Grus, Hydrus, Indus, Musca, Pavo, Phoenix, Triangulum Australe, Tucana and Volans. Bayer's most important legacy to astronomers was his system of identifying stars by Greek letters. In each constellation the brightest stars were assigned letters of the Greek alphabet, usually, but not always, in approximate order of brightness (Gemini, Orion and Sagittarius are notable examples of constellations in which the star labelled α (alpha) is not the brightest). Before Bayer's time, stars without proper names were identified by the cumbersome descriptions of the Greek astronomer Ptolemy such as 'in the left forearm of the advance twin', a reference to the 4th-magnitude star that we now term θ (theta) Geminorum. Clearly, to make sense of such descriptions astronomers had to know their constellation figures well, and even then there was considerable room for confusion. The system of so-called Bayer letters was a vast improvement, and remains in use to this day.

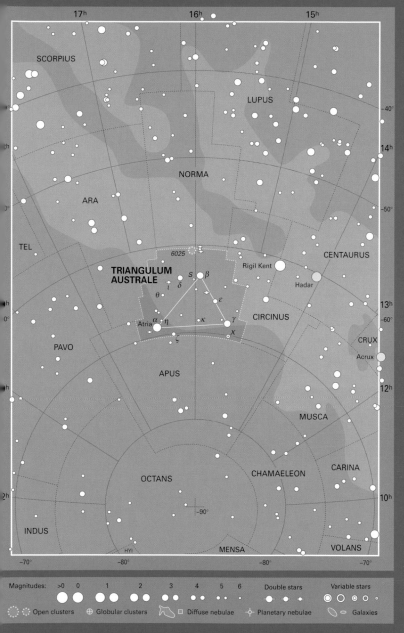

TUCANA   The Toucan

A constellation near the south pole of the sky, introduced in the late 16th century by the Dutch navigators Pieter Dirkszoon Keyser and Frederick de Houtman. It represents a toucan, the South American bird with the large beak. Its most notable features are the Small Magellanic Cloud, actually a neighbouring mini-galaxy, and the globular cluster known as 47 Tucanae.

α (alpha) Tucanae, 22h 19m −60°.3, mag. 2.9, is an orange giant star 199 l.y. away.

β (beta) Tuc, 0h 32m −63°.0, is a complex multiple star. Binoculars or small telescopes show that it consists of two almost identical blue-white stars, β^1 and β^2, mags. 4.4 and 4.5; β^2 is itself a close binary with a period of 44 years, requiring an aperture larger than 200 mm to divide. Nearby lies a mag. 5.1 white star. All three stars share the same proper motion through space, but their distances of 140, 172 and 152 l.y. suggest they are not truly connected.

γ (gamma) Tuc, 23h 17m −58°.2, mag. 4.0, is a white star 72 l.y. away.

δ (delta) Tuc, 22h 27m −65°.0, mag. 4.5, is a blue-white star 267 l.y. away with a 9th-mag. companion visible in small telescopes.

κ (kappa) Tuc, 1h 16m −68°.9, 67 l.y. away, is a double star consisting of components of mags. 5.1 and 7.3 visible in small telescopes. This pair moves through space with a wide mag. 7.2 star, itself a close binary with an orbital period of 86 years, just divisible in 150 mm aperture.

47 Tuc (NGC 104), 0h 24m −72°.1, is a prominent globular cluster the same apparent size as the full Moon, visible to the naked eye as a fuzzy 4th-mag. star; on early charts, it was actually catalogued as a star and given a stellar designation. Among globular clusters it is second only to ω (omega) Centauri in size and brightness. Telescopes of 100 mm aperture begin to resolve 47 Tuc, and even binoculars show the brilliant blaze of its star-packed core. It is among the closest globulars to us, 16,000 l.y. away.

NGC 362, 1h 03m −70°.8, is a 6th-mag. globular cluster visible in binoculars near the northern edge of the Small Magellanic Cloud, but not associated with it. NGC 362 actually lies 29,000 l.y. away, in our own Galaxy.

Small Magellanic Cloud (SMC), 0h 53m −73°, is a satellite galaxy of the Milky Way, as is its larger sibling, the Large Magellanic Cloud in Dorado. The Small Magellanic Cloud appears to the naked eye as a nebulous, tadpole-shaped patch 3½° across. Binoculars and small telescopes resolve star clusters and glowing gas clouds within it, although they are smaller and less impressive than those in the Large Magellanic Cloud. It lies about 200,000 l.y. away. (See the photograph on page 162.)

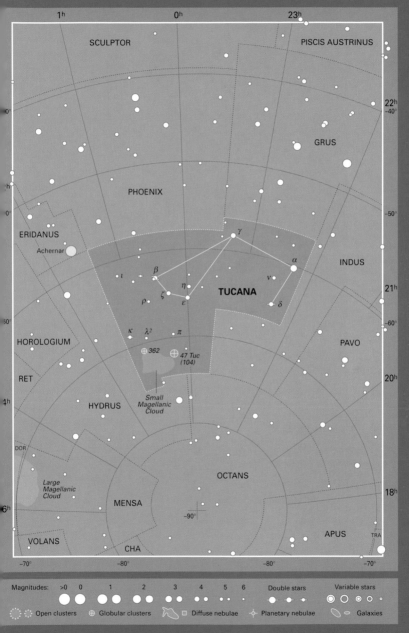

URSA MAJOR   The Great Bear

The third-largest constellation in the sky. Its central feature is the seven stars that make up the familiar shape variously called the Plough or the Big Dipper, the best-known of all star patterns, although why so many people, including the North American Indians, visualized this group as a bear remains a mystery. In Europe the pattern was seen as a wagon or chariot. Others, notably the Arabs, viewed the dipper shape not as a bear, but as a bier or coffin. In Greek mythology the bear represented Callisto, who was turned into a bear in punishment for her illicit love affair with Zeus. The two end stars in the dipper's bowl, Merak and Dubhe, are known as the Pointers, since they indicate the direction of Polaris, the north pole star in neighbouring Ursa Minor. The curving handle of the Big Dipper points to the bright star Arcturus in Boötes. At 11h 03.3m, +35° 58', lies the mag. 7.5 red dwarf Lalande 21185, which is the Sun's fourth-closest stellar neighbour, 8.3 l.y. away. Its designation comes from its number in a catalogue drawn up by the 18th-century French astronomer Joseph Lalande. With the exception of Alkaid and Dubhe, the stars of the Big Dipper are moving together through space, along with a number of other stars in the region; together, these stars make up the so-called Ursa Major moving cluster. Ursa Major contains numerous galaxies, but only a few of them are easily visible in amateur telescopes.

α (alpha) Ursae Majoris, 11h 04m +61°.8, (Dubhe, 'the bear'), mag. 1.8, is a yellow-orange giant 124 l.y. away. It has a close mag. 4.8 companion that orbits it in 44 years. The pair were closest in 2001 and should be divisible in 220 mm apertures after about 2010 as they move apart, being widest in 2020–2026.

β (beta) UMa, 11h 02m +56°.4, (Merak, 'flank'), mag. 2.3, is a blue-white star 79 l.y. away.

γ (gamma) UMa, 11h 54m +53°.7, (Phad or Phecda, 'thigh'), mag. 2.4, is a blue-white star 84 l.y. away.

δ (delta) UMa, 12h 15m +57°.0, (Megrez, 'root of the tail'), mag. 3.3, is a blue-white star 81 l.y. away.

ε (epsilon) UMa, 12h 54m +56°.0, (Alioth), mag. 1.8, is a blue-white star with a peculiar spectrum, 81 l.y. away.

ζ (zeta) UMa, 13h 24m +54°.9, (Mizar), mag. 2.2, is a celebrated multiple star. Good eyesight, or binoculars, reveals a mag. 4.0 companion, Alcor. Mizar is 78 l.y. from Earth and Alcor 81 l.y. away, too far apart to be a genuine binary. However, a small telescope reveals that Mizar has another mag. 4.0 companion closer to it, which definitely is related. This companion was first seen by the Italian astronomer Giovanni Riccioli in 1650, making Mizar the first double star to be discovered telescopically. Mizar itself was also the first star discovered to be a spectroscopic binary, by the American astronomer E. C. Pickering in 1889. The companion of Mizar is another spectroscopic binary, as is Alcor, making this a highly complex group. (See the diagram on page 250.)

η (eta) UMa, 13h 48m +49°.3, (Alkaid or Benetnasch, both from the Arabic for 'leader of the mourners'), mag. 1.9, is a blue-white main-sequence star 101 l.y. away. ▶

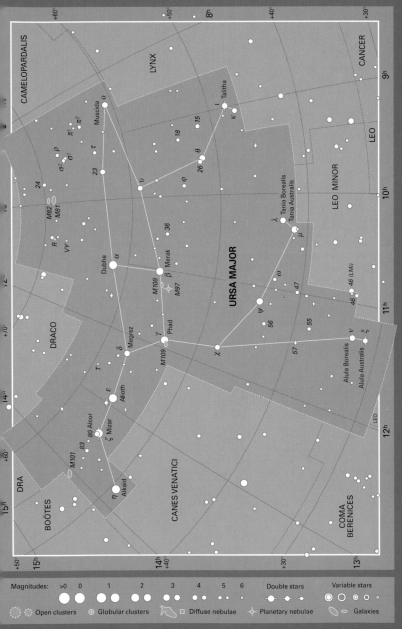

Magnitudes: >0 0 1 2 3 4 5 6   Double stars   Variable stars

Open clusters   Globular clusters   Diffuse nebulae   Planetary nebulae   Galaxies

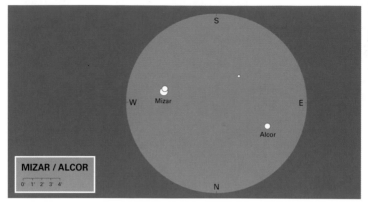

Mizar and Alcor as seen through a telescope. (Wil Tirion)

ξ (xi) UMa, 11h 18m +31°.5, 26 l.y. away, was the first binary star to have its orbit computed. Its two yellow components (both also spectroscopic binaries), of mags. 4.3 and 4.8, orbit each other with a period of 60 years. They are currently divisible in a 75-mm telescope and after 2015 will come within range of the smallest apertures, continuing to open out until 2035.

M81 (NGC 3031), 9h 56m +69°.1, is a beautiful 7th-mag. spiral galaxy, one of the brightest in the sky, visible in binoculars. A small telescope shows it as a roundish, softly glowing patch noticeably brighter towards the centre. Being tilted at an angle to us it appears somewhat elliptical in outline, covering over half a Moon's width at its longest. In the same telescopic field of view ½° to the north is M82 (see page 251); the two galaxies are about 10 million l.y. from us.

M97, the Owl Nebula, gets its popular name from the two dark patches, like an owl's eyes, either side of the 16th-magnitude central star. This star, a white dwarf, ejected the gas that forms the nebula and now illuminates it. An aperture of at least 100 mm is needed to see the eyes. Visually, the nebula appears grey rather than green. (AURA/NOAO/NSF)

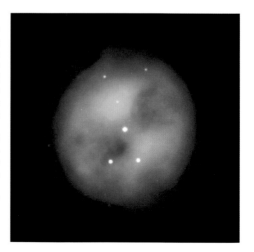

M82 (NGC 3034), 9h 56m +69°.7, is a neighbour galaxy of M81, a quarter as bright and less than half its size but still visible in binoculars. In a small telescope it appears as an elongated blur and can actually appear more prominent than M81 because of its higher surface brightness. Detailed studies show that M82 is actually an edge-on spiral galaxy mottled by dust clouds and experiencing a burst of star formation as a result of a recent interaction with M81.

M97 (NGC 3587), 11h 15m +55°.0, is an elusive 11th-mag. planetary nebula known as the Owl Nebula because of two dark patches like eyes that give it the appearance of an owl's face when seen through a large telescope. In moderate apertures it appears as only a pale disk over three times the size of Jupiter, and will probably need an aperture of at least 75 mm to be seen at all. The Owl lies about 1300 l.y. away.

M101 (NGC 5457), 14h 03m +54°.3, is a spiral galaxy visible in binoculars as a pale, rounded smudge about half the apparent size of the full Moon; because of its large size it is less prominent than its quoted 8th mag. would suggest. Long-exposure photographs show it as a face-on galaxy with widely spread spiral arms, but these are not apparent in small telescopes which tend to show only the elliptical central region. M101 lies 23 million l.y. away.

M101, a face-on spiral galaxy, displays an asymmetric shape in photographs, but in amateur instruments only the core and traces of the spiral arms will be seen. (George Jacoby, Bruce Bohannan, Mark Hanna/AURA/NOAO/NSF)

URSA MINOR   The Little Bear

A constellation said to have been introduced in about 600 BC by the Greek astronomer Thales. Ursa Minor currently contains the north celestial pole, within 1° of the conveniently placed 2nd-magnitude star we call Polaris, α (alpha) Ursae Minoris. Precession will carry the pole closest to Polaris (just under ½°) around AD 2100, after which it will start to move away again towards Cepheus, crossing the border in AD 2234 (see chart below). Ursa Minor is also termed the Little Dipper because its seven brightest stars outline a shape like a smaller version of the Big Dipper in Ursa Major. The stars β (beta) and γ (gamma) Ursae Minoris in the Little Dipper's bowl are called the Guardians of the Pole.

α (alpha) Ursae Minoris, 2h 32m +89°.3, (Polaris), mag. 2.0, is a yellow-white supergiant 431 l.y. away. It is a Cepheid variable (in fact, the nearest Cepheid to us) whose pulsations diminished during the 20th century, stabilizing by the 1990s at a few hundredths of a magnitude. Polaris is also a double star, with a mag. 8.2 companion visible in a small telescope. Binoculars and small telescopes show a ring of 8th- to 11th-mag. stars some ¾° across, on which Polaris appears to be strung like a lustrous pearl on a necklace.

β (beta) UMi, 14h 51m +74°.1, (Kochab), mag. 2.1, is an orange giant 126 l.y. away.

γ (gamma) UMi, 15h 21m +71°.8, (Pherkad), mag. 3.0, is a blue-white giant star 480 l.y. away. The mag. 5.0 orange giant 11 UMi that appears near it, as seen by the naked eye or in binoculars, is unrelated, lying 390 l.y. away.

ε (epsilon) UMi, 16h 46m +82°.0, mag. 4.2, 347 l.y. away, is a yellow giant eclipsing binary that varies in a period of 39.5 days by under 0.1 mag., not discernible to the naked eye.

η (eta) UMi, 16h 18m +75°.8, mag. 5.0, is a white main-sequence star 97 l.y. away. A wide mag. 5.5 companion, 19 UMi, is an unrelated background object.

The movement of the north celestial pole over 800 years as a result of precession. (Wil Tirion)

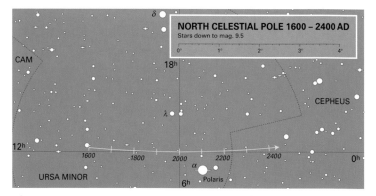

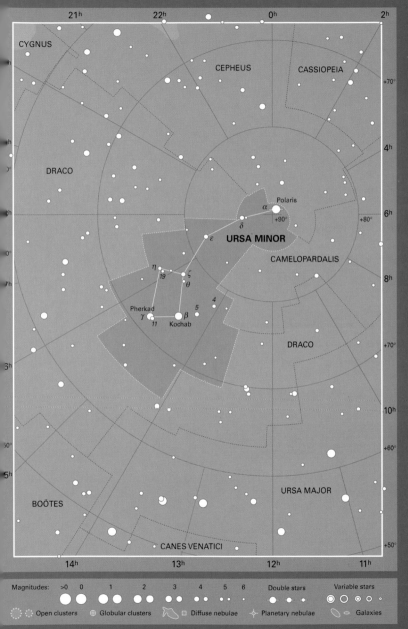

VELA   The Sails

Formerly part of the ancient constellation Argo Navis, the ship of Jason and the Argonauts, until made separate (along with Carina and Puppis) by Nicolas Louis de Lacaille in 1763. When Argo Navis was subdivided, its stars were not given new Greek letters; as a result, the labelling of the stars in Vela begins with γ (gamma). In fact, Argo Navis was so large that astronomers ran out of Greek letters to label its stars and turned to Roman letters; many of these Roman-lettered stars have ended up in Vela. The stars κ (kappa) and δ (delta) Velorum, in conjunction with ι (iota) and ε (epsilon) Carinae, form a shape known as the False Cross, sometimes mistaken for the real Southern Cross. Vela lies in a part of the Milky Way rich in faint nebulosity, visible on long-exposure photographs. This nebulosity is known as the Gum Nebula, after the Australian astronomer Colin S. Gum who drew attention to it in 1952. The Gum Nebula is believed to be the remains of one or more supernovae that occurred long ago. Another remnant of a supernova in the constellation is the Vela Pulsar, which flashes 11 times per second, one of the few pulsars that can be seen flashing at optical as well as radio wavelengths.

γ (gamma) Velorum, 8h 10m −47°.3, is an interesting multiple star. Binoculars or small telescopes show that it consists of two apparently unrelated blue-white components of mags. 1.8 and 4.3. The brighter of them is the brightest known Wolf–Rayet star, a rare class of stars with very hot surfaces which seem to be ejecting gas. It is 840 l.y. away, while the fainter star is estimated to lie 1600 l.y. away. There are also two wider companions, of 8th and 9th mags.

δ (delta) Vel, 8h 45m −54°.7, mag. 1.9, is a blue-white main-sequence star 80 l.y. away with a mag. 5.1 companion requiring 100 mm aperture to see.

κ (kappa) Vel, 9h 22m −55°.0, mag. 2.5, is a blue-white star 539 l.y. away.

λ (lambda) Vel, 9h 08m −43°.4, mag. 2.2, is an orange supergiant, irregularly variable (by less than 0.2 mag.), 573 l.y. away.

H Vel, 8h 56m −52°.7, 376 l.y. away, is a neat double of mags. 4.8 and 7.4, difficult in the smallest telescopes because of the magnitude contrast.

NGC 2547, 8h 11m −49°.3, 1400 l.y. away, is an open cluster of about 80 stars of mag. 6.5 and fainter, just visible to the naked eye and best seen in binoculars.

NGC 3132, 10h 08m −40°.5, is a relatively large and bright planetary nebula of 8th mag., known as the Eight-Burst Nebula. Through a small telescope it appears larger than Jupiter, with a central star of 10th mag. It lies 2600 l.y. away.

NGC 3228, 10h 22m −51°.7, is an open cluster of about 15 faint stars for binoculars and small telescopes, 1600 l.y. away.

IC 2391, 8h 40m −53°.1, is a large naked-eye cluster of 50 stars, 500 l.y. away, scattered around the mag. 3.6 blue-white star o (omicron) Vel, a small-amplitude variable of β Cephei type. About 1° from it is the binocular cluster NGC 2669.

IC 2395, 8h 41m −48°.2, is a binocular cluster of 40 stars, 3100 l.y. away. A mag. 5.5 star which seems to be the brightest member is probably a foreground star. Also visible ½° to the south is the 8th-mag. open cluster NGC 2670.

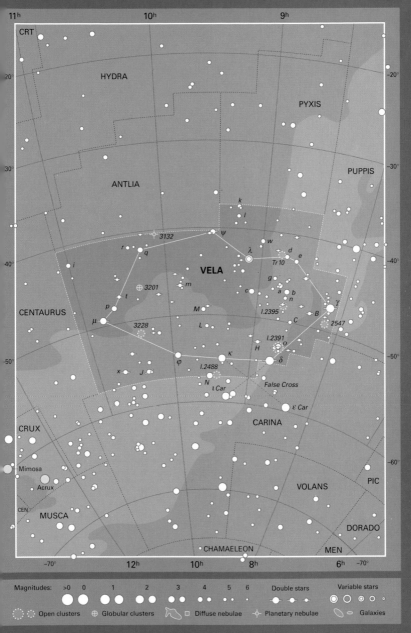

Magnitudes: >0 0 1 2 3 4 5 6   Double stars   Variable stars

Open clusters   ⊕ Globular clusters   Diffuse nebulae   ✧ Planetary nebulae   Galaxies

VIRGO   The Virgin

The second-largest constellation, and the largest of the zodiac. Virgo is usually identified as the goddess of justice, the scales of justice being represented by neighbouring Libra. But another legend sees her as Demeter, the corn goddess, and in the sky she is pictured holding an ear of wheat (the star Spica). The Sun passes through the constellation from mid-September to early November, and thus is within Virgo's boundaries at the time of the September equinox each year, when the Sun moves south of the celestial equator. Virgo contains the nearest major cluster of galaxies to us, which spills over into neighbouring Coma Berenices; the area is sometimes known as 'the realm of the galaxies'. The Virgo Cluster lies 55 million l.y. away and contains about 3000 members, several dozen of which are visible in apertures of 150 mm or so, although they appear as little more than hazy patches; some of the brightest members are mentioned below. Virgo contains the brightest quasar, 3C 273, located at 12h 29.1m, +2° 03′. It is unrelated to the Virgo Cluster of galaxies. Optically, 3C 273 appears as a 13th-mag. blue star. Its estimated distance is about 3000 million l.y.

α (alpha) Virginis, 13h 25m −11°.2, (Spica, 'ear of wheat'), mag. 1.0, is a blue-white main-sequence star 262 l.y. away. It is a spectroscopic binary, tidally distorted by its companion so that it varies slightly with a period of 4 days as it rotates, although only by 0.1 mag.

β (beta) Vir, 11h 51m +1°.8, (Zavijava), mag. 3.6, is a yellow-white main-sequence star 36 l.y. away.

γ (gamma) Vir, 12h 42m −1°.4, (Porrima), 39 l.y. away, is a celebrated double star. Together, the stars shine as mag. 2.7. But a small telescope reveals γ Vir to consist of a matching pair of white stars, each of mag. 3.5. They orbit each other every 169 years and are closest around the year 2005, when a 250-mm telescope is needed to split them. Rapidly moving apart thereafter, they become divisible in 100 mm by 2010, and in 60 mm after 2012, remaining within range of small telescopes for the rest of the 21st century.

δ (delta) Vir, 12h 56m +3°.4, mag. 3.4, is a red giant 202 l.y. away.

ϵ (epsilon) Vir, 13h 02m +11°.0, (Vindemiatrix, 'grape gatherer'), mag. 2.8, is a yellow giant 102 l.y. away.

θ (theta) Vir, 13h 10m −5°.5, mag. 4.4, is a blue-white star 415 l.y. away with a 9th-mag. companion visible in a small telescope.

τ (tau) Vir, 14h 02m +1°.5, mag. 4.3, a blue-white star 218 l.y. away, forms a wide optical double for small telescopes with a 9th-mag. companion.

φ (phi) Vir, 14h 28m −2°.2, 135 l.y. away, is a yellow giant of mag. 4.8 with a 9th-mag. companion, difficult to see in the smallest telescopes because of the brightness contrast.

M49 (NGC 4472), 12h 30m +8°.0, is an 8th-mag. elliptical galaxy visible as a rounded glow in a 75-mm telescope under low power. It is one of the largest and brightest members of the Virgo Cluster of galaxies.   ▶

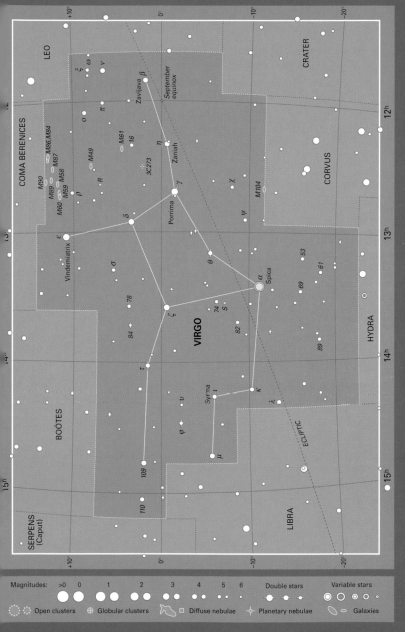

M58 (NGC 4579), 12h 38m +11°.8, is a 10th-mag. barred spiral galaxy with a noticeably brighter core.

M59 (NGC 4621), 12h 42m +11°.6, is a 10th-mag. elliptical galaxy with a starlike nucleus, lying about a quarter of the way from M60 to M58.

M60 (NGC 4649), 12h 44m +11°.5, is a 9th-mag. elliptical galaxy, one of the most prominent members of the Virgo Cluster, detectable in 75 mm aperture.

M84 (NGC 4374), 12h 25m +12°.9, and M86 (NGC 4406), 12h 26m +12°.9, are a pair of 9th-mag. elliptical galaxies appearing in the same telescopic field as fuzzy patches with noticeably brighter cores. M86 is slightly the larger of the two and noticeably elongated, whereas M84 appears round.

M87 (NGC 4486), 12h 31m +12°.4, is a celebrated giant elliptical galaxy. It is also a strong radio source, known as Virgo A, and an X-ray source. Photographs taken through large telescopes show a jet of matter emerging from M87. In amateur telescopes, M87 appears as a rounded, 9th-mag. glow with a noticeable nucleus.

M90 (NGC 4569), 12h 37m +13°.2, is a large 9th-mag. spiral galaxy, tilted at an angle to us so that it appears elongated.

M104 (NGC 4594), 12h 40m −11°.6, is an 8th-mag. spiral galaxy seen edge-on so that it appears elongated. It is popularly known as the Sombrero Hat because of its distinctive appearance on long-exposure photographs (see page 128). With its bulging nucleus ringed by tightly coiled spiral arms, its appearance is reminiscent of Saturn. Apertures of 150 mm reveal a dark lane of dust along its rim, silhouetted against the brighter arms and nucleus. The Sombrero is not in the Virgo Cluster but somewhat closer, about 35 million l.y. away.

VOLANS   The Flying Fish

This figure was invented at the end of the 16th century by the Dutch navigators Pieter Dirkszoon Keyser and Frederick de Houtman under the name Piscis Volans. None of its stars is particularly bright, but there are two fine double stars for small telescopes.

α (alpha) Volantis, 9h 02m −66°.4, mag. 4.0, is a blue-white star 124 l.y. away.

β (beta) Vol, 8h 26m −66°.1, mag. 3.8, is an orange giant 108 l.y. away.

γ (gamma) Vol, 7h 09m −70°.5, 142 l.y. away, is a pair of gold and cream stars of mags. 3.8 and 5.7, beautiful in a small telescope.

δ (delta) Vol, 7h 17m −68°.0, mag. 4.0, is a yellow-white giant 660 l.y. away.

ε (epsilon) Vol, 8h 08m −68°.6, 642 l.y. away, is a mag. 4.4 blue-white star with an 8th-mag. companion visible in a small telescope.

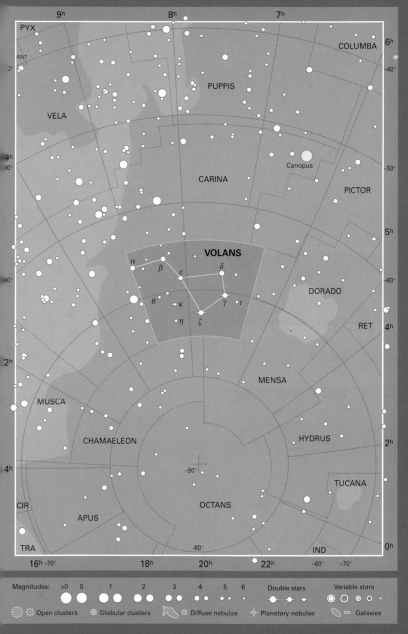

VULPECULA   The Fox

A faint constellation at the head of Cygnus. It was originated in 1687 by the Polish astronomer Johannes Hevelius, who called it Vulpecula cum Anser, the Fox and Goose. Since then the goose has fled, leaving the fox on its own. In 1967 this diminutive constellation was the site of an astounding discovery – the first pulsar, or flashing radio source, which was detected by radio astronomers in Cambridge, England, about 1½° north of the notable grouping called Brocchi's Cluster.

α (alpha) Vulpeculae, 19h 29m +24°.7, mag. 4.4, is a red giant 297 l.y. away. An unrelated mag. 5.8 orange giant, 8 Vul, 484 l.y. away, is seen nearby in binoculars.

T Vul, 20h 51m +28°.3, is a yellow-white supergiant Cepheid variable that ranges between mags. 5.4 and 6.1 every 4.4 days. It lies about 1700 l.y. away.

M27 (NGC 6853), 20h 00m +22°.7, the Dumbbell Nebula, is a large and bright planetary nebula, reputedly the most conspicuous of its kind, visible in binoculars as an elliptical misty glow. Its dumbbell shape is better seen through telescopes, which also reveal its greenish tinge. M27 is of 8th mag. and spans about one-quarter the apparent diameter of the full Moon. M27 is 1000 l.y. away. (For a photograph see page 218.)

Brocchi's Cluster (Collinder 399), 19h 25m +20°.2, is a striking binocular group of stars on the border with Sagitta, popularly known as the Coathanger because of its distinctive shape. The group consists of an almost straight line of six stars extending for three Moon diameters; from the centre of this line emerges a curve of four more stars, forming the Coathanger's hook. The brightest member of the group, in the hook, is 4 Vul, magnitude 5.1. Distances of the individual stars range from just over 200 l.y. to more than 1000 l.y., and all have different motions through space, so this is not a true cluster but a chance alignment.

The stars of Brocchi's Cluster form the shape of a coathanger. (Wil Tirion)

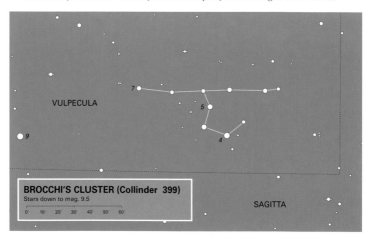

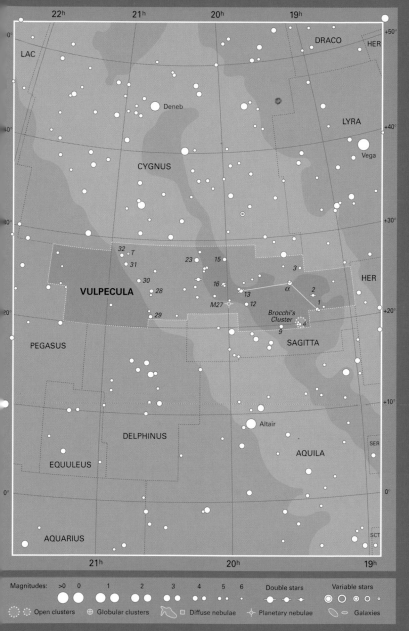

Magnitudes: >0 0 1 2 3 4 5 6   Double stars   Variable stars

Open clusters   Globular clusters   Diffuse nebulae   Planetary nebulae   Galaxies

The Pleiades star cluster in Taurus hovers above a winter landscape in this composite picture. (Robin Scagell)

SECTION II

Stars

Stars are balls of gas made incandescent by energy from nuclear reactions deep in their interiors. They come in a wide range of sizes and brightnesses, from faint dwarfs a hundredth of the Sun's diameter to dazzling supergiants hundreds of times the size of the Sun. They range in temperature from intensely hot blue-white stars (surfaces at more than 20,000°C) to cool red stars (3000°C). The Sun, which is a medium-temperature yellow star, turns out to be pretty average in all respects.

Stars are born from massive clouds of gas and dust within our Galaxy. An interstellar gas cloud is termed a *nebula* (plural: nebulae), from the Latin for 'cloud'. A nebula is not uniformly distributed in space, but contains denser knots – the seeds of future stars. If the knot is dense enough it begins to contract under the inward pull of its own gravity. As it gets smaller and denser it heats up, until the temperature and pressure at the centre of the shrinking blob become so great that nuclear reactions begin. The gas blob has switched on to become a true star, generating its own heat and light for millions of years.

Several star-spawning clouds are well within reach of observation by amateurs. Most famous is the Orion Nebula, marking the sword in the constellation of Orion the Hunter. This nebula is visible as a hazy green glow to the naked eye; binoculars show it more clearly. At the centre of the Orion Nebula is a star called θ^1 (theta[1]) Orionis, which small telescopes show to consist of four component stars. Energy emitted by the brightest of these four stars makes the nebula shine. But behind the bright, visible part of the cloud is an even larger, still-dark area where stars are being born at this moment. The Orion Nebula is estimated to contain enough matter to produce hundreds of stars: it is a star cluster in the making. Another famous stellar birthplace is the Tarantula Nebula in the southern constellation Dorado, which dwarfs the Orion Nebula and is in fact the largest nebula known.

One celebrated group of young stars is the Pleiades cluster, popularly known as the Seven Sisters, in the constellation Taurus, the Bull. At least five members of the Pleiades can be distinguished by normal eyesight; binoculars and small telescopes bring dozens more members into view. The whole cluster is estimated to contain about a hundred stars. The brightest and youngest of these formed no more than 2 million years ago, making them extremely youthful by astronomical standards.

The Pleiades is an example of a type of cluster referred to as an *open cluster* or *galactic cluster*. About a thousand are known to astronomers, and the most prominent of them are listed in this book. Near the Pleiades in Taurus is a larger and older open cluster, the Hyades, which is estimated to be about 500 million years old. Being older than the

Ghostly, glowing gas swirls of the Orion Nebula, M42, a cradle of star formation and one of the most-observed objects in the heavens. At its heart is the multiple star θ^1 (theta^1) Orionis, also known as the Trapezium; this lies near the tip of the dark intrusion known as the Fish's Mouth. Below it is the smaller, rounded knot of M43, actually part of the same great gas cloud. South is to the top in this picture. (Bill Schoening/AURA/NOAO/NSF)

Pleiades, the stars have had more time to drift apart. Eventually, most open clusters disperse completely. The Sun was probably a member of such a cluster when it was born 4600 million years ago. A different type of cluster is a globular cluster, described on page 283.

Much larger than open clusters are *stellar associations*, scatterings of young stars hundreds of light years across. It is no coincidence that most of the bright stars in Orion lie at similar distances from us (the most notable exception being Betelgeuse), for they are members of such an association, centred on the Orion Nebula about 1500 light years away. Three times closer to us is the extensive Scorpius–Centaurus association, which extends across 60° or more of sky from Scorpius via Lupus into Centaurus and Crux. Antares is the brightest member; other prominent members are β (beta) Centauri, α (alpha) and β (beta) Crucis, and the open cluster IC 2602 in Carina. Associations arise from particularly large clouds of gas and dust in the spiral arms of the Galaxy.

Nebulae are made of a 10:1 mixture of hydrogen and helium, the primary constituents of the Universe, so, naturally enough, stars have the same composition. Stars get their energy from nuclear reactions which transform hydrogen into helium. In the reactions, four hydrogen atoms are crushed together to make one atom of helium; an uncontrolled version of the same reaction occurs in a hydrogen bomb.

There are certain limits on the size of a star. A gas blob with less than about 8 per cent of the Sun's mass cannot become a star, because conditions in its interior will not become sufficiently extreme for nuclear reactions to begin. This 8 per cent limit may be considered the dividing line between a planet and a star. If the gaseous planet Jupiter in our Solar System had been about 80 times more massive than it actually is, it would have become a small star. At the other end of the scale, the largest stars have masses of about a hundred times that of the Sun. It was once thought that stars more massive than this would produce so much energy that they would literally disintegrate, but this may not be true in all cases. A few stars are known that seem to have masses greater than a hundred Suns, one example being η (eta) Carinae.

A star's most vital statistic is its mass, for this factor affects everything else about it: its temperature, its brightness and its lifetime. The stars with the least mass are, not surprisingly, the coolest; they are known as *red dwarfs*. A typical red dwarf such as Barnard's Star, the second-closest star to the Sun, has a mass about a tenth that of the Sun and glows a dull red with a surface temperature of about 3000°C. Even though Barnard's Star is only six light years away, it is too faint to be seen with the naked eye. Surprisingly enough, stars with the lowest mass live the longest. Their nuclear fires burn so slowly that they can survive for as much as a million million years, a hundred times as long as the Sun. The Sun itself, which by definition is of one solar mass, has a surface temperature of 5500°C, and is expected to live for about 10,000 million years. It is currently in the prime of its life.

The Horsehead Nebula in Orion, a dark cloud of cooler gas and dust seen in silhouette against a background of glowing hydrogen, looks like an intersellar chess piece. (Nigel Sharp/AURA/NOAO/NSF)

Moving up the scale, a star such as Sirius, which is twice the Sun's mass, can live for only about 1000 million years, a tenth of the Sun's age. The surface temperature of Sirius is a blue-white 11,000°C. Larger and hotter still, the star Spica in the constellation Virgo has a mass of about 11 Suns and a surface temperature of around 24,000°C. The lifetime of this intensely hot, highly luminous star is less than 1 per cent of the lifetime of the Sun.

A star's colour is a direct indicator of its temperature. The most precise way to measure a star's temperature is to study the spectrum of its light, which is done by splitting up the light in a device called a spectroscope. Stars are classified into a sequence of so-called *spectral types* according to their temperature (see the table on page 269). The bluest and hottest stars are classified as spectral types O and B. Prominent examples of B-type stars are α (alpha) and β (beta) Crucis, β (beta) Centauri and Spica; these are in fact the bluest of the first-magnitude stars. Then come the cooler blue-white A-type stars, which include Sirius, followed by F-type stars, which appear white or yellow-white; Procyon is an example. G-type stars are yellow; they include the Sun, α (alpha) Centauri and τ (tau) Ceti.

Cooler still are K-type stars, such as ε (epsilon) Eridani, which have orange hues. Coolest of all are the so-called red stars with an M-type spectrum, examples being Antares and Betelgeuse which are the reddest first-magnitude stars. Each spectral type is subdivided into ten steps from 0 to 9; on this more precise scale, the Sun ranks as a G2 star. The seemingly haphazard lettering sequence used for spectral types is the result of a previous classification scheme which was rearranged and shortened to produce the present system. The sequence of stellar spectral types can be remembered by the mnemonic: 'Oh Be A Fine Girl, Kiss Me'.

Columns of cooler hydrogen gas and dust up to 1 light year long protrude into the Eagle Nebula, M16, in Serpens. Ultraviolet light flooding from hot, young stars off the top of the picture evaporates gas from the surfaces of the columns, leaving behind denser knots where new stars are forming. In this image from the Hubble Space Telescope, hydrogen gas is depicted as green rather than the usual pink. (Jeff Hester and Paul Scowen, Arizona State University/NASA)

Star colours are necessarily subjective, given the individual characteristics of different people's eyes and the varying conditions under which stars are viewed. For example, astronomers define Vega, of spectral type A0, as pure white, yet to most eyes it has a distinctly bluish cast, as do the components of Castor, also of type A. At the other end of the scale, few so-called red giant or supergiant stars appear truly red, but more a deep orange or tan colour. An apparent paradox is that our Sun, whose light is usually thought of as white, is classified as a yellow star. In fact, what makes the daytime Sun appear white is its overpowering brilliance. If we could see it from a distance, so it was much fainter, it would indeed appear yellowish.

It turns out that a star of truly white appearance to the eye has a spectral type of around F0, similar to that of Canopus, 2000°C hotter than the Sun. The star colours given in the constellation notes in this book are an indication of how stars appear to an observer, although the real colours

NGC 2070 in the Large Magellanic Cloud, known as the Tarantula Nebula because of its spidery shape. At its heart is an open cluster, R 136, containing some of the hottest and most massive stars known. Such massive stars have short lifetimes and are likely to die in supernova explosions. (ESO)

STELLAR SPECTRAL TYPES

Type	Assigned colour	Temperature range (°C)	Examples
O	Blue	40,000–25,000	ζ Puppis (supergiant)
B	Blue	25,000–11,000	Spica (main sequence)
			Regulus (main sequence)
			Rigel (supergiant)
A	Blue-white	11,000–7500	Vega (main sequence)
			Sirius (main sequence)
			Deneb (supergiant)
F	White	7500–6000	Canopus (supergiant)
			Procyon (subgiant)
			Polaris (supergiant)
G	Yellow	6000–5000	Sun (main sequence)
			α Centauri (main sequence)
			τ Ceti (main sequence)
			Capella (giant)
K	Orange	5000–3500	ε Eridani (main sequence)
			Arcturus (giant)
			Aldebaran (giant)
M	Red	3500–3000	Barnard's Star (main sequence)
			Antares (supergiant)
			Betelgeuse (supergiant)

are subtle and your own visual impressions may well disagree. Indeed, you will find that the apparent intensity of the coloration varies from night to night with changing atmospheric transparency and clarity. For faint stars, you will be unlikely to see any colour at all without optical aid, for reasons to do with the sensitivity of the eye.

When the spectral type of stars is plotted against their actual luminosity (absolute magnitude), all stars that are in stable, hydrogen-burning middle age are found to lie in a well-defined band across the graph known as the *main sequence*. A star's position along the main sequence is fixed by its mass, with the least massive stars at the bottom and the most massive stars at the top. The Sun, as befits its middle-of-the-road nature, lies about halfway along the main sequence (see page 271). Such a plot of star brightness against spectral type is known as a Hertzsprung–Russell diagram, after the Danish astronomer Ejnar Hertzsprung and the American Henry Norris Russell who devised it in 1911–13.

Although most stars lie on the main sequence, a number of particularly bright stars lie above and to the right of it, while a few faint stars lie below and to the left of it. These stars are all in late stages of evolution. We can best understand what is happening to them by following the future evolution predicted for the Sun.

The Double Cluster in Perseus, NGC 869 and 884, is a pair of open clusters in a spiral arm of our Galaxy. NGC 884, at left, contains several red giants, suggesting that it is the older of the two. (Nigel Sharp/AURA/NOAO/NSF)

As mentioned earlier, the Sun formed about 4600 million years ago and is now about halfway through its expected lifespan. In a few thousand million years, though, it will start to run out of hydrogen at its core. In search of more hydrogen to use as fuel, the nuclear reactions inside the Sun will start to move outwards, releasing more energy. Eventually, when they are surrounded by a shell of burning hydrogen, even the helium atoms in the Sun's core will enter into nuclear reactions of their own, fusing together to form carbon atoms.

With all this extra energy being given off, the Sun will become much brighter than it is today and will start to swell alarmingly in size. But as the Sun's outer layers expand they will also cool, becoming redder in colour, and the Sun will turn into a *red giant*, similar to the bright stars Aldebaran and Arcturus. At its largest, the red giant Sun will grow to at least a hundred times its present diameter, engulfing Mercury, Venus and perhaps even the Earth within its outer layers. Needless to say, all life on our planet will long since have become extinct.

On the Hertzsprung–Russell diagram, the Sun's increase in brightness will move it upwards, off the main sequence, and its change in spectral

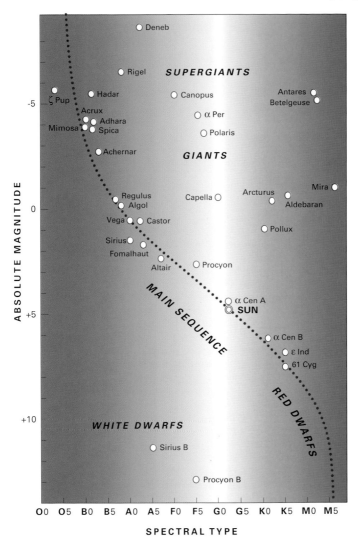

In the Hertzsprung–Russell diagram, the actual brightness of stars (their absolute magnitude) is plotted against their temperature (spectral type). Stars of different types fall in different areas of the diagram. The band running from top left to bottom right is the main sequence; the Sun falls approximately halfway along it. Giants and supergiants lie above the main sequence, while white dwarfs are below and to the left of it. (Wil Tirion)

STELLAR LUMINOSITY CLASSES

Ia0	Exceptionally bright supergiant
Ia	Bright supergiant
Iab	Less bright supergiant
Ib	Supergiant
II	Bright giant
III	Giant
IV	Subgiant
V	Main sequence
VI or sd	Subdwarf

type will in addition move it to the right. Stars at the top end of the main sequence, which are far more massive than the Sun, become so big and bright at this stage of their evolution that they are referred to not as mere giants but as *supergiants*. Prominent examples of red supergiants are Betelgeuse and Antares, both of which are hundreds of times larger than the Sun. Other stars which have not yet evolved enough to become red in colour but which are nevertheless firmly in the supergiant bracket are Rigel, Deneb and Polaris.

To distinguish whether a star is, say, a giant or supergiant, or lies on the main sequence, astronomers assign stars a luminosity class (see the table above) in addition to the spectral type. Here it should be noted that astronomers tend to think of stars as either dwarfs or giants, depending on whether they are on the main sequence or have evolved off it. Main sequence stars are sometimes referred to as dwarfs, even though the most massive of them may be several times larger than the Sun.

Taken together, the spectral type and luminosity class define the main properties of a star as it exists at present. But those properties change as the star ages. Stars spend only a few per cent of their total lifetime in the red giant phase, which in the case of stars like the Sun amounts to no more than a few hundred million years. A red giant is a star that has grown old and is about to die.

Once a red giant has swollen beyond a certain size its distended outer layers drift off into space, forming a stellar smoke ring known rather confusingly as a *planetary nebula*, even though it has nothing to do with planets. The name was first used in 1785 by William Herschel, because they looked like the small, rounded disks of planets as seen through his telescope. Probably the best-known of all planetary nebulae is the Ring Nebula in Lyra, though it is not the easiest to see. Much larger is the Dumbbell Nebula in Vulpecula, which can be picked up in binoculars on a clear, dark night. Two small but bright planetary nebulae for amateur telescopes are NGC 6826 in Cygnus and NGC 7662 in Andromeda.

At the centre of a planetary nebula, the core of the former red giant is exposed as a small, intensely hot star. Once the surrounding gases of the planetary nebula have dispersed, usually after a few thousand years, the central star remains as a so-called *white dwarf*. A white dwarf is only about the diameter of the Earth, but contains most of the matter of the original star; only about 10 per cent of the star's mass is lost in the planetary nebula stage. White dwarfs are therefore exceptionally dense bodies. A teaspoonful of white dwarf material would have a mass of thousands of kilograms. Over thousands of millions of years, white dwarfs slowly cool off and fade into oblivion.

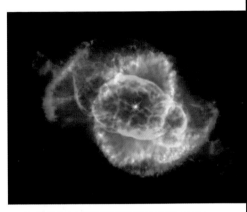

Two planetary nebulae photographed by the Hubble Space Telescope.
Right: NGC 6543, the Cat's Eye Nebula in Draco, consists of overlapping loops. Different colours represent gas at different temperatures. (Bruce Balick, U. of Washington/NASA)

Below: M57, the Ring Nebula in Lyra, is actually a cylinder of gas ejected by the central star; it appears ring-like because we are looking at it end-on. (AURA/STScI/NASA)

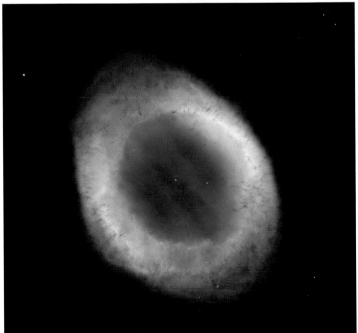

Being so small, white dwarfs are very faint. Not one is visible to the naked eye. Both the nearby bright stars Sirius and Procyon have white dwarf companions, but Procyon's companion is too close to its parent to be distinguishable in amateur telescopes, and the companion of Sirius can be glimpsed only under the most favourable conditions. The easiest white dwarf to see is a companion of the star o² (omicron²) Eridani (also known as 40 Eridani); a small telescope will show it. Of added interest is a fainter third member of this system, a red dwarf, which is also visible in amateur telescopes.

Our Sun, it seems, is destined to go through the stage of being a planetary nebula before fading away as a white dwarf. But stars with several times the Sun's mass, towards the top end of the main sequence, suffer a far more spectacular end. As we have seen, they first become dazzling supergiants rather than mere giants. They do not get a chance to

The Eskimo Nebula, NGC 2392 in Gemini, photographed with unprecedented clarity by the Hubble Space Telescope, gets its name from its resemblance to a face surrounded by a fur parka. In reality, the 'parka' is a disk of material thrown off the central star during its red giant phase, while the Eskimo's 'face' is a bubble of gas being blown into space by the intensely hot central star. (Andrew Fruchter, STScI/NASA/ESA)

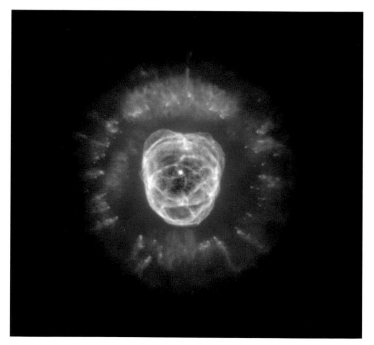

The η (eta) Carinae Nebula, NGC 3372, a large cloud of glowing hydrogen gas visible to the naked eye in the southern hemisphere, containing clusters of hot young stars. The nebula is named after η Carinae itself, a massive star that lies in the brightest central part, near a dark notch called the Keyhole (just above the V-shaped lane in this picture). η Carinae could erupt as a brilliant supernova within the next 10,000 years. (AURA/NOAO/NSF)

The Crab Nebula, M1, the remains of a star that exploded as a supernova. The blue colour comes from electrons spiralling in the Crab's magnetic field, while the red emission comes from hydrogen gas. The Crab Pulsar is the lowermost of the two stars near the nebula's centre; the other star is an unrelated foreground object. (Laird Thompson, University of Hawaii/CFHT)

reach the planetary nebula stage. So massive are they that the nuclear reactions at their centres continue in runaway fashion until the star becomes unstable and explodes. Such an explosion is known as a *supernova*.

In a supernova eruption a star's brightness increases millions of times, so that for a few days the star can rival the brilliance of an entire galaxy. The shattered outer layers of the star are thrown off into space at speeds of around 5000 km per second. In AD 1054 astronomers on Earth saw a star erupt as a supernova in the constellation Taurus. The star became brighter than Venus, and was visible in daylight for three weeks. It finally faded from naked-eye view more than a year after it had first appeared.

At the site of that explosion lies one of the most famous objects in the heavens: the Crab Nebula, the shattered remains of the star that erupted. The Crab Nebula is visible as a smudgy patch in amateur telescopes, but is best seen on long-exposure photographs taken with large instruments. Over the next 50,000 years or so the gases of the Crab Nebula will disperse into space, forming delicate traceries like those of the Veil Nebula in Cygnus, itself the remains of a former supernova.

The last supernova observed in our Galaxy was in 1604. This star, in the constellation Ophiuchus, reached a maximum magnitude of just over −2, as bright as Jupiter. It was studied by the German astronomer Johannes Kepler, and is often known as Kepler's Star. Hundreds of supernovae in other galaxies have been seen telescopically since then, but only one has become bright enough to be visible to the naked eye. That was Supernova 1987A, which erupted in the Large Magellanic Cloud, a near neighbour of the Milky Way. It was first spotted in 1987 on February 24 by astronomers in the southern hemisphere. It rose to a maximum magnitude of 2.9 in late May, eventually fading from naked-eye view by the end of that year. Another supernova in our Galaxy is long overdue. When it comes, it should be a spectacular sight. Many astronomers dream of seeing it outshining the other stars in the night sky, dazzling the eye and casting shadows.

A star might not blow itself entirely to bits in a supernova explosion. Sometimes the central core of the exploded star is left as an object even smaller and denser than a white dwarf, known as a *neutron star*. In a neutron star, the protons and electrons of the star's atoms have been crushed by the tremendous forces of the supernova so that they combine to form the particles known as neutrons. A typical neutron star is a mere 20 km in diameter, but contains as much mass as one or two Suns. Being so minute, neutron stars can spin very rapidly without flying apart. Each time they spin we receive a flash of radiation like a lighthouse beam. Astronomers have detected radio pulses from several hundred such sources, which they term *pulsars*; one lies at the centre of the Crab Nebula. The Crab Pulsar flashes 30 times per second; others pulse more slowly, down to once every four seconds. Most neutron stars are too faint to be seen optically, but the pulsar in the Crab Nebula has been seen flashing in step with the radio pulses.

If the core of the exploded star has a mass of more than three Suns, then even a neutron star is not the end for it. Instead, it becomes something still more bizarre: a *black hole*. No force can shore up a dead star weighing more than three solar masses against the inward pull of its own gravity. It continues to shrink, becoming ever smaller and denser until its gravity becomes so great that nothing can escape from it, not even its own light. It has dug its own grave – a black hole. Since a black hole is by definition invisible, it is of only academic interest to amateur observers. However, professional astronomers have detected X-ray emissions from various locations in space which they believe come from hot gas plunging into the bottomless pits of black holes. The best-known candidate for a black hole is Cygnus X-1; it lies near a visible 9th-magnitude star in the constellation Cygnus.

Double and Multiple Stars

To the naked eye, stars appear as solitary, isolated objects. But the great majority of them – over 75 per cent – actually have one or more companion stars, too faint or too close to be seen separately with the naked eye but which can be distinguished telescopically. Many attractive double and multiple stars are within the range of binoculars and small telescopes.

There are two sorts of double star. In one type, the two stars are not actually related but happen to lie in the same line of sight by chance; in this arrangement, termed an *optical double*, one star may be many times farther from us than the other. Optical doubles are comparatively rare. Most double stars are physically linked to each other by gravity, forming a genuine *binary* system. The two stars in an optical binary orbit their mutual centre of gravity, which can take centuries. In systems of more than two stars, the orbits can become very complicated.

Twins or triplets are the most common, but some stellar families can be much larger. One celebrated multiple star is the so-called Double Double, ε (epsilon) Lyrae. Binoculars show it to be a wide double, but modest-sized amateur telescopes reveal that each of these stars is itself double, making a quadruple system. Even more remarkable is Castor, a system of six stars all linked by gravity. Amateur telescopes show that Castor consists of two bright blue-white stars close together, with a fainter third star some way off. Professional observations have revealed that each of these three stars is a *spectroscopic binary*. In a spectroscopic binary the two stars are too close together to be seen individually in any telescope, but analysis of the light from the star reveals the companion's presence. The two brightest stars of Castor take nearly five centuries to orbit each other, whereas their spectroscopic companions whirl around them in only a few days. Sometimes the members of a binary system eclipse each other as seen from Earth. These eclipsing binaries are dealt with in the next section, on variable stars.

For amateurs, the attraction of observing double stars lies in separating (or 'splitting') close doubles and in comparing their brightnesses and colours. Some pairs have a beautiful colour contrast, as with the yellow and blue-green components of Albireo, β (beta) Cygni. Star colours can become more noticeable if you slightly defocus the telescope, or tap it to make the image vibrate. The closer together two stars are, the larger the aperture of telescope needed to split them. If one star is much fainter than the other it will be swamped by the primary star's light, and so will be more difficult to see than a companion that is of similar brightness to the main star. In the case of binaries with relatively short orbital periods, the separation can change markedly over a few years as the stars move relative to each other. The constellation notes in this book give guides to the likely aperture needed to split various double stars. But there can be no hard and fast rules. So much depends on the observer's eyesight, the quality of telescope and the observing conditions. The only way to find out for sure is to look yourself.

Variable Stars

Certain stars vary in brightness, and are known as *variable stars*. Amateur astronomers can make valuable observations of them. The observer estimates the brightness of the variable star by comparing it with nearby stars of known constant magnitude, and the observations are plotted on a graph to form what is known as a *light curve*, demonstrating how the star's brightness varies with time. Such a graph can reveal much about the nature of the star under study.

As might be expected, the usual reason that a star appears to vary is because of actual changes in its light output, but that is not the only possible explanation. In some cases the star is a member of a binary system in which one star periodically eclipses the other. For this to happen, of course, the stars' orbits must be oriented virtually edge-on to us. The first such *eclipsing binary* star to be noticed, and still the most famous, is Algol in the constellation Perseus. Algol consists of a blue dwarf, from which most of the light comes, orbited by a fainter yellow subgiant. Every 2.87 days the magnitude of Algol drops from 2.1 to 3.4 as the fainter star eclipses the brighter one; the eclipse lasts about 10 hours. Compare it at maximum with α (alpha) Persei, magnitude 1.8, and at minimum with δ (delta) Persei, magnitude 3.0. There is also a secondary minimum when the brighter star eclipses the fainter one, but the drop in light is too small for the eye to notice. Most variable-star observers begin by following the changes in Algol's brightness; regular predictions of its eclipses are issued by astronomical societies and in astronomy magazines.

Another eclipsing binary of note is β (beta) Lyrae, which varies between magnitudes 3.3 and 4.4 with a period of 12.9 days. Compare it with nearby γ (gamma) Lyrae, constant at magnitude 3.2, and κ (kappa) Lyrae, magnitude 4.3. In the case of β Lyrae and binaries like it, the stars are so close together that their mutual gravity distorts them into elongated shapes. As a result, the amount of light that we receive from such pairs is not constant, even outside eclipses.

Of the stars that vary intrinsically in brightness, most do so because of physical changes in their size. These are known as *pulsating variables* (not to be confused with pulsars). Of particular importance to astronomers are the so-called *Cepheid variables*, named after their prototype δ (delta) Cephei. Cepheid variables are yellow supergiant stars that go through one cycle of pulsations in periods ranging from about 2 to 40 days, varying in brightness by up to one magnitude as they do so. δ Cephei itself varies between magnitudes 3.5 and 4.4 every 5.4 days, and is thus an easy object for amateur observation. Good comparison stars are ε (epsilon) Cephei, magnitude 4.2, and ζ (zeta) Cephei, magnitude 3.4.

The importance of Cepheid variables is that the length of their light cycle is directly related to their absolute magnitude: the brighter the Cepheid, the longer it takes to complete one cycle of variation. Astronomers therefore have an easy way of finding these stars' absolute magnitude, simply by observing their period of fluctuation. When the absolute

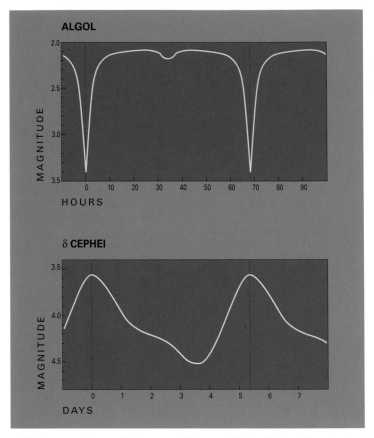

*Light curves of two well-known variable stars: Algol, or β (beta) Persei, and
δ (delta) Cephei. Their brightness rises and falls in a predictable fashion over
the course of a few days. Note that the rise to maximum of δ Cephei is quicker
than its subsequent decline. The small secondary minimum of Algol is
undetectable to the naked eye. (Wil Tirion)*

magnitude is compared with the apparent magnitude, the star's distance
is easily computed. Cepheid variables thus make important distance
indicators in astronomy.

Related pulsating stars are the *RR Lyrae variables*. These are old blue
stars frequently found in globular clusters, and they vary by about 0.5 to
1.5 magnitudes in less than a day. The prototype, RR Lyrae itself, varies
from magnitude 7.1 to 8.1 and back in 0.57 days. Minor additional types
of pulsating variable include β (beta) Cephei stars and δ (delta) Scuti

stars, both of which have short periods of only a few hours and ranges of variation too small to be noticeable to the naked eye.

Each type of variable star falls in a specific area on the Hertzsprung–Russell diagram, and represents stars of different mass at various stages of their evolution. Variability of one kind or another seems to be an inevitable consequence of the ageing process of a star.

Red giants and red supergiants are old stars that frequently prove to be variable. They pulsate, but not with anything like the same regularity as the types of stars mentioned above. The most abundant variables known are the Mira stars, also known as *long-period variables*, which have periods ranging from about three months to two years, and amplitudes of several magnitudes. Their prototype is Mira, o (omicron) Ceti, in the constellation Cetus, the Whale, a red giant whose period averages 332 days during which it ranges between about magnitudes 3 and 9; the exact period and amplitude differ slightly from cycle to cycle. Another prominent example of the same type is χ (chi) Cygni.

More erratic are the *semi-regular variables*, typically with periods of about 100 days and amplitudes of 1 or 2 magnitudes. The *irregular variables* have little or no discernible pattern at all to their fluctuations. All these stars are red giants and red supergiants that have reached a stage of instability, oscillating in size and brightness. Sometimes it is difficult to decide which class a star should be placed in. Examples of semi-regular and irregular variables are Antares, Betelgeuse, α (alpha) Herculis and μ (mu) Cephei.

Most spectacular of all variable stars are the *novae*, which suddenly and unexpectedly erupt by perhaps 10 magnitudes (10,000 times) or more, sometimes becoming visible to the naked eye where no star appeared before. Their name comes from the Latin meaning 'new', for they were once thought to be genuinely new stars. Now we know that they are merely old, faint stars undergoing a temporary outburst. Despite their name, they are not related to supernovae, which explode for different reasons, as mentioned on page 276.

According to current theory, novae are close double stars, one member being a white dwarf. Gas spilling from the companion star onto the white dwarf is thrown off in an eruption. The star does not disrupt itself in a nova outburst. In fact, some novae have undergone more than one recorded outburst, notably RS Ophiuchi and T Pyxidis, and perhaps all novae recur, given time.

A nova rises to maximum brightness in a few days. Amateur astronomers are often the first to spot such eruptions and to notify professional observatories. After a few days or weeks at maximum comes a slow decline over several months, as the nova slowly sinks back to its previous obscurity, sometimes punctuated by minor additional outbursts. Following the progress of a nova is one of the most fascinating aspects of variable star observation. Prominent naked-eye novae occur only about once a decade, but many fainter ones are visible through binoculars.

M81, a near-perfect spiral galaxy in Ursa Major, is tilted at a slight angle to our line of sight. (Nigel Sharp/AURA/NOAO/NSF)

Milky Way, Galaxies and the Universe

Our Sun and all the stars visible in the night sky are members of a vast aggregation of stars known as the Galaxy (given a capital G to distinguish it from any other galaxy). Our Galaxy is spiral in shape, with arms composed of stars and nebulae winding outwards from a central bulge of stars. It is about 100,000 light years in diameter; the Sun lies in a spiral arm 30,000 light years from the Galaxy's centre. Astronomers estimate that the Galaxy contains some 250,000 million stars.

Most of the stars in the Galaxy lie in a disk about 2000 light years thick. Seen from our position within the Galaxy, this disk of stars appears as a faint, hazy band crossing the sky on clear, dark nights. We call this band the Milky Way, and the name Milky Way is often used for our entire Galaxy. The starfields of the Milky Way are particularly dense in the region of Sagittarius, which is the direction of the Galaxy's centre. Note that the plane of the Milky Way is tilted at 63° with respect to the celestial equator. This results from a combination of the tilt of the Earth's axis, and the fact that the plane of the Earth's orbit is tilted with respect to the plane of the Galaxy.

Dotted around our Galaxy in a spherical halo are more than a hundred ball-shaped conglomerations of stars known as *globular clusters*. These contain from a hundred thousand to several million stars, all bound together by gravity. Several globular clusters are visible with the naked

M22 in Sagittarius is one of the most prominent of the globular clusters that surround our Galaxy. (Nigel Sharp, REU program/AURA/NOAO/NSF)

M51, the beautiful Whirlpool Galaxy in Canes Venatici, and its small companion galaxy NGC 5195. This companion has apparently twisted the outstretched spiral arm of M51 during a recent encounter, and it now lies behind the arm. (Todd Boroson/AURA/NOAO/NSF)

eye or binoculars. The brightest are ω (omega) Centauri and 47 Tucanae, both in the southern hemisphere; in the northern hemisphere the best example is M13 in Hercules. To the naked eye and in binoculars, these objects appear as softly glowing patches of light. Moderate-sized telescopes start to resolve some of the individual red giant stars, giving the clusters a speckled appearance. Globular clusters formed early in the history of the Galaxy and contain some of the most ancient stars known, 10,000 million years or more old, over twice the age of the Sun.

Our Galaxy has two small companion galaxies called the Magellanic Clouds. To the naked eye they appear like detached portions of the Milky Way, in the southern constellations Dorado and Tucana. The Large Magellanic Cloud contains about a tenth the number of stars in our Galaxy and lies about 170,000 light years from us. The Small Magellanic Cloud has only about a fifth as many stars as the Large Magellanic Cloud, and lies somewhat farther off, about 200,000 light years. Both Clouds contain numerous star clusters and bright nebulae, and are rich territories for sweeping with instruments of all sizes. One can but imagine the magnificent view of our Galaxy that any astronomers living in the Magellanic Clouds would have.

Countless other galaxies are dotted like islands in the Universe as far as the largest telescopes can see. Most galaxies are members of clusters containing up to thousands of galaxies. Our own Milky Way is the second-largest member of a small cluster of some three dozen galaxies known as

Galaxies are classified according to shape: elliptical, type E; spiral, type S; barred spiral, type SB; and irregular. (Wil Tirion)

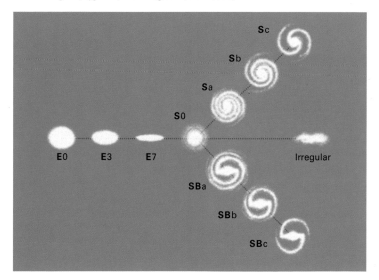

Right: NGC 5850 is an 11th-mag. barred spiral in the Virgo Cluster. Its arms are wound tightly around the central bar, forming the shape of a Greek letter theta (θ). (W. M. Keck Observatory)

Below: NGC 4013 in Ursa Major is a spiral galaxy seen edge-on, crossed by a dark lane of dust in its central plane. (Blair Savage and Chris Howk, U. Wisconsin/ Nigel Sharp, NOAO/ WIYN/NSF)

the Local Group. The largest galaxy in the Local Group is visible to the naked eye as a fuzzy, elongated patch in the constellation Andromeda. The Andromeda Galaxy is estimated to contain about twice as many stars as our own Galaxy, and to be about 25 per cent greater in diameter. It lies about 2.5 million light years away. Long-exposure photographs reveal that the Andromeda Galaxy is a spiral, but tilted so that we see it almost edge-on. If our own Galaxy were viewed from outside, it would look much like this. Amateur telescopes show that the Andromeda Galaxy has two small companion galaxies, its equivalent of the Magellanic Clouds.

The only other member of the Local Group within easy reach of amateur instruments is M33 in Triangulum, another spiral galaxy, somewhat farther from us than the Andromeda Galaxy and containing considerably fewer stars. It can be picked up in binoculars under clear, dark skies. The nearest rich cluster of galaxies to the Local Group lies in the constellation of Virgo, part of it spilling over into neighbouring Coma Berenices. Of its 3000 or so known members, dozens are within the reach of amateur telescopes.

Astronomers classify galaxies into three main types: elliptical, spiral and barred spiral. *Elliptical galaxies* range in shape from virtually spherical, designated E0, to flattened lens shapes, designated E7. They include both the largest and the smallest galaxies in the Universe. Supergiant ellipticals composed of up to 10 million million stars are the most luminous galaxies known. An example is M87 in the Virgo Cluster. At the

Right: M82 in Ursa Major is an edge-on spiral galaxy ploughing through a cloud of dust. (Nigel Sharp/AURA/ NOAO/NSF)

Below: M87 is a massive elliptical galaxy in the Virgo Cluster, surrounded by a swarm of hundreds of globular clusters, visible here as hazy dots. (AURA/NOAO/NSF)

other end of the scale, dwarf ellipticals resemble large globular clusters. Dwarf ellipticals may be the most abundant type of galaxy in the Universe, but their faintness makes them difficult to see.

Spiral galaxies (type S), such as the Andromeda Galaxy, have arms winding out from a central bulge. There are usually two arms, but sometimes more. In *barred spirals* (type SB) the arms emerge from the ends of a bar of stars that runs across the galaxy's centre. Spiral and barred spiral galaxies are subdivided according to how tightly their arms are wound: types Sa and SBa have the tightest-wound arms, while types Sc and SBc have the loosest-wound arms. M31 in Andromeda is type Sb. Until recently, our own Galaxy was thought to fall midway between Sb and Sc, but there is growing evidence that it might instead be a barred spiral.

Most of the galaxies we see in the Universe are large, bright spirals but, as mentioned above, they may actually be outnumbered by dwarf galaxies that are too faint to detect over great distances. In addition to these three main types, there are certain galaxies classified as *irregular*; the Magellanic Clouds are usually reckoned to fall into this class, although there is some semblance of spiral structure in the Large Magellanic Cloud.

Being faint, fuzzy objects, galaxies are best seen where skies are clear and dark, away from city haze and stray light. Low magnification is best, to increase the contrast of the galaxy against the sky background. Through a telescope you should see the nucleus of a galaxy as a starlike point surrounded by the misty halo of the rest of the galaxy. Do not

M77 in Cetus is the most prominent example of a Seyfert galaxy, a type of spiral with an unusually bright and active nucleus. (AURA/NOAO/NSF)

NGC 5128, also known as Centaurus A, is a supergiant elliptical galaxy with a prominent dust band, probably the result of a merger of two galaxies. (ESO)

expect to see prominent spiral arms, as on a long-exposure photograph, except through larger apertures at dark sites. Galaxies present themselves to us at all angles, so even spirals can seem to be elliptical in shape when viewed edge-on.

Peculiar things are going on within some galaxies. For example, certain galaxies give off vast amounts of energy as radio waves; these *radio galaxies* include the supergiant ellipticals M87 in Virgo and NGC 5128 (pictured above) in Centaurus. Some spiral galaxies have unusually bright nuclei. These are termed *Seyfert galaxies*, after the American astronomer Carl Seyfert, who was the first to draw attention to them, in 1943. M77 in Cetus (left) is the brightest Seyfert galaxy.

Most peculiar of all are the objects known as *quasars*, which emit as much energy as hundreds of normal galaxies from an area less than a light year in diameter. Despite their exceptional nature, quasars are not at all exciting visually, which is why they were overlooked until 1963. The brightest quasar, 3C 273, appears as an unimposing 13th-magnitude star in Virgo. Observations with the Hubble Space Telescope have shown that quasars are actually the highly luminous centres of galaxies far off in the Universe. Radio galaxies, Seyferts and quasars are probably all related; perhaps they are all young galaxies seen at different stages of their evolution. Their central powerhouse is thought to be a massive black hole that gobbles up stars and gas from the surrounding galaxy, and its activity can be boosted by infalling gas when two galaxies collide.

In 1929 the American astronomer Edwin Hubble made the most significant of all discoveries in cosmology: the galaxies are moving apart from one another as though the Universe is expanding, like a balloon being inflated. (But clusters of galaxies such as the Local Group do not themselves expand – they are held together by their mutual gravitational attraction.) Hubble's discovery that the Universe is expanding came from a study of the spectrum of each galaxy's light. This revealed that the light from the galaxies was being lengthened in wavelength as a result of high-speed recession (this is called the Doppler effect). Such a lengthening of wavelength is called a *redshift*, because the light from the galaxy is moved towards the red (longer-wavelength) end of the spectrum. Incidentally, the redshift does not make galaxies actually look redder, because the blue end of the spectrum is filled in by light that was formerly at ultraviolet wavelengths.

Hubble found that the amount of redshift in a galaxy's light is directly related to its distance, the most distant galaxies having the greatest

The redshift in the light from a distant galaxy is measured by the change in position of lines in its spectrum. Here, the change in position is shown in the visual window, the range of wavelengths to which the human eye is sensitive. All light from the galaxy is shifted by the same amount, so while some of the galaxy's light moves out of sight beyond the red end of the visual window, other light moves into the visible window from the ultraviolet. (Wil Tirion)

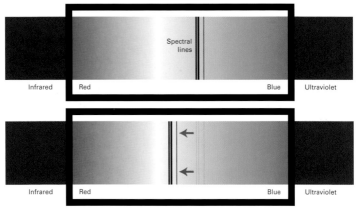

VISUAL WINDOW

Spectral lines

Infrared   Red   Blue   Ultraviolet

Infrared   Red   Blue   Ultraviolet

Stephan's Quintet in Pegasus is a group of five galaxies: NGC 7317 (top left), NGC 7318A and B (centre), NGC 7319 (bottom right) and NGC 7320 (lower left). However, NGC 7320, which appears larger than the others, has a lower redshift than the others, which suggests it is a foreground object and not a true member of the cluster at all. (Nigel Sharp/AURA/NOAO/NSF)

redshifts. Therefore, by measuring the redshift of a galaxy astronomers can tell how far away it is. Quasars, for instance, exhibit such enormous redshifts that they must be the most distant objects visible in the Universe, over 10,000 million light years away.

Since the Universe is expanding, it is logical to conclude that it was once smaller and more densely packed than it is now. According to the most widely accepted theory, the entire Universe was originally a compressed, superdense blob which, for some unknown reason, exploded in a cataclysm known as the Big Bang. The galaxies are the fragments from that explosion, still flying outwards. As far as anyone can tell at present, the Universe will continue to expand for ever. According to the best estimates the Big Bang took place around 13,000 million years ago; that is the age of the Universe as we know it. It is impossible to tell what, if anything, happened before the Big Bang.

The Sun

Our Sun is a glowing sphere of hydrogen and helium gas 1.4 million km in diameter, 109 times the diameter of the Earth and 745 times as massive as all the planets put together. The Sun is vital to all life on Earth because it provides the heat and light that makes our planet habitable. To astronomers, the Sun is important because it is the only star that we can observe in close-up. Studying the Sun tells us a lot that we could not otherwise know about stars in general.

Whereas most celestial objects pose problems for observers because they are faint, the Sun presents entirely the opposite problem: it is so bright that it is dangerous to look at. Anyone glancing for an instant at the Sun through any form of optical instrument, be it binoculars or telescope, risks instant blindness. Even staring with the naked eye at the Sun

The Sun on May 19, 2000, during the rise to solar maximum showing several sunspot groups, each large enough to swallow the Earth many times over. Note the effect of limb darkening, against which brighter faculae may be seen. (Mees Solar Observatory, University of Hawaii)

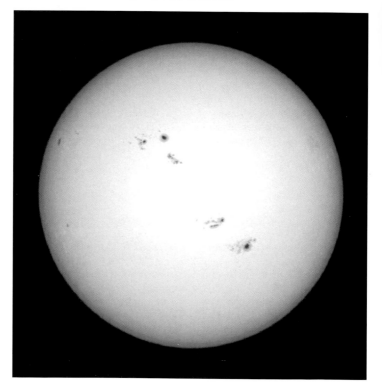

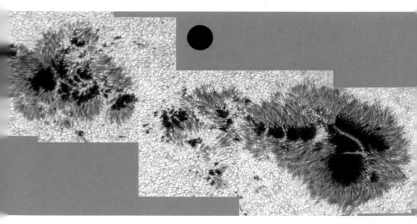

A complex group of sunspots some 200,000 km long as it appeared on September 4, 1998. The black circle shows the Earth to the same scale. Filamentary structure in the outer penumbra of the spots gives them a sunflower-like appearance. Note also the granulated appearance of the surrounding solar photosphere. (Kiepenheuer-Institut für Sonnenphysik)

for more than a few seconds can permanently damage your eyesight. There is a safe and easy way to observe the Sun, and that is by projecting its image onto a piece of white card.

Some telescopes come equipped with dark Sun filters to place behind the eyepiece. These should never be used, for the focused light and heat of the Sun can crack them, with inevitably disastrous results. Safe filters for telescopic solar observation do exist, made either of glass or plastic film with a metallic coating. These filters fit over the front of the telescope tube, reducing the amount of incoming light and heat to safe levels. Filters made from a thin plastic film called Mylar give a bluish image, whereas other types give yellow or orange images.

A telescopic view of the Sun, seen either by projection or through a safe filter, displays the brilliant solar surface, the *photosphere* ('sphere of light'), consisting of seething gases at a temperature of 5500°C. Although intensely hot by terrestrial standards, this is cool by comparison with the Sun's core, where the energy-generating nuclear reactions take place; there, the temperature is calculated to be about 15 million °C.

The photosphere exhibits a mottled, rice-grain effect termed *granulation*, caused by cells of hot gas bubbling up in the photosphere like water boiling in a pan. Granules range from about 300 km to 1500 km in diameter. Look carefully at a projected image of the Sun, and you will notice that the edges of the Sun appear fainter than the centre of the disk, an effect known as *limb darkening*. This is caused by the fact that the gases of the photosphere are somewhat transparent, so that at the centre

293

of the disk we are looking more deeply into the Sun's interior than at the limbs. Brighter patches known as *faculae* may be visible against the limb darkening. Faculae are areas of higher temperature on the photosphere. A number of prominent dark markings known as *sunspots* will probably also be visible. They are areas of cooler gas that appear dark by contrast against the brighter photosphere.

Sunspots are temporary features that arise where magnetic lines of force burst through the photosphere from within the Sun. Apparently, the presence of a strong magnetic field blocks the outward flow of heat from inside the Sun, producing the cooler spot. Sunspots have a dark centre known as the *umbra*, of temperature about 4000°C, surrounded by a brighter *penumbra* at about 5000°C. (The temperature of a sunspot's umbra is similar to that of the surface of a red giant such as Aldebaran.)

Sunspots range in size from small pores no bigger than a large granule to enormous, complex patches many times larger than the Earth. Spots of that size are visible to the naked eye when the Sun is dimmed by the atmosphere, shortly after sunrise or before sunset, or through special solar filters. A major sunspot takes about a week to develop to full size, then slowly dies away again over the next fortnight or so.

As the Sun rotates, new spots are brought into view at one limb while others disappear around the opposite limb. The largest sunspots tend to form in groups large enough to span the best part of the distance from the Earth to the Moon. Spot groups can persist for one or two solar rotations (a month or two). Usually, a spot group contains two main components, aligned east–west. The spot that leads across the disk as the Sun rotates is termed the p (preceding) spot, and is usually larger than its follower, the f spot. The p and f spots have opposite magnetic polarities, like the ends of a horseshoe magnet. The horseshoe is completed by invisible lines of force looping between the spots.

Sometimes the intense magnetic fields in a complex sunspot group become entangled, releasing a sudden flash of energy known as a *flare* that may last from a few minutes to an hour. Atomic particles are spewed out into space by the eruption of a flare. These particles reach Earth about a day later, causing effects in the upper atmosphere such as radio interference and the ethereal displays known as *aurorae*. In an aurora, the sky glows with patches of green and red light that can take the form of folded drapery or arches, shimmering and changing shape for hours on end. Aurorae occur near the Earth's magnetic poles and extend to lower latitudes only at times of extreme solar activity, so they are seldom visible outside far northerly or southerly latitudes.

Until the 1970s, flares were the only known cause of aurorae, but it is now recognized that another form of activity called coronal mass ejections (CMEs) actually has a greater effect. CMEs are huge bubbles of hot gas thrown off from the Sun, sometimes at the same time as a flare but often without flares. The existence of CMEs was not recognized earlier because they can be detected only from spacecraft.

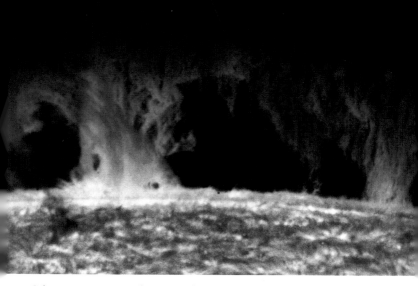

Solar prominences extending 65,000 km into space look like enormous burning trees at the edge of the Sun. This photograph was taken at the wavelength of light emitted by hydrogen gas. (Big Bear Solar Observatory)

By observing the progress of spots as they are carried across the Sun's face, we can measure the rotation period of the Sun. Being gaseous and not solid, the Sun does not rotate at the same rate at all latitudes. It spins most quickly at its equator, in 25 days; this falls to about 28 days at latitude 45°; and near the poles it is slowest of all, 34 days. The average figure usually quoted, 25.38 days, refers to the rotation rate at latitude 17°. As a result of the Earth's orbital motion around the Sun, a spot takes about two days extra to return to the same place on the Sun as seen from Earth.

The number of sunspots on view waxes and wanes in a cycle lasting 11 years on average, although the length of individual cycles has been as short as 8 years and as long as 16 years. At times of minimum activity the Sun may be spotless for days on end, whereas at solar maximum over a hundred spots may be visible at any one time. However, the level of activity varies considerably from cycle to cycle; the average number of sunspots visible at maximum has ranged from 40 to 180, and even when the Sun is supposedly quiet some large spots and flares can break out. Solar activity is notoriously unpredictable, which adds to its fascination for observers.

A few general rules can be deduced, however. The first spots of each new cycle appear at latitudes of about 30–35° north and south of the Sun's equator. As the cycle progresses, spots tend to form closer to the equator. Sunspot numbers build up to a peak, then start to decay again. As minimum approaches, the last spots of the cycle are found at latitudes between 5° and 10° north and south of the equator. At the same time – that is, around solar minimum – the first spots of the next cycle are forming at higher latitudes. Sunspots are seldom found on the Sun's equator, and spots at latitudes greater than 40° are exceedingly rare.

Above the photosphere is a tenuous layer of gas known as the *chromosphere*, about 10,000 km deep. So faint is the chromosphere that it is normally invisible except with special instruments. It can, however, be seen for a few seconds at a total eclipse, when it appears as a pinkish crescent just before and after the Moon completely covers the face of the Sun. This pinkish colour is caused by light from hydrogen gas, and gives rise to its name, which means 'colour sphere'.

Also visible around the edge of the Sun at total eclipses are huge clouds of gas extending from the chromosphere into space, known as *prominences*. They have the same characteristic rosy-pink colour as the chromosphere, caused by emission from hydrogen. Like so many features of the Sun, prominences are controlled by magnetic fields. So-called *quiescent prominences* extend for 100,000 km or more across the Sun, sometimes forming graceful arches tens of thousands of kilometres high. When seen silhouetted against the brighter background of the photosphere, they are termed *filaments*. Quiescent prominences can last for months. At the other end of the scale are *eruptive prominences*, with lifetimes of only a few hours. They are flares seen at the edge of the Sun, ejecting material into space at speeds of up to 1000 km per second. All forms of solar activity – sunspots, flares and prominences – follow the 11-year solar cycle.

The Sun's crowning glory is its *corona*, a faint halo of gas that comes into view only when the brilliant photosphere is blotted out at a total solar eclipse. The corona is composed of exceptionally rarefied gas at a temperature of 1 to 2 million °C. Petal-like streamers of coronal gas extend from the Sun's equatorial zone, while shorter, more delicate plumes fan out from the polar regions. The shape of the corona changes during the solar cycle: at solar maximum, when there are more active areas on the Sun, the corona appears more rounded in shape than at solar minimum.

Gas from the corona is continually flowing away from the Sun out into the Solar System, forming what is known as the *solar wind*. Atomic particles of the solar wind are detected streaming past the Earth at a speed of about 400 km per second. The most obvious effect of the solar wind is to push comet tails away from the Sun. The solar wind extends outwards beyond the orbit of the most distant planet, finally merging with the thin gas between the stars. In a sense, therefore, all the planets of the Solar System can be said to lie within the outer reaches of the Sun's corona.

Right: The spectacular corona of the Sun springs into view at total solar eclipses. This was its appearance at the eclipse of July 11, 1991, near the time of solar minimum, photographed from Mexico. (Armagh Planetarium)

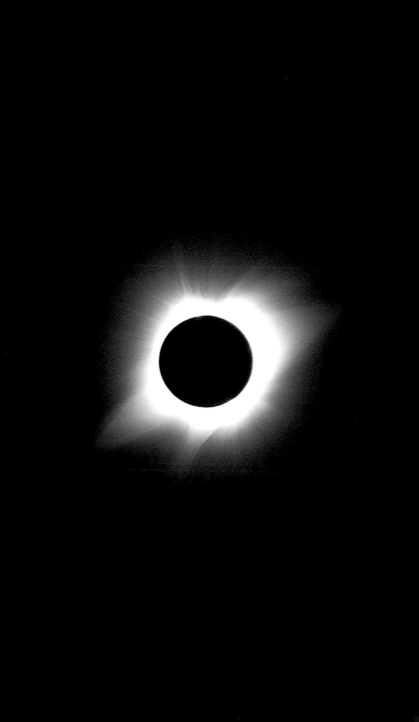

The Solar System

In one sense our Sun is unusual, for it is not accompanied by another star, as are most stars, but has a family of nine planets, assorted moons and countless smaller lumps of debris. This retinue of bodies, all held captive by the Sun's gravity, is called the Solar System.

All the objects in the Solar System shine by reflecting the light from the Sun. Several of the planets when at their brightest can equal or outshine the brightest stars. All the planets orbit the Sun in approximately the same flat plane, so they are always to be found near the ecliptic. A bright 'star' that disturbs a familiar constellation pattern along the ecliptic is therefore almost certainly a planet (although it could, just possibly, be a nova).

As seen from above the Sun's north pole, the planets orbit the Sun in an anticlockwise direction. Their orbits are slightly elliptical in shape, so that each planet's distance from the Sun varies somewhat during one orbit. In order of average distance from the Sun, the nine planets are: Mercury, Venus, Earth, Mars, Jupiter, Saturn, Uranus, Neptune and Pluto. The four inner planets are rocky, relatively small bodies. Then come four giant planets made mostly of gas and liquid. Pluto is a small, frozen oddity on the outer limits.

Venus is the planet with the most circular orbit; its distance from the Sun varies by little more than 1.5 million km. The planet with the most elliptical orbit is distant Pluto, which at times actually strays across the orbit of Neptune, as it last did between 1979 and 1999. Some of the minor bodies, or asteroids, in the Solar System have orbits that are more elliptical than this, and the comets have highly distended orbits that can take them from the innermost part of the Solar System out to way beyond Pluto. The closest point of a celestial body's orbit to the Sun is known as the *perihelion*; the farthest point is the *aphelion*.

The basic unit of distance in the Solar System is the *astronomical unit*, the average distance of the Earth from the Sun; it is equivalent to 149,597,870 km. Light takes 499 seconds (8.3 minutes) to cross this distance, so we on Earth see the Sun as it actually appeared 8.3 minutes ago. Clearly, although the astronomical unit is large compared with everyday distances on Earth, it is insignificant in comparison with a light year. There are 63,240 astronomical units in a light year.

How long a planet takes to orbit the Sun depends on its distance: the closest planets orbit the quickest, and the farthest orbit the slowest. Each planet's orbital period is technically its 'year', although the period is usually expressed in terms of Earth days and years. The orbital periods of the planets range from 88 days for Mercury to 248 years for Pluto. These orbital periods are known technically as *sidereal periods*, for they are measured with respect to the distant stars.

As the two inner planets, Mercury and Venus, move around their orbits they periodically come between us and the Sun; when that happens, they are said to be at *inferior conjunction*. That is also the time when Mercury

SUN

Mercury

Venus

Earth

Mars

Jupiter

Saturn

Uranus

Neptune

Pluto

Sun
Mercury
Venus
Earth
Mars

Asteroids

Jupiter

Saturn

Uranus

Neptune

Pluto

Above left: The sizes of the planets to scale, compared with a section of the Sun. Above right: The orbits of the planets to scale. (Wil Tirion)

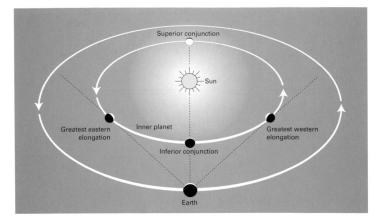

When Venus or Mercury lies directly between the Earth and the Sun it is said to be at inferior conjunction; on the far side of the Sun it is at superior conjunction. Maximum angular separation from the Sun is termed greatest elongation. (Wil Tirion)

and Venus are closest to the Earth, although this fact is of little use to observers since the planets are lost in the Sun's glare and their illuminated hemispheres are in any case turned away from the Earth.

The orbits of Mercury and Venus are inclined by 7° and 3.4° respectively to the orbit of the Earth, but that is sufficient to ensure that they usually pass above or below the disk of the Sun at inferior conjunction. But occasionally Mercury or Venus does actually cross the face of the Sun as seen from Earth. At such an event, called a *transit*, the planet appears as a tiny dark dot like a small sunspot against the glaring solar surface. Transits of Mercury are more common than those of Venus: the next are due in 2003, 2006, 2016 and 2019, whereas the transits of Venus in 2004 and 2012 will be the only ones for over a century.

When Mercury or Venus lies on the far side of the Sun from us it is said to be at *superior conjunction*. When Mars or any of the planets beyond it lies on the far side of the Sun it is simply said to be at conjunction – being farther from the Sun than we are, outer planets can never come between Earth and the Sun, so there is no ambiguity about which sort of conjunction is meant. Planets at conjunction are invisible in the Sun's glare.

The best time to view the two inner planets is when they are at their maximum possible angular separation from the Sun, known as *greatest elongation*. At greatest elongation Venus appears almost exactly half-phase, although in the case of Mercury the phase can vary noticeably from 50 per cent because of the planet's elliptical orbit. Eastern elongation means that the planets are setting after the Sun in the evening sky; at western elongation, the planets rise before the Sun in the morning sky.

The time between one greatest eastern elongation of Mercury and the next, or one greatest western elongation and the next, is 116 days. With Venus, greatest western elongations or greatest eastern elongations recur every 584 days, although Venus is so prominent that it can be adequately observed well away from greatest elongation. Incidentally, the interval between one given appearance of a planet and the next – be it conjunction, elongation or whatever – is known as its *synodic period*. This differs from the sidereal period because our observation platform, the Earth, is moving around the Sun.

The best opportunities to see Mars and the outer planets is when they are directly opposite the Sun in the sky; this is termed *opposition* (Mercury and Venus, being between the Earth and the Sun, cannot come to opposition, of course). A planet at opposition appears due south for northern hemisphere observers (due north for observers in the southern hemisphere) at midnight local time, or 1 a.m. if daylight-saving time is in operation. Planets lie closest to the Earth when at opposition, so this is when they appear at their biggest and brightest.

Other Solar Systems?

Planets are thought to be born from a disk of gas and dust left over after the formation of a star, a process that happened some 4.6 billion years ago around our Sun. Several stars are known to have such disks around them, including Vega, Fomalhaut and β (beta) Pictoris, so in these cases we may well be catching the process of planet formation in action. Astronomers have also detected signs of fully formed planetary systems, in most cases from the slight wobble in position of a star as a large planet orbits it. The first star with a planet detected in this way was 51 Pegasi, in 1995. Four years later, not just one but three planets were found around υ (upsilon) Andromedae, the first multi-planet system known. Other techniques have since confirmed the existence of extrasolar planets, such as tiny dips in a star's light as a planet crosses in front of it and the detection of a star's light reflected from the surface of a planet. With large telescopes of the future, it may be possible to obtain direct images of planets around other stars.

A disk of dust around β (beta) Pictoris, photographed by the Hubble Space Telescope and depicted in false colour, presents persuasive evidence of planet formation. Beta Pictoris itself is obscured by the black circle at the centre. The dust disk is aligned almost edge-on to us, and is larger than the orbit of Pluto around the Sun. (NASA)

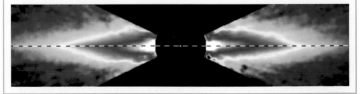

The Moon

The Moon, the Earth's natural satellite and nearest celestial neighbour, is an object of perennial fascination for observation with instruments of all sizes. Despite its small size – 3475 km in diameter, roughly a quarter that of the Earth – it is so close to us, on average 384,400 km, that even ordinary binoculars reveal a wealth of detail on its cratered surface. Some of the most interesting objects to look out for are shown in the maps on pages 322–333.

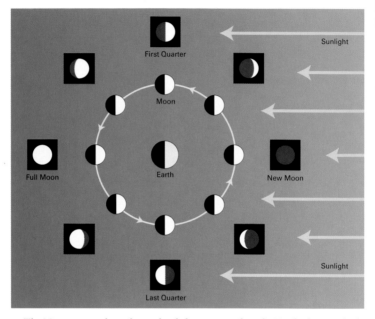

The Moon passes through a cycle of phases as it orbits the Earth, during which we see differing proportions of its illuminated side. (Wil Tirion)

In the course of a month the Moon goes through a cycle of phases from new (unilluminated) through waxing half (first quarter), full, waning half (last quarter), back to new again. Strictly speaking, there are two types of month. The first lasts 27.3 days, the time the Moon takes to complete one orbit of the Earth relative to a fixed point such as the distant stars; this is known as a *sidereal month*. But the Earth also moves relative to the Sun in that time, so the Moon must complete rather more than one orbit to return to the same phase as seen from Earth. The time the Moon takes to complete one cycle of phases, 29.5 days, is called a *synodic month*.

With the exception of a few small areas near the poles, each spot on the Moon is subjected to two weeks of daylight, during which surface

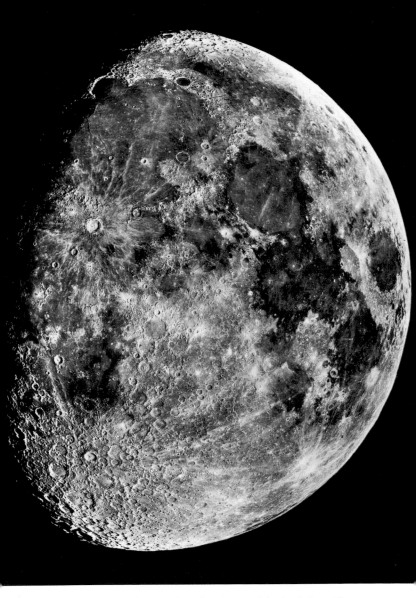

The Moon at gibbous phase, 10 days after new. Dark lowland plains (the maria), relatively devoid of craters, contrast with bright, rugged highlands. The large mare towards the top is Mare Imbrium, with the horseshoe-shaped bay Sinus Iridum on its shore. South of Mare Imbrium is the prominent rayed crater Copernicus. Most of the features in this photograph can be picked out with small telescopes or even binoculars, if firmly mounted. (Lick Observatory)

temperatures reach over 100°C, i.e. higher than the boiling point of water, followed by a two-week night when temperatures plummet to −170°C or less. At the poles, though, some crater floors remain permanently shaded from the Sun and hence temperatures never rise above freezing. Here, ice deposited by cometary impacts could exist beneath the surface. Evidence for such lunar ice fields was received from NASA's Lunar Prospector spacecraft, which detected signs of hydrogen in the polar regions, presumably released by water molecules, as it orbited the Moon in 1998 and 1999.

The Moon spins on its axis in 27.3 days, the same time it takes to complete one orbit of the Earth, so that the same face of the Moon is always turned towards us; this is known as a *captured rotation*. In practice, though, we can see slightly more than half the Moon's surface. The Moon's equator is tilted at about 6½° to the plane of its orbit, so that at times we can see as much as 6½° over the Moon's north or south pole; this is known as *libration in latitude*. In addition, the Moon's speed of motion along its elliptical orbit changes rhythmically as it approaches and recedes from the Earth, while its axial rotation remains uniform. The Moon therefore seems to rock slightly to west and east as it orbits the Earth, so that we can see up to 7¾° or so around each edge; this is known as *libration in longitude*. The net effect of these librations is that we can see 59 per cent of the Moon's surface at one time or another.

The line dividing the lit and unlit portions of the Moon is known as the *terminator*. Objects near the terminator are thrown into sharp relief by the low angle of illumination, so that craters and mountains appear particularly rugged. As the Sun rises higher over the moonscape, details

Partners in space: the Moon and Earth photographed in 1992 by the Galileo space probe, on its way to Jupiter. The Moon is in the foreground, moving from left to right. On the Earth, Antarctica is visible through clouds at the bottom.
(NASA)

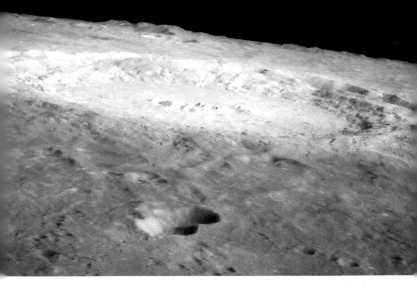

An oblique view of the giant impact crater Copernicus, taken from orbit by the Apollo 12 astronauts in 1969, demonstrates the relative shallowness of such large craters. The keyhole-shaped feature in the foreground is Fauth. (NASA)

become more washed out. Near full Moon many individual formations are difficult to pick out. An exception is those craters that have bright ray systems, apparently made of pulverized rock thrown out from the crater during its formation; the rays become more prominent under high illumination. It is easiest to pick out the difference in contrast between the bright highlands and the dark lowland plains at full Moon.

After looking at the brilliant full Moon it comes as a surprise to realize that the lunar surface rocks are actually dark grey in colour; on average, the Moon's surface reflects only 12 per cent of the light that hits it. If the Moon were, for instance, covered in clouds like those of Venus, it would be over five times brighter.

Formations on the Moon bear a variety of curious names. The dark lowland plains are termed *maria* (singular: mare), Latin for 'seas', because the first observers imagined them to be stretches of water; the name persists, even though it has been clear for centuries that the Moon possesses neither air nor liquid water. Thus we have, for instance, Oceanus Procellarum (Ocean of Storms), Mare Imbrium (Sea of Rains) and Mare Tranquillitatis (Sea of Tranquillity). Less prominent lowlands are termed bays (Sinus), marshes (Palus) or lakes (Lacus). Mountains on the Moon are named after terrestrial ranges; hence we have the lunar Alps (Montes Alpes) and the lunar Apennines (Montes Apenninus). The craters have been named after philosophers and scientists of the past, although it is true to say that the selection has been somewhat arbitrary.

We owe the modern system of lunar nomenclature to an Italian astronomer, Giovanni Riccioli, who in 1651 published a map bearing many of the names now universally accepted. Riccioli allocated himself a prominent crater near one edge of the Moon, and named a large neighbouring formation after his pupil, Francesco Grimaldi. Less favoured colleagues did not fare so well. Galileo, one of the greatest scientists of all time, but who fell out with the Church, is given an insignificant 16-km-diameter crater in Oceanus Procellarum.

Riccioli's system has been followed to the present day but, now that the near and far sides of the Moon have been mapped in detail by spacecraft,

Mare Humorum, a lowland plain in the southwestern sector of the Moon. The large, partially flooded crater on its northern shore is Gassendi, noted for the complex clefts on its floor. Glows known as transient lunar phenomena have been reportedly seen in this area. (European Southern Observatory)

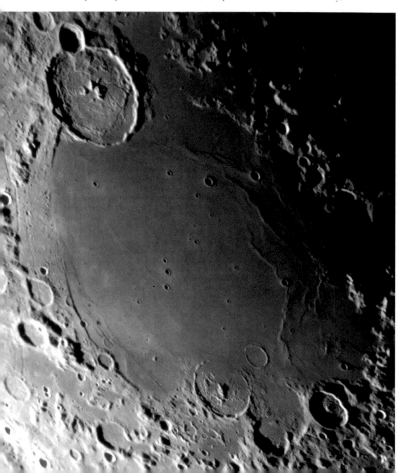

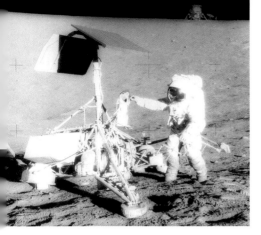

Apollo 12 made a precision landing in November 1969 alongside the Surveyor 3 probe that had arrived 2½ years earlier. Here, astronaut Pete Conrad prepares to remove part of the Surveyor's camera, which was brought back to Earth. In the background is the lunar module. (NASA)

the ground rules have been extended to include names from areas of human endeavour other than astronomy. Thus we now find Freud, H. G. Wells and Montgolfier commemorated on the Moon. Finding suitable names for features on celestial bodies has become a pressing problem now that so many planets and their moons have been surveyed by space probe.

Lunar observation began in 1609 when Galileo turned his first telescope towards the Moon. His drawings, published in 1610, were crude by modern standards but served to show that the Moon's surface is rugged and mountainous. Debate raged for the next three and a half centuries over the origin of the Moon's features. One school of thought held that they were caused by volcanic action; the opposition believed the craters and maria to have been blasted out by massive impacts of meteorites and asteroids. There is no point in recapitulating the arguments here. Suffice it to say that the impact theory emerged as the undisputed winner, although it is accepted that there has also been a certain amount of volcanic activity on the Moon.

Not until the late 1960s, when space probes and astronauts first reached the Moon, was the controversy resolved. In 1964 and 1965, three American probes called Rangers 7, 8 and 9 zoomed towards the Moon, sending back a stream of photographs showing features down to a metre across, a hundred times smaller than Earth-based telescopes could see. The lesson was that even the apparently flattest parts of the Moon's surface were pitted with small craters caused by aeons of meteorite bombardment. Landing sites for future manned spacecraft would therefore have to be chosen with particular care, to prevent the landing craft from toppling into a crater or hitting a boulder (this nearly happened on the Apollo 11 descent).

This initial cursory examination by the Rangers was followed by a two-pronged attack: the Surveyors, a series of craft which soft-landed automatically to give an astronaut's-eye view of the lunar surface (the

Rangers had simply crashed), and the Lunar Orbiters which, as their name implies, photographed the Moon from close orbit. Between 1966 and 1968, these two series of probes revolutionized our knowledge of the Moon, and paved the way for the manned Apollo landings. The Surveyors showed that the Moon's topsoil is made of compacted dust, firm enough to bear the weight of astronauts and their spacecraft. From Lunar Orbiter photographs, astronomers compiled their most detailed maps of the near and far sides of the Moon.

The Moon's far side had first been glimpsed by Luna 3, a Soviet probe, in October 1959. Its photographs were poor by modern standards, but they did at least reveal the main difference between the two hemispheres of the Moon: there are scarcely any mare areas on the Moon's far side. Instead, heavily cratered bright uplands dominate the scene. The reason for this asymmetry is that the Moon's crust is some 25 km thicker on the far side.

Large lowland basins like Mare Imbrium do exist on the far side of the Moon, but they have not been flooded by dark lava. Volcanic lavas from the Moon's interior found it easier to leak out through the thinner crust of the Earth-facing hemisphere. The most prominent dark area on the Moon's far side is not a true mare at all, but a deep crater called Tsiolkovsky, 180 km in diameter (almost as large as Sinus Iridum on the visible hemisphere).

Once the Lunar Orbiters had spied out potential landing sites, the stage was set for astronauts to follow. On July 20, 1969, the Apollo 11 lunar module called Eagle carried Neil Armstrong and Edwin Aldrin to the first manned lunar landing, in the southwestern Sea of Tranquillity. The two astronauts spent two hours exploring the Moon's surface, setting up experiments and collecting samples for geologists to study. For the first time, humans had touched another world in space.

Tsiolkovsky, a 180-km-diameter crater on the far side of the Moon with a dark floor and a prominent central peak, photographed by Lunar Orbiter 3 in 1967. Lunar Orbiter pictures appear banded because they were transmitted to Earth in strips. (NASA)

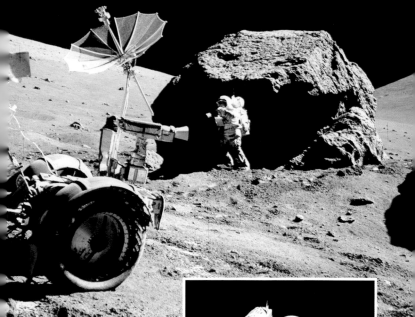

On the Apollo 17 mission, the last in the series, astronauts Eugene Cernan and Harrison Schmitt spent three days exploring southeastern Mare Serenitatis.

Above: Schmitt is dwarfed by a rock that has rolled downhill. In the foreground is the electric Moon car, the lunar rover.
Right: Schmitt collects pebbles from the lunar soil with a rake. (NASA)

By the time Apollo 17 concluded the series of manned landings in December 1972, astronauts had brought back to Earth over 380 kg of Moon samples, most of which is stored in Houston, Texas. When divided into the overall cost of Apollo, one kilogram of Moon rocks is worth 100 million US dollars. In addition to the Apollo samples, three Soviet automatic lunar probes have returned to Earth with a few hundred grams of Moon soil.

What have we learned from these precious specimens? The most astounding fact about the Moon rocks is their immense age. The Apollo 11 samples, for instance, proved to be 3700 million years old, older than virtually any rocks on Earth – yet the site from which they came, Mare Tranquillitatis, is one of the youngest areas on the Moon. The youngest of all the rocks brought back, from the Apollo 12 site in Oceanus Procellarum, are 3200 million years old. As expected, the lunar maria turned out to be covered with lava flows similar in composition to volcanic basalt on Earth. They do not resemble a rugged lava field on Earth because, over the thousands of millions of years since the lavas were laid down, the sandblasting effect of micrometeorites has eroded the surface rocks to form a layer of soil several metres deep, called the *regolith*.

By contrast, the highlands, sampled by later Apollo missions, consist of a paler rock called anorthosite, rare on Earth. Rocks from the highlands proved older than those from the maria, mostly dating back 4000 million years or more. Their jumbled, fragmented nature bears witness to the violent bombardment from meteorites which the Moon suffered early in its history.

Despite scientists' hopes, the rich haul of Moon rocks did not conclusively answer the question of the Moon's origin. All three existing theories – that the Moon split from our planet shortly after its formation, that it was once a separate body that was captured by the Earth's gravity, or that the Earth and Moon formed side by side, much as they are now – had drawbacks.

Following the Apollo landings, a fourth theory emerged which combined aspects of the previous three. According to this now widely accepted view, a stray body the size of Mars struck the young Earth a glancing blow, spraying debris into orbit around the Earth where it coalesced into the Moon. Precise dating of Moon samples suggests this collision happened about 4.5 billion years ago, some 50 million years after the Earth formed. By then, iron had sunk to the Earth's centre, producing a core, so that the material thrown off the outer layers in the collision was predominantly rocky. This is consistent with the finding, derived from tracking the motion of the Lunar Prospector probe as it orbited the

Tycho, 85 km wide, in the Moon's southern highlands, is the youngest large crater on the Moon. Its bright rays extend across the visible face of the Moon. Terraces on its inner slopes, also seen on other large impact craters, are caused by slumping. This photograph was taken by Lunar Orbiter 5. (NASA)

Letronne, a crater flooded by lava on the southern shore of Oceanus Procellarum. One wall has been destroyed, although parts of its central peak remain. The low angle of illumination throws the wrinkles of solidified lava into sharp relief. This photograph was taken from orbit during the Apollo 16 mission. (NASA)

Moon in 1998–99, that the Moon has only a small metallic core, a few per cent of its total mass. By contrast, the Earth's iron core accounts for about 30 per cent of its mass.

While many details of the Moon's origin remain obscure, we now have a much clearer picture of its subsequent history. Heat released by its rapid accumulation in orbit around the Earth (a process termed *accretion*) melted its outer layers. A scum of less dense rock formed a primitive crust which was then battered for several hundred million years by the infall of other debris left over from the formation of the Solar System. Other rocky bodies in the Solar System, notably lunar-like Mercury, also bear the scars of this same mopping-up operation. This heavy bombardment created the Moon's jumbled highlands and carved out the mare basins, in the process removing much of the crust on one hemisphere – the side that was to become the Earth facing hemisphere, when the Moon's rotation became tidally locked.

About 4000 million years ago, the storm of meteoric debris abated. Then, slowly, molten lava began to seep out from inside the Moon, solidifying to form the dark lowland maria. Wrinkles of solidified lava are visible through binoculars and small telescopes when the maria are under low illumination. In particular, look out for the Serpentine Ridge in eastern Mare Serenitatis when the Moon is five to six days old, or six days past full. Mare Tranquillitatis and Mare Imbrium also have prominent wrinkle ridges. In fact, one 'crater' on Mare Tranquillitatis, called Lamont, consists of nothing more than low ridges of solidified lava. In places, particularly in western Oceanus Procellarum, a number of blister-like domes have been produced by the upwelling of molten lava. But volcanic cones like Vesuvius are absent on the Moon, evidently because the lunar lavas were too runny to build up into mountains.

By 1000 million years ago the volcanic outpourings had ceased, leaving the Moon cold and dead. Since then it has remained virtually unchanged, save for the arrival of an occasional meteorite to punch a new crater in its surface. For instance, the crater Copernicus was formed roughly 1000 million years ago; Tycho was blasted out about 300 million years ago.

Yet the Moon may not be completely inactive today. From time to time observers have reported transient events, such as glows and obscurations, around the edges of the maria and in certain craters – Aristarchus seems to be a favourite spot for these *transient lunar phenomena* (TLPs). Most TLPs have been observed by amateur astronomers. Their reality remains controversial, but if they do occur they are probably caused by the release of gas from the Moon's interior.

Other features to look out for when observing the Moon are grooves in the surface, known as *rilles*, apparently caused by faulting. For instance, the crater Hyginus near the Moon's centre lies in the middle of a long, rimless cleft along which smaller craters have been formed by subsidence. Another type of rille, called a sinuous rille, snakes over mare surfaces like a meandering river. Apollo 15 landed at the edge of one of these sinuous rilles, called Hadley Rille. Such winding rilles are not dried-up river beds but are probably collapsed tunnels through which underground lava streams once flowed.

Meandering near the foothills of the Apennine Mountains at the eastern rim of Mare Imbrium, Hadley Rille is a dried-up lava channel that was visited by the Apollo 15 astronauts in 1971. (NASA)

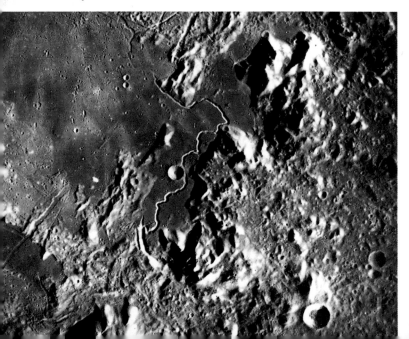

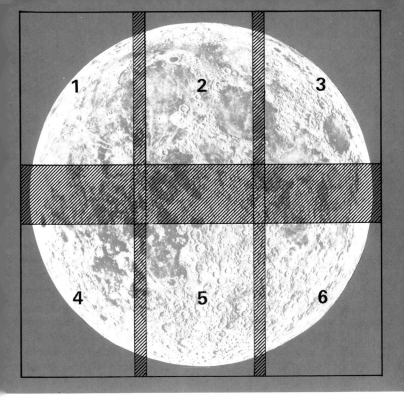

Key to the Moon maps on pages 322–333. These maps show the Moon in conventional orientation, with north at the top, as it appears to the naked eye and in binoculars. Through an astronomical telescope, south will be at the top, so telescope users must invert the maps. West is at the left, as on Earth, although in the rest of the sky west is in the opposite direction, towards the western horizon.

Moon Map 1

Aristarchus Brilliant young crater 40 km in diameter with multiple terracing on its inner walls. The brightest area on the Moon, and centre of a major ray system. Dark bands are visible on the inner walls under high illumination. Aristarchus has been the location of numerous red glows known as transient lunar phenomena (TLPs), possibly caused by outgassing from the surface. Mountains to the north exhibit fantastic sculpturing.

Copernicus Diameter 93 km. One of the most magnificent craters on the Moon. Centre of a major ray system. Terraced walls and numerous hummocky central peaks. Rays from Copernicus are splashed for more than 600 km across the Mare Imbrium and Oceanus Procellarum.

Encke Low-walled crater 28 km in diameter covered by rays from nearby Kepler. Brilliant under high illumination.

Euler   Small, sharp crater, 28 km across, in southwestern Mare Imbrium. Centre of a minor ray system.

Fauth   Keyhole-shaped crater south of Copernicus, nearly 2 km deep. Striking under low illumination.

Harpalus   Crater 39 km in diameter on Mare Frigoris, bright under high illumination. The smaller crater Foucault, 23 km in diameter, lies between it and Sinus Iridum.

Herodotus   Companion to Aristarchus, similar in size (diameter 35 km) but different in structure – it has a dark, lava-flooded floor and it is not a ray centre. The W-shaped Vallis Schröteri (Schröter's Valley), 150 km long, starts at a craterlet outside the northern wall of Herodotus.

Hevelius   Large, bright crater 115 km in diameter on western shore of Oceanus Procellarum, with clefts crossing the floor. Smaller, sharper crater Cavalerius (diameter 60 km) adjoins to the north.

Kepler   Major ray centre in Oceanus Procellarum. Brilliant crater, 31 km in diameter, with central peak and heavily terraced walls.

Mairan   Prominent crater in highlands west of Sinus Iridum. Diameter 40 km.

Marius   Dark, flat-floored ring 41 km in diameter on Oceanus Procellarum, notable because it lies on a wrinkle ridge. There are many dome-like structures in this area where lava has bubbled up through the surface.

Oceanus Procellarum   Vast dark plain with no clear-cut borders extending from Mare Imbrium southwards to Mare Humorum. Oceanus Procellarum has a maximum width of about 2000 km and occupies an area of more than two million square km. It is dotted with numerous craters and bright rays.

Prinz   U-shaped formation 52 km in diameter, remains of a half-destroyed crater flooded by lava from Oceanus Procellarum. Notable because of sinuous rilles to the north.

Pythagoras   Magnificent large crater 128 km in diameter with terraced walls rising nearly 5 km and prominent central peak, near the northwestern limb of the Moon.

Reiner   Sharp crater 30 km in diameter west of Kepler in Oceanus Procellarum. Particularly notable is a tadpole-shaped splash of brighter material on the dark plain to the west and north.

Reinhold   Prominent crater 42 km in diameter southwest of Copernicus, with terraced walls. A smaller, lower ring to the northeast is Reinhold B.

Rümker   Remarkable formation on northwestern Oceanus Procellarum, visible only under low illumination. Rümker is an irregular, lumpy dome 70 km wide.

Sinus Iridum   Beautiful large (diameter 250 km) bay on the Mare Imbrium. Its seaward wall has been broken down by invading lava and is reduced to a few low wrinkle ridges. Its remaining walls, known as Montes Jura (the Jura Mountains), are brilliantly illuminated by the morning Sun. The deep, 38-km-diameter crater on the northern rim is Bianchini.

Moon Map 2

Agrippa   Oval-shaped crater 46 km in diameter with central peak, forming a neat pair with Godin.

Anaxagoras   Crater 51 km in diameter near the north pole of the Moon, centre of an extensive ray system.

Archimedes   Distinctive, flooded ring 83 km in diameter in eastern Mare Imbrium, notable for its almost perfectly flat floor. Apollo 15 landed southeast of here, at the foot of the Montes Apenninus, in 1971.

Aristillus   Prominent crater 55 km in diameter in eastern Mare Imbrium with terraced walls, numerous surrounding ridges and a complex central peak 900 m high. Under high illumination it is seen to be surrounded by a faint ray system. Forms a pair with Autolycus to the south.

Aristoteles   Magnificent, partially lava-flooded crater 87 km in diameter, touching the smaller crater Mitchell to the east. Numerous ridges radiate from its outer walls. Makes a pair with Eudoxus to the south.

Autolycus   Prominent crater 39 km in diameter, south of Aristillus. High illumination shows it to be the centre of a faint ray system.

Cassini   Flooded ring of unusual appearance with partially destroyed walls, 56 km in diameter. Contains a bowl-shaped crater, Cassini A, 15 km wide.

Eratosthenes   Prominent, deep crater on the edge of Sinus Aestuum at the southern end of Montes Apenninus (the Apennine Mountains). Terraced walls and craterlet on central peak. Diameter 58 km.

Eudoxus   Rugged crater 67 km in diameter with small central peak and terraced walls. South of Aristoteles.

Godin   Smaller but deeper companion to Agrippa; 35 km in diameter with central peak, bright walls and faint ray system.

Hyginus   Rimless crater 9 km in diameter, centre of a cleft or rille 220 km long, visible in small telescopes, evidently formed by collapse of the surface. To the east is another cleft, Rima Ariadaeus.

Lambert   Crater 30 km in diameter in Mare Imbrium, with central craterlet. Situated on a wrinkle ridge. Under low illumination a larger 'ghost' ring, Lambert R, is seen to the south.

Linné   Bright spot on Mare Serenitatis, best seen under high illumination. At its centre is a small, young crater 2.4 km in diameter.

Manilius   Bright crater 39 km in diameter in Mare Vaporum, with terraced walls and central peak. Develops a ray system as illumination increases, as does its neighbour Menelaus (diameter 27 km) on the rim of Mare Serenitatis.

Mare Imbrium   Enormous circular plain 1150 km in diameter, dominating this section. Bounded by the Montes Alpes, Caucasus, Apenninus and Carpatus, but open on the southwest to Oceanus Procellarum. Mare Imbrium has a double structure: traces of a smaller inner ring are visible, marked by a few isolated mountains and wrinkle ridges. Note the different shades of lava on its surface. Isolated mountains protruding from its dark floor include Pico and Piton, plus the ranges known as Montes Recti, Spizbergen and Teneriffe.

Mare Serenitatis   A major rounded lunar sea 660 × 600 km, bounded on the northwest by Montes Caucasus and on the southwest by Montes Haemus. A bright ray from Tycho crosses the dark lava plain, passing through the crater Bessel, 16 km in diameter. Under high illumination, Mare Serenitatis is seen to be rimmed with darker lava. Note also a major wrinkle ridge (the Serpentine Ridge) on the east side. Apollo 17 landed at the southeastern edge of Mare Serenitatis in 1972.

Plato   Unmistakable large, dark-floored crater in the highlands north of Mare Imbrium, 101 km in diameter; prominent under all conditions of illumination. Tiny craterlets pockmark its flat floor. Temporary obscurations of the surface, presumed to be caused by outgassing, have been observed in this area. Landslips appear to have detached part of the inner western wall.

Pytheas   Small (diameter 20 km) but deep and prominent crater on Mare Imbrium, rhomboidal in outline. Becomes brilliant under high illumination.

Stadius   'Ghost' ring to the east of Copernicus, outlined only by a few ridges and craterlets. Visible only under low illumination. Diameter 69 km.

Timocharis   Bright crater 34 km in diameter on Mare Imbrium, with terraced walls and distinctive central crater. Faint ray centre.

Triesnecker   Crater 26 km in diameter surrounded by a system of clefts.

Vallis Alpes (Alpine Valley)   Flat-floored valley 180 km long through the lunar Alps, connecting Mare Imbrium with Mare Frigoris.

Moon Map 3

Atlas   Large crater, 87 km in diameter, with terraced walls and complex floor. A ruined ring abuts to the northwest. Atlas and Hercules form one of the many crater pairs in this region.

Burckhardt   Complex crater 57 km across, overlapping older formations (Burckhardt E and F) on either side.

Bürg   Prominent crater despite its moderate size (diameter 40 km). Central peak. Lies in the centre of Lacus Mortis. Note nearby rilles.

Cleomedes   Large, irregular-shaped crater 126 km in diameter with partially flooded floor, to north of Mare Crisium. West wall is interrupted by 43-km-diameter crater Tralles.

Endymion   Large, dark-floored enclosure 125 km in diameter, with walls up to 4900 m high.

Franklin   Crater 56 km in diameter. Forms a pair with the 40-km-diameter Cepheus to the northwest.

Geminus   Prominent crater 86 km in diameter with central peak. Bright rays emanate from two smaller craters nearby, Messala B and Geminus C.

Hercules Flat-floored enclosure 67 km in diameter containing sharp, bright bowl crater Hercules G.

Le Monnier Old and flooded crater with its western wall washed away by lava from Mare Serenitatis. Diameter 61 km.

Mare Crisium Unmistakable dark lowland plain ringed by high mountains like an oversized crater, 420 × 550 km. Main feature on its flat, lava-flooded floor is the crater Picard, 23 km across, with smaller crater Peirce to its north.

Mare Tranquillitatis An irregular-shaped lowland, 540 × 780 km. Wrinkle ridges attest to numerous lava flows over the plain. Main crater on the mare is the distorted Arago, on the western side, 26 km in diameter. Apollo 11 made the first manned lunar landing in southwest Mare Tranquillitatis in 1969.

Plinius Distinctive crater 42 km in diameter with complex central peak combined with a central craterlet. Stands between Mare Serenitatis and Mare Tranquillitatis, along with Dawes to the northeast, 18 km in diameter.

Posidonius Large (diameter 100 km) partially flooded and ruined crater on the northeast shore of Mare Serenitatis. Its floor contains a curving ridge, several rilles and a small bowl crater. Ruined structure Chacornac, 51 km in diameter, abuts to the southeast.

Proclus Small (28 km wide) but brilliant crater on the western edge of Mare Crisium. High illumination shows it to be the centre of a fan-shaped ray system.

Taruntius Low-walled crater in northwestern Mare Fecunditatis, 56 km in diameter, with concentric inner ring. Crater Cameron (formerly Taruntius C) interrupts the northwestern wall. Centre of a faint ray system.

Thales Bright ray crater 32 km in diameter, northeast of Mare Frigoris.

Moon Map 4

Bullialdus Handsome crater in Mare Nubium with terraced walls and complex central peak. Diameter 59 km. Two smaller craters, Bullialdus A and B, form a chain extending south.

Flamsteed Small (diameter 21 km) crater on Oceanus Procellarum with much larger ring of eroded hills to its north, Flamsteed P.

Gassendi Large, 110-km-diameter, partially flooded ring on northern border of Mare Humorum. Complex internal pattern of clefts, ridges and hillocks. A rich area for transient lunar phenomena (TLPs). Deeper crater Gassendi A interrupts the north wall, with the smaller Gassendi B lying farther north.

Grimaldi Vast, 220-km-diameter, dark-floored enclosure at the western limb of the Moon with broad, crater-strewn walls. Nearer the limb is a smaller dark patch, marking the floor of Riccioli, 140 km in diameter.

Hainzel Curious keyhole-shaped formation, composed of three craters fused together; the two smaller ones are Hainzel A and C.

Hippalus Lava-flooded bay 58 km in diameter on the shores of Mare Humorum, in an area with many rilles.

Lansberg   Prominent crater 40 km in diameter with massive walls and a central peak, on Oceanus Procellarum. Apollo 12 landed southeast of it in 1969.

Letronne   Large bay, 120 km wide, on the south side of Oceanus Procellarum. Its seaward-facing wall has evidently been washed away by invading dark lava.

Mare Humorum   Rounded lowland plain 370 km across with Gassendi on its northern rim. Ringed by clefts and wrinkle ridges. On the south it invades the rings Doppelmayer and Lee, although Vitello escapes destruction. On the east is the ringed bay Hippalus, associated with much surface faulting.

Schickard   Major dark-floored enclosure 227 km across. South of Schickard are the overlapping craters Nasmyth (diameter 77 km) and Phocylides (114 km). Adjoining it to the southwest is the extraordinary plateau Wargentin, 84 km wide, evidently a crater filled to the brim with solidified lava.

Schiller   Curious footprint-shaped enclosure, 165 × 65 km.

Sirsalis and Sirsalis A   Twin craters 42 and 49 km wide near a 280-km-long cleft, Rima Sirsalis, which stretches towards 130-km Darwin.

Moon Map 5

Abulfeda   Prominent smooth-floored crater with sculptured inner walls, 65 km in diameter. Apollo 16 landed in the highlands north of here in 1972.

Albategnius   Large (diameter 136 km) walled enclosure with central peak. The prominent crater Klein, 44 km in diameter, breaks the southwest wall.

Aliacensis   Prominent crater with irregular outline, 80 km in diameter. Forms a pair with Werner.

Alpetragius   Bowl-shaped crater 3900 m deep on outer slopes of Alphonsus, with large central dome. Diameter 40 km.

Alphonsus   Large enclosure 118 km in diameter with complex walls and a central ridge. Numerous craterlets and clefts cover the floor; several dark patches are visible under high illumination. Alphonsus is the site of reported obscurations believed to be caused by the release of gas from the surface.

Arzachel   Magnificent crater 96 km in diameter with terraced walls and prominent central peak. To its east is a noticeable crater 31 km in diameter with central peak, like a smaller version of Alpetragius, called Parrot C.

Barocius   Large formation southeast of Maurolycus, diameter 82 km. Its northeast wall is broken by Barocius B. To the southwest is Clairaut, 75 km in diameter, between Barocius and Cuvier.

Birt   Sharp, bright crater 17 km in diameter on eastern Mare Nubium. Telescopes reveal a smaller crater (diameter 7 km), Birt A, on the east wall, and under low illumination a rille to the west.

Blancanus   Crater 110 km in diameter, south of Clavius.

Clavius   Magnificent walled plain 225 km in diameter. Note the distinctive arc of smaller craters across its convex floor. Its south wall is interrupted by the 50-km crater Rutherfurd, and its northeast wall by 52-km Porter.

Delambre   Prominent crater 53 km in diameter with irregular interior, south-west of Mare Tranquillitatis.

Deslandres   Huge, low and eroded formation 235 km in diameter, southeast of Mare Nubium. The partially ruined ring Lexell, diameter 63 km, opens onto its southern side, and the prominent crater Hell (diameter 33 km) lies on its western floor.

Fra Mauro   Largest member, 94 km wide, of an old, eroded crater group north of Mare Nubium, also including Bonpland (diameter 60 km), Parry (47 km) and Guericke (60 km). Apollo 14 landed just north of Fra Mauro in 1971.

Heraclitus   Curious elongated formation 90 km long with central ridge, south of Stöfler. Its southern end is rounded off by the crater Heraclitus D. Between Heraclitus and Stöfler is the crater Licetus, diameter 75 km. Touching Heraclitus to the east is Cuvier, also 75 km in diameter.

Herschel   Deep (3900 m) crater with elongated central peak, north of Ptolemaeus; diameter 41 km. Farther north is the slightly smaller Spörer, a partially filled ring.

Hipparchus   Large, eroded enclosure 150 km in diameter north of Albategnius. Its central 'peak' is actually a small ruined crater. On its northeastern floor is Horrocks, 31 km in diameter and 2800 m deep. Between Hipparchus and Albategnius lies the crater Halley (diameter 36 km), east of which is Hind, 29 km wide and 2800 m deep.

Longomontanus   Large walled plain in the rugged southern uplands of the Moon. Diameter 145 km. A ridge to the east forms a crescent-shaped enclosure known as Longomontanus Z.

Maginus   Major walled plain 185 km in diameter north of Clavius. Convex floor with small peaks in centre. Southwest wall is interrupted by smaller crater, Maginus C.

Mare Nubium   Irregular dark lowland plain, covered with numerous wrinkle ridges and ghost craters. On its southern shore is the flooded crater Pitatus, and in the southwest are the dark-floored pair Campanus and Mercator (48 km and 47 km in diameter). Its most celebrated feature is a 120-km-long fault on the eastern side called the Rupes Recta, or Straight Wall, between the craters Birt and Thebit. The Rupes Recta appears to run north–south through the remains of an old flooded crater of which only the eastern half remains. At the southern end of the Rupes Recta are the Stag's Horn Mountains, apparently the remains of a flooded crater.

Maurolycus   Distinctive crater, 114 km wide, with twin central peak. Its walls rise to 5000 m. It partially obliterates a smaller, unnamed formation to the north.

Moretus   Crater with a central peak in the jumbled uplands southeast of Clavius. Diameter 114 km. Closer to the south pole are the craters Short (diameter 71 km) and Newton (diameter 79 km), heavily foreshortened.

Pitatus   Large, dark-floored ring 105 km across on the southern shore of Mare Nubium. Invading lava from Mare Nubium has partly destroyed the crater's walls and left only the vestige of a central peak. Note the rilles around its inner walls. A smaller, similarly flooded ring adjoining it to the east is Hesiodus.

Ptolemaeus Vast walled plain 153 km in diameter, hexagonal in shape. Its ancient floor is heavily pockmarked with smaller craters, the most prominent being Ptolemaeus A.

Purbach Battered but still prominent large crater, 115 km in diameter. Its floor contains ridges, and its north wall is interrupted by the oval crater Purbach G, while its south wall intrudes into Regiomontanus.

Regiomontanus Flooded crater, 124 km across. Its central peak has a summit craterlet. Forms a pair with Purbach, both noticeably hexagonal in outline.

Scheiner Crater 110 km in diameter southwest of Clavius. The largest of three craterlets on its floor is Scheiner A.

Stöfler Large, flat-floored formation 126 km wide, west of Maurolycus. Its eastern wall is destroyed by the intrusion of several craters, the largest being Faraday, diameter 69 km. The southern wall of Faraday is disturbed by Faraday C, which itself intrudes into Stöfler P.

Thebit Fascinating triple crater on southeastern Mare Nubium. The main crater, 55 km in diameter, is broken by the 20-km Thebit A, which in turn is broken by the still-smaller Thebit L.

Tycho Magnificent crater in the Moon's southern uplands, 85 km in diameter. Prominent at all angles of illumination, and brilliant under high lighting. Massively terraced walls rising up to 4500 m, imposing central peak and rough floor. Tycho is the major ray crater on the Moon. Rays from Tycho extend for 1500 km or more in all directions. Note the dark 'collar' around Tycho under high illumination. Probably the youngest of the Moon's major features.

Walter Large crater 128 km in diameter, considerably modified by landslips on the inner walls and by several interior craterlets. Appears almost square in outline.

Werner Prominent crater 70 km in diameter with walls 4200 m high, notably sharper and more rounded than neighbouring craters. The floor of Werner is dotted with several hills.

Moon Map 6

Capella Prominent crater north of Mare Nectaris, 45 km in diameter, deformed by a surface fault. Large central peak. Isidorus, 42 km in diameter, adjoins it to the west.

Catharina One of a curving trio of craters around western Mare Nectaris. Diameter 104 km. A faint ring, Catharina P, covers much of its northern floor.

Cyrillus Crater 95 km in diameter, with complex terraced walls, multiple central peak and rugged floor. Overlapped by Theophilus.

Fracastorius Horseshoe-shaped bay 124 km in diameter on the southern shore of Mare Nectaris. Dark lava has breached its northern wall and flooded its interior.

Janssen   Vast, irregular enclosure, 180 × 240 km across, heavily bombarded. In the north it is interrupted by Fabricius, 78 km wide, which has a central peak. On the west wall is a smaller (diameter 34 km) sharp crater, Lockyer. Southeast of Janssen are the twin craters Steinheil and Watt (67 km and 66 km in diameter respectively). Farther north of Fabricius is Metius, diameter 88 km.

Langrenus   Magnificent, bright-walled plain on eastern Mare Fecunditatis with terraced walls, outer ridges and a complex central peak. Its diameter is 133 km. Langrenus is a ray centre. Northwest of it, on Mare Fecunditatis, is a trio of smaller craters called Langrenus F, B and K in order of decreasing size. South of Langrenus is the large, flooded formation Vendelinus, 155 km across.

Mädler   Prominent crater (diameter 28 km) on northwest Mare Nectaris, with a central ridge.

Mare Fecunditatis   Irregularly shaped 820 × 660 km dark lowland area, connecting with Mare Tranquillitatis. On its western border it invades several craters, notably Gutenberg (diameter 71 km) and Goclenius (diameter 55 × 75 km). Note the numerous clefts in this area.

Mare Nectaris   Rounded lowland plain 350 km wide, bordered by several large craters, notably Theophilus, Cyrillus, Catharina and Fracastorius. An outer mountain ring, Rupes Altai (the Altai Scarp), surrounds Mare Nectaris.

Messier and Messier A   Elliptical pair of craters on Mare Fecunditatis, prominent despite their small sizes, 11 km and 13 km. Two bright rays extend from the western member, Messier A. Both craters appear brilliant under high illumination.

Palitzsch   A crater and valley on the eastern side of Petavius. The crater itself, 41 km wide, is at the southern end of the valley, which is 110 km long.

Petavius   Magnificent walled enclosure 177 km in diameter. A prominent rille runs across the floor from the massive, complex central peak to the terraced walls, which appear double in parts. Ridges radiate from its outer walls. West of Petavius is Wrottesley, 57 km in diameter, and with a central peak. Southwest of it is Snellius, 83 km in diameter, which lies across the cratered Vallis Snellius.

Piccolomini   Beautiful crater on Rupes Altai (the Altai Scarp). Diameter 89 km, with a broad central peak and terraced walls.

Theophilus   Imposing crater 110 km in diameter on the northwestern rim of Mare Nectaris, with a massive 2200-m central mountain. Terraced walls rise over 5000 m above the floor, with many external ridges.

Vallis Rheita (Rheita Valley)   Crater chain northeast of Janssen and Fabricius. It can be traced for a total length of about 500 km. The crater Rheita itself, 70 km in diameter and with a small central peak, lies at the northern end of the valley.

Map 1

322

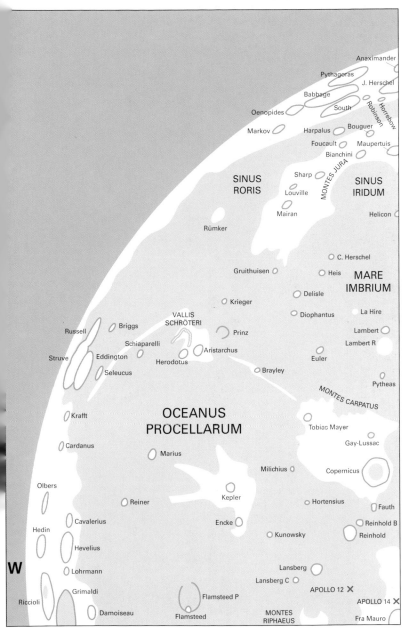

Anaximander

Pythagoras

J. Herschel

Babbage

Horrebow

Robinson

Oenopides

South

Bouguer

Markov

Harpalus

Foucault

Maupertuis

Bianchini

Sharp

MONTES JURA

SINUS
RORIS

Louville

SINUS
IRIDUM

Mairan

Helicon

Rümker

C. Herschel

Gruithuisen

Heis

MARE
IMBRIUM

Delisle

Krieger

Diophantus

VALLIS
SCHRÖTERI

La Hire

Briggs

Prinz

Lambert

Russell

Schiaparelli

Aristarchus

Lambert R

Struve

Eddington

Herodotus

Euler

Seleucus

Brayley

Pytheas

MONTES CARPATUS

Krafft

OCEANUS
PROCELLARUM

Tobias Mayer

Cardanus

Gay-Lussac

Marius

Milichius

Copernicus

Olbers

Kepler

Reiner

Hortensius

Fauth

Cavalerius

Encke

Reinhold B

Hedin

Kunowsky

Reinhold

Hevelius

W

Lansberg

Lohrmann

Lansberg C

APOLLO 12 ✕

Grimaldi

Flamsteed P

APOLLO 14 ✕

Riccioli

Damoiseau

Flamsteed

MONTES
RIPHAEUS

Fra Mauro

Map 2

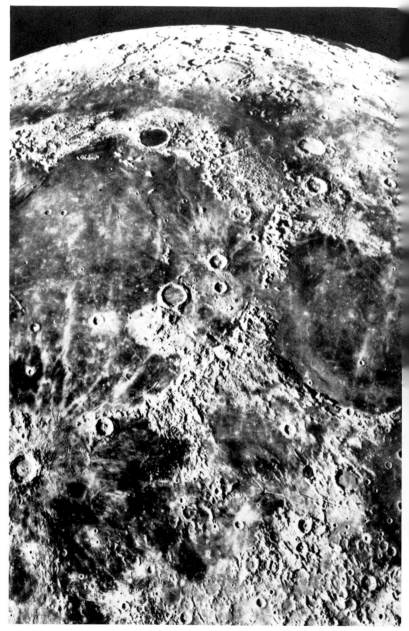

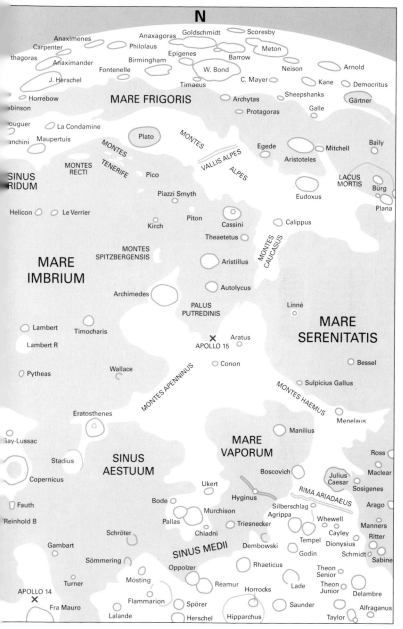

Map 3

Thales
Strabo
MARE HUMBOLDTIANUM
rtner

Endymion

Baily
Atlas
Mercurius
ACUS
ORTIS
Chevallier
Bürg
Hercules
Shuckburgh
Zeno
Mason
Oersted
Schumacher
Plana
Williams
Grove
Cepheus
Hooke
Messala
LACUS SOMNIORUM
Franklin
Gauss
Daniell
Berzelius
Bernoulli
Geminus
Berosus
Posidonius
Burckhardt
Chacornac
Debes
Hahn
Newcomb
Tralles
Delmotte
Le Monnier
Römer
Cleomedes
Eimmart
Plutarch
MARE SERENITATIS
Macrobius
MARE ANGUIS
Littrow
Hill
Tisserand
✕
APOLLO 17
Maraldi
Carmichael
Peirce
Dawes
Vitruvius
Franz
Proclus
MARE CRISIUM
Alhazen
Plinius
Yerkes
Picard
Hansen
Jansen
Lyell
Glaisher
Lick
Ross
Maclear
Cauchy
da Vinci
Condorcet
Sosigenes
Auzout
MARE TRANQUILLITATIS
Arago
Firmicus
MARE UNDARUM
Manners
Lamont
Cameron
Taruntius
Apollonius
Dionysius
Ritter
Maskelyne
Secchi
Dubyago
Sabine
MARE FECUNDITATIS
MARE SPUMANS
Schmidt
✕ APOLLO 11
Censorinus
Delambre
Webb
Messier A ⃝⃝ Messier
Taylor
Alfraganus
Torricelli
Langrenus F
Kästner
MARE MARGINIS
MARE SMYTHII
E

327

Map 4

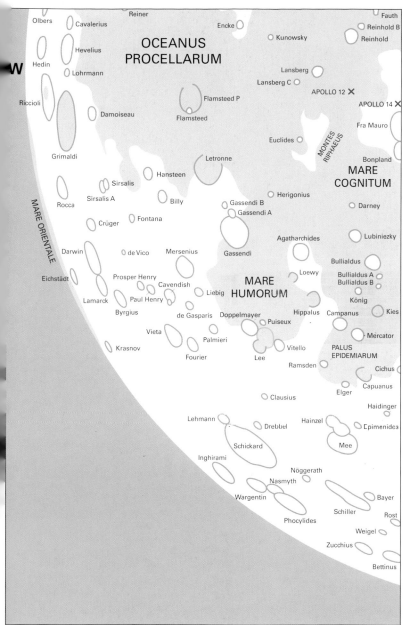

Olbers
Cavalerius
Reiner
Encke
Fauth
Reinhold B
Reinhold
Kunowsky

OCEANUS PROCELLARUM

Hevelius
Hedin
Lohrmann

Lansberg
Lansberg C
APOLLO 12 ✕
APOLLO 14 ✕

Riccioli
Damoiseau
Flamsteed P
Flamsteed
Euclides
Fra Mauro

MONTES RIPHAEUS

Grimaldi
Letronne
Hansteen
Bonpland

MARE COGNITUM

Sirsalis
Sirsalis A
Rocca
Billy
Herigonius
Darney

Crüger
Fontana
Gassendi B
Gassendi A
Agatharchides
Lubiniezky

Darwin
de Vico
Mersenius
Gassendi
Bullialdus
Loewy
Bullialdus A
Bullialdus B

Eichstädt
Prosper Henry
Cavendish
MARE HUMORUM
König

Lamarck
Paul Henry
Liebig
Hippalus
Campanus
Kies

Byrgius
de Gasparis
Doppelmayer
Puiseux
Mercator

Vieta
Palmieri
PALUS EPIDEMIARUM

Krasnov
Fourier
Lee
Vitello
Ramsden
Cichus
Capuanus

MARE ORIENTALE

Elger
Clausius
Haidinger

Lehmann
Drebbel
Hainzel
Epimenides

Schickard
Mee

Inghirami
Nöggerath
Nasmyth
Bayer

Wargentin
Schiller
Rost

Phocylides
Weigel
Zucchius
Bettinus

W

Map 5

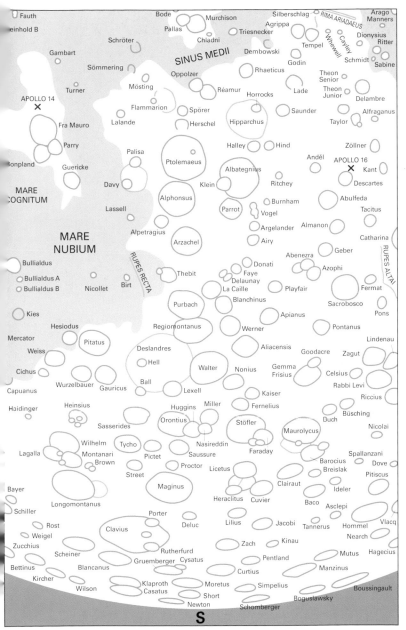

Fauth
einhold B
Reinhold B
Gambart
APOLLO 14 ✕
Fra Mauro
Parry
Bonland
Bonpland
Guericke
MARE COGNITUM
COGNITUM
Davy
Lassell
MARE NUBIUM
Bullialdus
Bullialdus A
Bullialdus B
Nicollet
Kies
Hesiodus
Mercator
Weiss
Cichus
Capuanus
Haidinger
Lagalla
Bayer
Schiller
Rost
Weigel
Zucchius
Bettinus
Kircher
Wilson
Turner
Sömmering
Mösting
Flammarion
Lalande
Palisa
Alpetragius
Pitatus
Wurzelbauer
Gauricus
Heinsius
Sasserides
Wilhelm
Montanari
Brown
Longomontanus
Scheiner
Blancanus
Klaproth
Casatus
Bode
Pallas
Schröter
Chladni
SINUS MEDII
Oppolzer
Spörer
Herschel
Ptolemaeus
Arzachel
RUPES RECTA
Thebit
Birt
Purbach
Regiomontanus
Deslandres
Hell
Ball
Walter
Lexell
Huggins
Orontius
Tycho
Pictet
Street
Maginus
Porter
Deluc
Clavius
Rutherfurd
Gruemberger
Cysatus
Moretus
Short
Newton
Murchison
Agrippa
Triesnecker
Dembowski
Godin
Rhaeticus
Réamur
Horrocks
Saunder
Hipparchus
Halley
Hind
Albategnius
Klein
Ritchey
Parrot
Vogel
Burnham
Argelander
Almanon
Airy
Donati
Faye
Delaunay
La Caille
Playfair
Blanchinus
Apianus
Werner
Aliacensis
Goodacre
Nonius
Gemma Frisius
Kaiser
Fernelius
Miller
Stöfler
Maurolycus
Nasireddin
Faraday
Saussure
Proctor
Licetus
Clairaut
Heraclitus
Cuvier
Lilius
Jacobi
Zach
Kinau
Pentland
Curtius
Simpelius
Schomberger
Boguslawsky
Silberschlag
RIMA ARIADAEUS
Arago
Manners
Whewell
Cayley
Dionysius
Ritter
Tempel
Schmidt
Sabine
Theon Senior
Theon Junior
Delambre
Lade
Alfraganus
Taylor
Zöllner
Andĕl
APOLLO 16 ✕
Kant
Descartes
Abulfeda
Tacitus
Catharina
RUPES ALTAI
Geber
Abenezra
Azophi
Fermat
Sacrobosco
Pons
Pontanus
Lindenau
Zagut
Celsius
Rabbi Levi
Riccius
Büsching
Duch
Nicolai
Spallanzani
Barocius
Breislak
Dove
Pitiscus
Ideler
Baco
Asclepi
Tannerus
Hommel
Vlacq
Nearch
Mutus
Hagecius
Manzinus
Boussingault

S

Map 6

Southeast Section

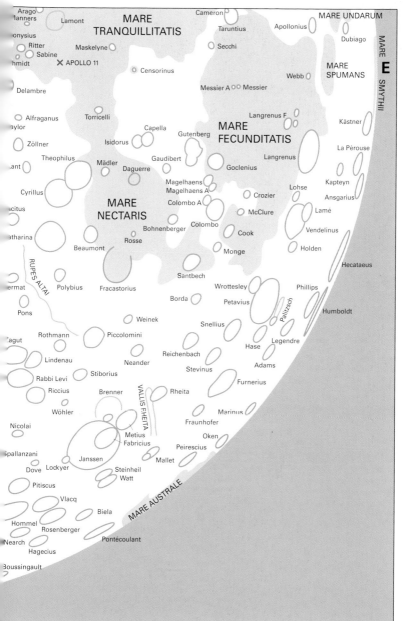

Arago
Lamont
Manners
MARE TRANQUILLITATIS
Cameron
Taruntius
Apollonius
MARE UNDARUM
Dubiago
MARE SMYTHII
E
Dionysius
Ritter
Sabine
Maskelyne
Secchi
Schmidt
X APOLLO 11
Censorinus
Webb
MARE SPUMANS
Delambre
Messier A Messier
Alfraganus
Torricelli
Langrenus F
Kästner
Taylor
Capella
Gutenberg
MARE FECUNDITATIS
Zöllner
Isidorus
La Pérouse
Theophilus
Mädler
Gaudibert
Langrenus
Daguerre
Goclenius
Kapteyn
Kant
Cyrillus
Magelhaens
Magelhaens A
Crozier
Lohse
Ansgarius
Tacitus
MARE NECTARIS
Colombo A
McClure
Lamé
Catharina
Bohnenberger
Colombo
Cook
Vendelinus
Rosse
Beaumont
Monge
Holden
RUPES ALTAI
Polybius
Fracastorius
Santbech
Hecataeus
Fermat
Wrottesley
Phillips
Pons
Borda
Petavius
Palitzsch
Humboldt
Weinek
Snellius
Rothmann
Piccolomini
Legendre
Tagut
Hase
Lindenau
Reichenbach
Neander
Stevinus
Adams
Rabbi Levi
Stiborius
Furnerius
Riccius
Brenner
Rheita
VALLIS RHEITA
Wöhler
Marinus
Nicolai
Fraunhofer
Metius
Fabricius
Oken
Spallanzani
Peirescius
Janssen
Mallet
Dove
Lockyer
Steinheil
Watt
Pitiscus
Vlacq
Biela
MARE AUSTRALE
Hommel
Rosenberger
Pontécoulant
Nearch
Hagecius
Boussingault

Eclipses of the Sun and Moon

Occasionally the Sun, Moon and Earth line up exactly to cause an eclipse of either the Sun or the Moon. The Sun is eclipsed when the Moon passes directly between it and the Earth; the Moon's shadow falls on part of the Earth, and from within the shadow part or all of the Sun is obscured from view. Eclipses of the Moon occur when the Moon is on the opposite side of the Earth to the Sun; the Moon can then enter the Earth's shadow and be darkened.

If the Moon orbited the Earth in the same plane that the Earth orbits the Sun, a solar eclipse would occur at each new Moon and a lunar eclipse at each full Moon. But the Moon's orbit is inclined at 5° to that of the Earth, just enough to ensure that the alignment of the three bodies is seldom exact. Only on those occasions when the Moon crosses the Earth's orbit at the time of new or full Moon does an eclipse occur.

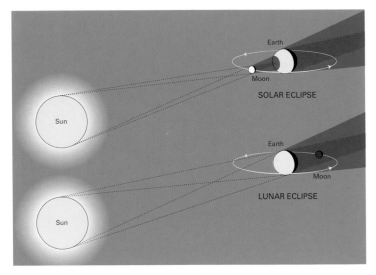

When the Moon passes in front of the Sun, a solar eclipse occurs. When the Moon enters the Earth's shadow, a lunar eclipse occurs. (Wil Tirion)

Each year at least two eclipses of the Sun are visible somewhere on Earth, and there can be as many as five, while there can be up to three eclipses of the Moon. The maximum number of eclipses possible in one year, both solar and lunar, is seven. Whereas a lunar eclipse is visible from anywhere on Earth that the Moon is above the horizon, an eclipse of the Sun can be seen only from within the narrow band along which the Moon's shadow passes. Hence, from any one place, lunar eclipses are about twice as frequent as solar ones.

Scientifically, total eclipses of the Sun are by far the more important. At a total solar eclipse the Moon completely blots out the brilliant disk of the Sun, allowing astronomers to observe the Sun's faint outer halo of gas, the corona. To see the Sun totally eclipsed we must be in the darkest central portion, or umbra, of the Moon's shadow as it sweeps over the Earth. The band of totality is usually no more than a few hundred kilometres across, but outside this is a much wider area that experiences a partial eclipse. Astronomers travel across the globe for the few precious moments that totality affords. The longest that a total eclipse of the Sun can last is 7 minutes 31 seconds, but the usual duration is around 2–4 minutes.

Total solar eclipses demonstrate one of the most extraordinary coincidences in nature: the Sun and Moon appear virtually the same size in the sky. This results from the fact that the Moon, despite being 400 times smaller than the Sun, is also 400 times closer. Sometimes, though, when the Moon is at its farthest from us in its elliptical orbit, it appears slightly too small to cover the Sun completely, and a ring of bright sunlight remains visible around the Moon's obscuring disk. Such an event is termed an *annular* eclipse (from the Latin word 'annulus' meaning a ring, not because they occur yearly). Partial and annular eclipses are of curiosity value only; they have none of the scientific importance of a total eclipse.

A solar eclipse gets underway at *first contact*, when the edge of the Moon begins its progress across the face of the Sun. Totality is still 1½ hours away. The partial phases of the eclipse can be observed by looking

Safe ways to watch a solar eclipse

Suitable filters for looking at the Sun with the naked eye are now produced commercially, so there is no excuse for risking your eyesight by using unsafe materials. One popular type of filter specifically intended for solar observation is made of aluminized plastic film, known by the tradename Mylar, while a second type consists of a thicker, black plastic. Mylar filters give a bluish image but black polymer filters give a more natural-looking orange image.

A welder's glass of shade 13 or 14 makes a safe solar filter for naked-eye observing. Two or three layers of heavily fogged black-and-white photographic negative film are also safe, because the silver in the film absorbs the Sun's heat as well as its light. Unsuitable materials include sunglasses, neutral-density photographic filters, colour film and compact discs (CDs); although these and other materials dim the light from the Sun, the Sun's heat will still get through to damage your eyes.

A simple arrangement not involving filters is to make a pinhole in a piece of card and allow the Sun's light to shine through this hole onto a white surface. You are in fact using a pinhole camera to observe the Sun. Drawbacks with this arrangement are that the image so formed is small and faint and requires strong sunlight, undimmed by clouds, to work successfully.

The 'diamond ring' effect, plus pink prominences visible against the pearly light of the Sun's inner corona, at the end of the total solar eclipse of July 11, 1991, viewed from Mexico. For a view of the full corona at the same eclipse, see page 297. (Armagh Planetarium)

at the Sun through a special filter (see the panel on page 335), or by projecting the Sun's image through binoculars or a telescope onto white card. When you do this, compare the jet-black outline of the Moon with the umbra of sunspots. This will confirm that sunspots are not totally black; instead their colour appears somewhat brownish.

Not until about 20 minutes before totality, when the Sun's disk is over two-thirds covered, does the sky start to darken noticeably. An eerie half-light covers the landscape; animals act as though night is falling. Totality itself rushes up as the last crescent sliver of sunlight is blotted out by the Moon. In the final seconds, chinks of sunlight peep between the mountains at the Moon's rugged edge, producing *Baily's beads*, named after the English astronomer Francis Baily who described them after the eclipse of 1836. Often one bead shines brighter than the others, producing the effect of a diamond ring.

Then, second contact: the Moon completely covers the Sun, and the pearl-coloured corona springs into view. Now it is safe to watch without eye protection. Feather-like plumes and streamers of the corona extend outwards from the polar and equatorial regions of the Sun for several solar diameters. Pinkish-red prominences can be seen looping out from the Sun's chromosphere around the dark outline of the Moon. Bright stars are visible in the darkened sky. All too soon, the beautiful spectacle is over. The diamond ring flashes out at third contact, signalling the end of totality. At fourth contact, the Moon moves completely clear of the Sun. The eclipse is over.

Eclipses of the Moon are far less spectacular. The Moon takes several hours to move completely through the dark inner part of the Earth's shadow, the umbra. The outer part of the shadow, the penumbra, is so light that it produces little noticeable darkening of the Moon's surface.

A total eclipse of the Moon can last up to 1¾ hours, but even when totally eclipsed the Moon seldom completely disappears. The reason is that light is bent into the Earth's shadow by the Earth's atmosphere, giving the eclipsed Moon a coppery-red colour. Dark eclipses occur when there are a lot of clouds and dust in the Earth's atmosphere, which block the light. Although of interest as a natural spectacle, a lunar eclipse is of little scientific importance.

Unusual colour effects were seen at the very dark total lunar eclipse of December 9, 1992, in which most of the Moon became almost invisible to the naked eye, apart from a brighter crescent with a bluish tinge. (Eric Hutton)

Mercury

Mercury is a disappointing object for observers. Small telescopes show its phases as it orbits the Sun every 88 days, but through even the largest instruments only a few smudgy markings can be seen on its surface, less distinct than the Moon's features to the naked eye. Most observers must therefore be content with simply catching a glimpse of this elusive object during one of its periodic appearances in the evening or morning sky.

Being the innermost planet, Mercury never appears far from the Sun in the sky. Circumstances dictate that there are two good times to look for Mercury. One is when it is setting after the Sun during evenings in spring (around March/April in the northern hemisphere or September/October in the southern hemisphere). The second is when it is a morning object rising before the Sun in the autumn (September/October in the northern hemisphere, March/April in the southern hemisphere).

An additional complication is that its orbit is markedly elliptical, ranging from 46 to 70 million km from the Sun, so that even on these occasions there are times when Mercury is easier to see than at others. Even when Mercury is best placed, a low, clear horizon will be needed to see it; binoculars help to pick the planet out of the twilight glow, for Mercury can never be seen against a truly dark sky. With all these complications, it is little wonder that many town dwellers have never seen the planet. Nevertheless it is worth looking for, because at its best it can become almost as bright as Sirius.

The difficulty in observing Mercury led to a long-standing misconception about the time it takes to rotate on its axis. Towards the end of the 19th century, the Italian astronomer Giovanni Schiaparelli proposed, after a long series of observations, that the planet spins on its axis in 88 days, the same time as it takes to orbit the Sun. It would therefore keep one face turned permanently towards the Sun, as the Moon does to the Earth. In the 1920s, the Greek-born astronomer Eugène Antoniadi compiled a map showing smudgy markings on the surface of Mercury based on an assumed 88-day rotation period. This map seemed to settle the matter once and for all.

Then, in 1965, came a surprise. At Arecibo Radio Observatory, astronomers Rolf Dyce and Gordon Pettengill bounced radio waves off the surface of Mercury. From the change in frequency of the reflected radio waves, they deduced that Mercury spins once every 59 days, two-thirds of the time it takes to orbit the Sun. Therefore the Sun does rise and set on Mercury, but very slowly. For the Sun to go once around the sky as seen from the surface of the planet – say, from one noon to the next – takes 176 Earth days, during which time Mercury orbits the Sun twice, spinning three times on its axis.

In the skies of Mercury the Sun appears two and a half times as large as it does from Earth. The daytime side of Mercury is continually blasted by lethal doses of high-energy solar radiation. The Sun's intense heat roasts the surface rocks to over 400°C at noon on the equator, hot enough to

The lunar-like landscape of Mercury, photographed by the American space probe Mariner 10 in March 1974. The bright ray crater just above centre is called Kuiper. The largest craters are about 200 km across. (USGS)

melt tin and lead. Without an atmosphere to hold in the heat, the planet's surface cools to a frigid −180°C during the long night.

With a diameter of 4879 km, Mercury is only 50 per cent larger than our Moon, and smaller than every other planet in the Solar System except Pluto. Astronomers had long assumed that Mercury resembled our Moon in appearance, but it took the space probe Mariner 10 in 1974 to show how remarkably similar-looking Mercury and the Moon really are. As Mariner 10 flew past Mercury, its cameras photographed a surface heavily pockmarked with craters of all sizes, similar to the lunar highlands. Mariner 10 photographed less than half Mercury's surface, yet, if that portion is typical of the rest of the planet, it is enough to tell us much of the previously unknown story of Mercury.

Craters on Mercury look almost identical to their lunar counterparts. There are deep, young craters, eroded ancient craters, craters with terraced walls, central peaks and bright rays. Many of the features on Mercury have been named after artists, composers and writers, thereby breaking the near-monopoly previously held by astronomers' names on Solar System bodies. For instance, we now find on Mercury memorials to Bach, Mozart, Van Gogh and Chekhov.

Sometimes it is difficult to distinguish at a glance between a picture of Mercury and one of the Moon. Almost certainly, the craters on both bodies have been formed in the same way, by the impact of large meteorites early in the history of the Solar System. One noticeable difference on Mercury is that material ejected from the craters has not travelled as far as on the Moon, because Mercury's surface gravity is stronger – over twice that of the Moon, but still only 38 per cent of the Earth's. Another result of the higher gravity of Mercury is that craters tend to be shallower for a given diameter than on the Moon.

Meandering cliffs called *lobate scarps*, several hundred kilometres long and a kilometre or so high, are a feature of Mercury's surface unlike anything on the Moon. These are believed to have been caused by a shrinking of the planet as its core cooled early in its history, leading to compression and faulting in the crustal rocks. The surface rocks of Mercury are actually slightly darker in tone than those of the Moon, reflecting a mere 11 per cent of the sunlight hitting them, compared with an average 12 per cent for the Moon; Mercury in fact has the darkest surface of any planet in the Solar System overall, although the Moon's maria are darker than Mercury.

Between many of the large craters in the highlands of Mercury are areas peppered only with small craters. These areas, termed the *intercrater plains*, have no real counterpart on the Moon. They clearly pre-date the large craters, but whether they were formed by volcanic action or are deposits of ejecta from large impacts is uncertain. Surveys from a future Mercury orbiter should resolve the question. Other areas of particular interest for first-hand study are deep craters near the poles whose interiors are permanently shaded from the Sun, thereby preserving in

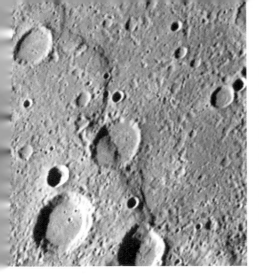

Santa Maria Rupes, a surface fold of the type known as a lobate scarp, runs nearly vertically for 200 km through this picture from Mariner 10. It cuts across old craters and intercrater plains, evidence that Mercury has shrunk slightly in size since these features were formed. (NASA/JPL/ Northwestern University)

deep freeze any gases that have seeped from the planet over its history as well as ice from cometary impacts.

Of all the features seen by Mariner 10, the most prominent is an enormous bull's-eye structure, partly hidden by shadow, named the Caloris Basin. It is 1300 km across, similar to Mare Imbrium on the Moon and fully one-quarter the diameter of Mercury. It was presumably formed by the impact of an asteroid after most of the surface had already been cratered. The Caloris Basin contains several concentric rings of mountains, and is surrounded by a number of radial ridges and grooves. Most importantly from a geological point of view, its interior, and much of the low-lying land around it, has been flooded by lava. Geological activity died out on Mercury over 3000 million years ago, as it did on the Moon. Since then, little has changed except for the random arrival of a stray meteorite.

Despite its outward resemblance to the Moon, inwardly Mercury is believed to be more like the Earth. Mercury has a relatively large mass for its small diameter, which implies that it has a large iron core three-quarters of the planet's diameter. A core that size would be as big as the Moon. The existence of an iron core was directly confirmed when Mariner 10 measured a magnetic field around the planet, albeit only 1 per cent the strength of the Earth's magnetic field – but that is still far stronger than the magnetic fields of Venus and Mars.

One explanation for such a disproportionately large core is that Mercury was originally much bigger, but had most of its rocky outer layers blasted off in a collision with a stray body of similar size to our Moon. This collision could also have knocked Mercury's orbit into its elliptical shape. In its own way, Mercury turns out to be a fascinating world containing many clues about the origin and development of the Solar System.

Venus

Many people have seen Venus without realizing it. The planet appears as the brilliant evening or morning 'star', the most prominent object in the twilight, outshining every genuine star with its cold white light. So striking is Venus at its best that it is frequently reported as a hovering UFO.

Venus orbits the Sun every 225 days at a distance of 108 million km; it can pass closer to Earth than any other planet, within 40 million km. With a diameter of 12,100 km, only 650 km smaller than the Earth, it is almost a twin of our own planet in size. But the main reason for the brilliance of Venus in the sky is not its size or proximity but its cloak of unbroken clouds that reflect two-thirds of the light hitting them. While making Venus so prominent, these clouds also prevent astronomers from seeing its surface.

The clouds of Venus circulate around the planet every 4 days, spiralling from equator to pole and creating a V- or Y-shaped pattern visible in ultraviolet light, as in this photograph from the Pioneer Venus orbiter. (NASA/Ames)

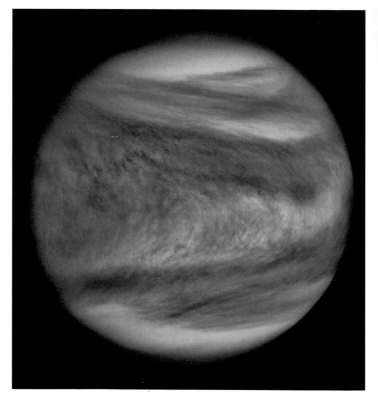

Through a telescope, Venus appears like a white billiard ball that goes through a cycle of phases as it orbits the Sun. At half phase, around the time of greatest elongation, it appears about half the size of Jupiter, but as Venus approaches the Earth its apparent size grows to equal or exceed that of Jupiter. One complete cycle of phases of Venus seen from Earth (its synodic period) takes 584 days, 2½ times longer than one orbit (the sidereal period); this is a consequence of the rapid relative movement of the Earth and Venus in their respective orbits around the Sun.

When it is a crescent, Venus is close enough to the Earth for its phase to be picked out in modest binoculars; some people even claim to have seen Venus as a crescent with the naked eye. The planet appears most brilliant when 28 per cent of its disk is illuminated as seen from Earth, this being the most favourable combination of distance and phase. Venus can reach a maximum magnitude of −4.7, nearly seven times brighter than the next most prominent planet, Jupiter. Being so bright, Venus is best observed against a twilight sky to reduce the dazzle.

Only the vaguest markings can be made out in the clouds of Venus with Earth-based telescopes; some dusky shadings are visible, and often the clouds appear to be brighter at the poles, with a surrounding darker collar. Cloud patterns are often shaped like a sideways V or Y. Observers usually draw the disk of Venus to a diameter of 50 mm, and estimate the brightness of features on a scale from 0 (extremely bright) to 5 (unusually dark). The terminator (the edge of the illuminated portion) can appear irregular, due not so much to differences in the height of the clouds as to differences in brightness. These effects are caused by the corkscrew circulation of clouds around Venus from its equator to its pole, as has become clear from space-probe photographs.

Unable to see the planet's surface because of the enveloping clouds, astronomers could only guess at the rotation period of Venus until the 1960s – and they guessed incorrectly. As with Mercury, radar observations provided the surprising truth. It turns out that Venus rotates on its axis from east to west, the opposite direction from the Earth and other planets, and it does so very slowly: once every 243 days, longer than the 225 days it takes to orbit the Sun. Its clouds, though, rotate every four days, also retrograde (east to west), a result of high-speed winds in the upper atmosphere.

Before space probes arrived there, theories about the nature of the planet's surface abounded. Since Venus was so similar in size to the Earth, it was tempting to speculate that conditions there might be Earth-like. One charming notion was that Venus resembled our planet as it had been in Carboniferous times, with steaming jungles and even dinosaurs. Some astronomers proposed that the planet was covered with water, while others imagined it to be a world of deserts. None of the theories came close to anticipating the uniquely hostile conditions on Venus.

Radio astronomers provided the first clue in the late 1950s when they detected radio-wave emissions from the planet, which implied that it was very hot – even hotter than boiling water. By comparison, the deserts of the Earth are only mildly warm. These readings, doubted at the time, were confirmed by the American probe Mariner 2 which scanned the planet as it flew past it in 1962.

Conditions on Venus were experienced directly for the first time by a Soviet probe, Venera 4, when it parachuted into the atmosphere in October 1967. It found that Venus's atmosphere is made almost entirely of unbreathable carbon dioxide gas, but the probe was destroyed by the intense heat and crushing pressure long before it reached the surface. Venera 7 was the first probe to land intact on the surface, in 1970 on December 15. It registered a temperature of 475°C and an atmospheric pressure 90 times that on Earth. Venera 7 landed on the night side of the planet; in 1972 its successor Venera 8 landed on the day side, finding identical conditions.

The dense atmosphere of Venus traps heat like a blanket, keeping the temperature constant over the entire planet. Under such pressure-cooker conditions, the atmosphere of Venus behaves more like a liquid than a gas. To explore Venus is like trying to explore a scalding hot ocean, and requires a refrigerated spacecraft reinforced like a submarine.

Large meteorites can penetrate the thick atmosphere of Venus to produce impact craters. In the foreground is Howe, 37 km wide, surrounded by a blanket of bright ejecta. This image, created from Magellan radar data, has been coloured to match the hues recorded by Soviet Venera landers. (NASA)

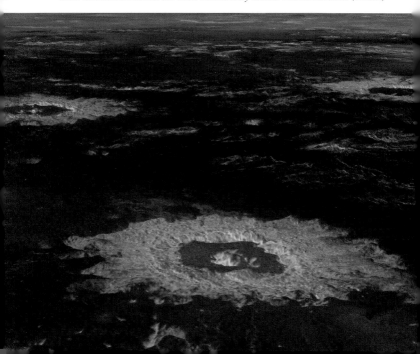

Maat Mons, a volcano on Venus 8 km high, in a view reconstructed from Magellan radar imagery. Lava flows extend across the plains in the foreground, where an impact crater can also be seen. The vertical scale has been exaggerated ten times, so the terrain seems rougher than it really is. (NASA)

Why should Venus be so hot – hotter even than the day side of Mercury, despite the fact that its clouds reflect over three-quarters of the incoming sunlight? The answer lies in the *greenhouse effect*, which works far more effectively on Venus than on Earth. About 1 per cent of the incoming sunlight penetrates to the planet's surface, so that it is as gloomy there as on a heavily overcast day on Earth. That incoming sunlight is absorbed by the surface and is re-radiated at longer wavelengths, in the infrared. Although the carbon dioxide of the atmosphere is transparent to visible light, it traps infrared; since infrared is heat energy, the temperature of the atmosphere rises.

It turns out that Venus and the Earth have similar amounts of carbon dioxide, but on Earth most of it is locked away in rocks such as limestone. Whereas the amounts of carbon dioxide on both planets are similar, Venus is almost entirely devoid of water – whatever water it originally possessed has long since evaporated and been lost to space. Only a trace of water vapour remains, but that is sufficient to boost the effect of the carbon dioxide in causing the greenhouse effect of the atmosphere of Venus.

A final contribution to the greenhouse effect is provided by the clouds of Venus. These are made not of water vapour, as are the clouds of Earth, but of sulphuric acid of 80 per cent concentration, stronger than in a car battery. Sulphuric acid, too, absorbs infrared. Taken together, these three factors of carbon dioxide, water vapour and sulphuric acid turn Venus into a perfect suntrap. The clouds add to the nastiness of Venus in another way: from them descend showers of corrosive sulphuric acid rain. Despite its heavenly name, Venus turns out to be an incarnation of hell.

In December 1978 a flotilla of five American space probes called Pioneer plummeted into the atmosphere of Venus. They found that the uppermost layer of sulphuric acid clouds, the one which observers view through telescopes, starts about 65 km above the planet's surface and is a few kilometres thick. Around 55 km altitude is a thin haze layer apparently consisting of sulphuric acid particles, which give the clouds their yellow tinge. The densest cloud layer of all is at about 50 km altitude, and it is from this layer that the rain of sulphuric acid droplets falls. Below the clouds, the gloom is broken by flashes of lightning, while thunder reverberates in the atmosphere.

Although the clouds of Venus mask its surface from view, astronomers have nevertheless been able to map the planet's features by radar, which penetrates the clouds. Radar observations from Earth during the 1970s first revealed some features, and detailed maps of the entire planet have since been made by spacecraft, notably the Magellan probe that went into orbit around Venus in 1990.

Venus is mostly rolling plains, but there are three main continental areas. One, called Ishtar Terra, the size of the United States, has a mountain range, Maxwell Montes, which towers 12 km above the mean surface level, higher than Mount Everest on Earth. The largest continental area of all, Aphrodite Terra, the size of South America, is cut by a system of rift valleys that extends for thousands of kilometres.

Magellan's radar 'eye' spotted impact craters ranging in size from over 100 km across down to 3 km, demonstrating that large meteorites can get through the dense atmosphere without burning up. Most exciting of all were volcanic mountains with fresh-looking lava flows on their flanks, notably Maat Mons, at 8½ km high the second-highest peak on the planet, which lies near the equator in Aphrodite Terra. Surrounding it are lava flows estimated to have been no more than ten years old when Magellan observed them. Clearly, Venus is still an active planet, with highlands formed by volcanic action.

Other types of volcanic feature on Venus include flattened, pancake-like domes, evidently formed by the outpouring of sticky lava, and ringlike formations of cracks and ridges termed coronae, hundreds of kilometres across, apparently caused by subsidence after upwelling of magma from below.

Photographs from Soviet lander probes sitting on the planet's surface show a rocky wasteland bathed in a sulphurous orange glow. Chemical analyses made by these probes confirm that the surface rocks of Venus are similar in composition to volcanic basalts on Earth, as would be expected from the volcanic activity on the planet. Venus is a tantalizing vision of an Earth that might have been – and a terrifying demonstration of what the Earth itself might have become had it been born closer to the Sun.

Mars

Mars is distinguishable by its intense reddish-orange hue, stronger than the colour of any star and the cause of its association with the god of war. At its best Mars can shine at magnitude −2.8, rivalling Jupiter. But the most favourable appearances of the planet are few and far between, the main reason being the marked ellipticity of its orbit, which takes it from 206 to 249 million km from the Sun (average distance 228 million km).

If the Earth passes Mars when Mars is closest to the Sun, only about 55 million km separates the two bodies and astronomers get their best views of the red planet; at such times a magnification of ×75 will show the planet the same size as the full Moon to the naked eye. But when farthest from the Sun, Mars lies 100 million km from Earth at opposition, almost twice the distance, and appears unimpressive even in powerful telescopes. The closest approaches of Mars occur at intervals of about 15 years, as in 2003 and 2018, so astronomers do not waste their chances of observing it at its best. Positions of Mars for a five-year period are shown in the charts on pages 356–365.

Mars at its close approach of 1997, photographed by the Hubble Space Telescope. The dark feature at the centre is Syrtis Major. South of it is the lowland basin Hellas, filled with frost and clouds. At right, white clouds also lie around the volcano Elysium Mons. (Steve Lee, University of Colorado/ Jim Bell, Cornell University/Mike Wolff, Space Science Institute/NASA)

Mars has a diameter of 6790 km, just over half that of the Earth. Its day is just over half an hour longer than our own – 24 hours 37 minutes – but its year is nearly twice as long, 687 Earth days. Its orbit lies outside the Earth's, so Mars can never appear as a crescent. But at times it displays a distinctly gibbous phase, like the Moon a couple of days from full.

Binoculars show Mars as nothing more than an orange dot of light; a telescope is needed to bring into view the main features of the planet. Among the most obvious are the white polar caps, which stand out in stark contrast to its ochre-coloured deserts. Occasionally, violent winds whip up dust storms in the thin atmosphere, obscuring the dusky surface markings. Frustratingly for observers, the worst dust storms tend to blow up when Mars is at its closest to the Sun, thus ruining the best observing opportunities. The most prominent dark surface marking, a large triangular area named Syrtis Major, was first noted by the Dutchman Christiaan Huygens in 1659. Syrtis Major, along with the polar caps, should be visible in a modest amateur telescope.

Observers usually draw Mars with a disk size of 50 mm, the same as Venus, although some prefer a diameter of 42 mm, to correspond to the planet's diameter of 4200 miles. Features are estimated on an intensity scale from 0 (the polar caps) to 10 (the black sky). On this scale, the bright deserts rank about 2 and the darkest surface shadings about 8. Different surface features will be seen as the planet rotates, of course, turning through nearly 15° every hour. Since the rotation period of Mars is slightly slower than that of Earth, the same features will be on view just over half an hour later each night.

The temptation to assume too many similarities between Mars and the Earth led early astronomers astray. The dark areas, which range in colour from brown to grey-green, were termed seas and lakes in the belief that they really were filled with water, while the orange areas were named after places on Earth – there is an Arabia, Libya, Syria and Sinai on Mars. Towards the end of the 19th century astronomers realized that there were no oceans on Mars after all, but this opened the way for a much more in-triguing explanation for the dark areas: that they were covered with primitive vegetation, such as moss or lichen. In support of this view, observers noted that when the polar caps melted in the Martian summer, the surface markings became larger and darker; this was interpreted as vegetation growing in the milder, wetter conditions.

The most extreme proponent of the life-on-Mars idea was an American astronomer, Percival Lowell. His ideas were inspired by an Italian observer, Giovanni Schiaparelli, who in 1877 reported seeing long, straight lines that appeared to criss-cross the planet's surface. Schiaparelli called these lines *canali*, a word which in Italian means 'channels'; but inevitably it was translated as 'canals', implying that they were artificial, although Schiaparelli himself kept an open mind about their true nature.

For Lowell, there was no doubt: the canals were evidence of an advanced civilization on Mars. His beliefs sparked off a whole generation

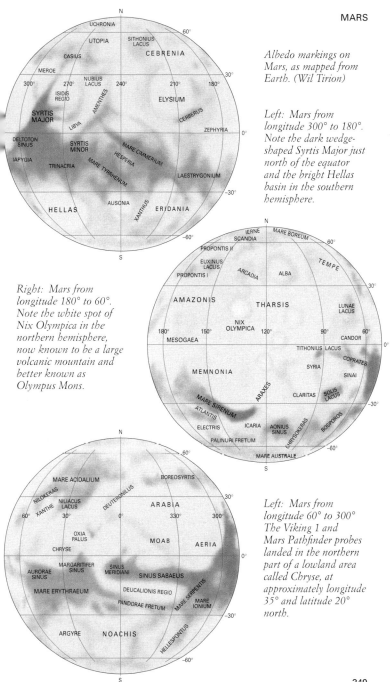

Albedo markings on Mars, as mapped from Earth. (Wil Tirion)

Left: Mars from longitude 300° to 180°. Note the dark wedge-shaped Syrtis Major just north of the equator and the bright Hellas basin in the southern hemisphere.

Right: Mars from longitude 180° to 60°. Note the white spot of Nix Olympica in the northern hemisphere, now known to be a large volcanic mountain and better known as Olympus Mons.

Left: Mars from longitude 60° to 300° The Viking 1 and Mars Pathfinder probes landed in the northern part of a lowland area called Chryse, at approximately longitude 35° and latitude 20° north.

of science fiction, including the famous *War of the Worlds* by H. G. Wells. Lowell set up his own observatory at Flagstaff, Arizona, in 1894 specifically to study Mars. There he produced fanciful maps of the canal network and wrote books such as *Mars as the Abode of Life*, published in 1908, in which he expounded his theory of a Martian civilization clinging to existence on an arid planet, reliant on the canals to bring meltwater from the polar caps to irrigate crops at the equator.

Most other astronomers failed to see the canals, or in their place could detect only broad, irregular smudges. After Lowell's death in 1916 a few devoted followers kept the canal theory alive, but the idea of Martian civilization was doomed to oblivion in the light of new knowledge about conditions on the planet.

By the 1950s it was clear that the atmosphere on Mars was far too thin for a human to breathe. Under so thin an atmosphere, temperatures would be frigid and dangerous amounts of ultraviolet radiation would penetrate to the ground. To make matters worse, no oxygen could be detected in the planet's atmosphere, although carbon dioxide was known to be present. No advanced life form could exist under such conditions, although hardy vegetation was still not ruled out.

Mars is the subject of an ongoing programme of exploration by space probes, which started in July 1965 when the American craft Mariner 4 swooped past at a distance of 10,000 km, sending back the first close-up photographs. Its most important revelation was that there are craters on Mars, looking similar to those on the Moon only somewhat more eroded because of the effects of the planet's atmosphere.

Unfortunately, a lunar-like Mars seemed an unpromising abode for life of even the lowliest form. The image of Mars as a dead world, in both the geological and biological sense, was reinforced in 1969 when Mariners 6 and 7 photographed more craters, apparently all of impact origin. But attitudes changed when Mariner 9 went into orbit around Mars and made the first survey of the entire planet in 1971–2. It revealed major formations that previous Mariners had, by ill luck, completely missed.

For a start, there was a chain of three volcanic mountains atop a highland area known as the Tharsis Bulge which straddles the equator; these volcanoes are named Arsia Mons, Pavonis Mons and Ascraeus Mons ('Mons' means mountain). Other types of feature on Mars and the names given to them include lowland plains (Planitia), highland plateaus (Planum), valleys (Vallis), canyons (Chasma), and eroded craters (Patera). Northwest of the chain of Tharsis volcanoes is an even bigger volcanic mountain, Olympus Mons, originally seen from Earth as a white ring and then known as Nix Olympica ('the snows of Olympus'). Olympus Mons, 600 km wide and 27 km high, is the largest volcano in the Solar System, larger even than the volcanic islands of Hawaii on Earth. The whole Tharsis area is noted for the frequent appearance of white clouds, often described as being shaped like a letter W. The presence of mountains explains the preference of clouds for this region.

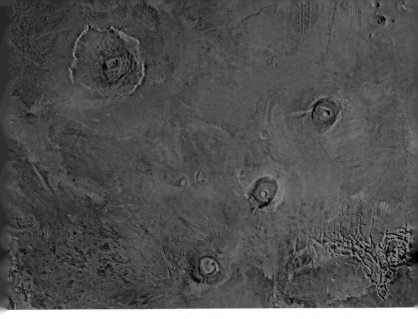

Olympus Mons (top left) and the chain of three volcanoes of the Tharsis Bulge on Mars, Ascraeus Mons (top), Pavonis Mons (middle) and Arsia Mons, in a mosaic of images from the Viking orbiter. The cracked terrain at bottom right is Noctis Labyrinthus, the start of the Valles Marineris fault complex. (USGS)

Another dramatic and unexpected feature was an immense system of canyons, up to 500 km wide and 4 km deep, extending east from the Tharsis Bulge. This massive rift in the surface of Mars, now called the Valles Marineris (Mariner Valleys, named in the plural because of their multiplicity), not merely dwarfs the Earth's Grand Canyon, but at 4000 km long could span the entire United States. Valles Marineris corresponds to a broad, fuzzy 'canal' seen from Earth, called Coprates; it is visible from Earth mainly because dark dust collects on its floor.

Coprates is one of the few supposed canals that has any corresponding surface feature. A few other canals, notably one called Cerberus, do coincide with dark streaks on the Martian surface, caused by darker dust or rock; but all of these are broad and irregular markings, unlike the thin, straight canals that Lowell drew. Close comparison of Lowell's canal maps with space probe pictures reveals few correlations at all. There seems no real explanation for the canal network that Lowell and his followers drew, other than the fallibility of human observers straining at, or past, the limits of visibility.

Although the Lowellian canals were not in evidence, Mariner 9 found convincing signs that water had once flowed on Mars, an impression amply confirmed by subsequent probes. Sinuous channels, looking like dried-up river beds, snake across parts of the planet's surface. Some of

the lowlands appear to have once been inundated by flash floods. Liquid water cannot exist on Mars today because the atmospheric pressure is too low. The existence of ancient watercourses imply that in the past the atmosphere was denser – and a denser atmosphere would keep the planet warmer. Perhaps the volcanoes of Mars gushed out enough gas to change the climate temporarily. If that were so, life might have had a chance to arise on Mars after all, and microscopic organisms such as bacteria might still cling to existence among the red sands of the planet.

The first search for Martian life was carried out in 1976 by two American space probes called Viking. Each probe came in two halves: a lander and an orbiter. The Viking 1 lander touched down on a lowland plain called Chryse in the northern hemisphere, over which water appeared to have flowed during the wet times on Mars. The other lander descended on the opposite side of the planet in an area known as Utopia, which is encroached upon by the outer fringes of the north polar cap during winter.

The landers carried colour cameras as well as instruments to analyse the soil and the atmosphere. Each Viking lander found itself on a rock-strewn desert without any visible signs of life – no plants, insects or animal tracks. A mechanical arm picked up samples of the soil and tipped them into an on-board biological laboratory which set to work in search of Martian microorganisms. To the disappointment of many, neither Viking found life in the soil of Mars, despite exhaustive tests.

That is not to say that Mars is totally sterile; maybe life lies in places that the Vikings were unable to sample, such as deeper underground or inside rocks. Such a possibility was strengthened in 1996 when organic molecules, and even apparent fossils of simple organisms like bacteria, were reported to have been found in a meteorite from Mars that had fallen to Earth in Antarctica. But those claims are highly controversial – and will probably remain so until samples are brought back automatically from Mars by robotic landers.

Above: The rocky red surface of Mars, photographed by the Mars Pathfinder lander in July 1997. On the horizon are the hills nicknamed Twin Peaks. (Timothy Parker, JPL/NASA)

Left: Sojourner, the small rover carried by Pathfinder, examines a rock called Yogi. (Peter Smith, University of Arizona/NASA)

A lifeless Mars, though, should not be too surprising in the light of readings from instruments on landers such as the Vikings, which experienced just how hostile conditions on the planet really are. The Viking 1 lander recorded a maximum summer afternoon air temperature of −29°C, whereas farther north at the Viking 2 landing site temperatures dropped well below −100°C as winter approached and the carbon dioxide in the atmosphere began to freeze, forming patches of white frost on the surface. Atmospheric pressure was a mere 7.5 millibars (750 pascals), equivalent to the pressure at a height of 35 km above the Earth's surface.

Although the Viking lander results were a disappointment for the biologists, there was plenty to interest the geologists. As expected, the rusty redness of Mars turned out to be caused by large amounts of iron oxide in the surface rocks. Mars may be the richest source of iron ore in the Solar System. Even the sky is pink, a result of fine particles of dust suspended in the wispy atmosphere. In fact, the surface of Mars is drier and dustier than any desert on Earth.

While the Viking landers studied Mars from the surface, the orbiters continued the surveying job begun by Mariner 9. Mars turns out to be a planet of two contrasting halves. The northern hemisphere is the lower and smoother of the two, flooded by lava from volcanoes. The youngest eruptions occurred within the past few hundred million years. The southern hemisphere – the area that the first Mariners photographed – is higher and has been heavily cratered by meteorite impacts, giving it a broad similarity to the lunar highlands.

In the southern hemisphere of Mars are two large impact basins, Hellas and Argyre. Argyre, 900 km in diameter, is similar in size to the Moon's Mare Imbrium, while Hellas is nearly three times larger, equivalent in size to Oceanus Procellarum. Unlike the lunar maria, however, Argyre and Hellas appear to be filled with light-coloured dust. Many of the dust storms that periodically obscure the surface of Mars after perihelion start in Hellas.

Remarkable new views of the surface came in 1997 from the Mars Pathfinder spacecraft which landed in the northern hemisphere in Chryse Planitia, about 800 km east of the Viking 1 lander. It was a lowland area that looked from orbit to have been flooded in the distant past, an impression that was borne out by the pictures from Pathfinder which

Phobos (right) and Deimos, the moons of Mars, are thought to be captured asteroids. Here they are compared with the asteroid Gaspra, top, all at the same scale and under similar lighting conditions. Gaspra was photographed by the Galileo probe, Phobos and Deimos by the Viking orbiter. (NASA)

showed a rocky plain littered with boulders and smaller rounded pebbles similar to those deposited by immense floods on Earth (see the panorama on pages 352–3). Pathfinder carried a mini rover called Sojourner which roamed the surface in the vicinity of the lander, testing the composition of various rocks. A variety of compositions were found, consistent with the rocks having been swept there from various places by an ancient flood. However, judging from the density of impact craters in the region, that flood must have taken place 2 billion years or more ago.

It comes as a surprise to learn that Mars today is not short of water, although most of that water is in frozen form, both in the polar caps and in a subsurface permafrost layer polewards of about 30° latitude north and south. This offers another explanation for the ancient water channels of Mars: the melting of subsurface ice by volcanic heating or meteorite impacts. Whether this, rather than a more Earth-like climate, was the real source of the ancient floods, or whether a combination of both factors was responsible, remains unresolved.

The polar caps of Mars are made of water ice a few metres thick, augmented each winter by carbon dioxide. Carbon dioxide freezes out of the atmosphere to produce a thin sprinkling of frost that can extend more than halfway to the equator. During each Martian year the atmospheric pressure changes by 20 per cent or more as carbon dioxide evaporates from one polar cap with the coming of spring, migrates to the opposite hemisphere and then freezes out again as winter arrives at the other pole.

One final puzzle from the pre-space-probe era of observation remained to be explained: what causes the seasonal changes in the dark areas if there is no vegetation on Mars? Wind-blown dust provides the answer. Orbiting probes have recorded many examples of changes in surface markings caused by light and dark dust being blown by seasonal winds. Syrtis Major, for example, is a gently sloping area of darker volcanic rock which periodically becomes partly covered by paler dust and is later swept clear again. Dust blown by the winds on Mars, which can reach 200 km per hour, is also a powerful erosive agent, sandblasting the surface features of the planet.

No discussion of Mars would be complete without reference to its two tiny moons, Phobos and Deimos. They were discovered in 1877 by Asaph Hall using the 66-cm (26-inch) refractor at the US Naval Observatory in Washington, DC, and both are beyond the range of normal amateur telescopes. Space probe photographs show that both are cratered chunks of rock shaped like lumpy potatoes. Phobos, larger and closer to Mars, measures about 27 × 18 km; Deimos is about 15 × 10 km.

Phobos is remarkable in that it orbits Mars three times a day. It also lies closer than any other moon to its parent – a mere 6000 km above the surface of Mars. Probably they are asteroids which strayed too close to Mars and were captured by the planet's gravity. They will make a fascinating sight in the sky for the first astronauts to set foot on Mars, the planet of red, frozen deserts which was so nearly suitable for life of its own.

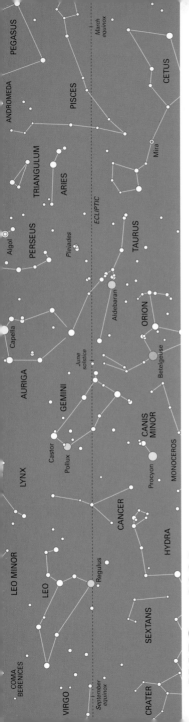

MARS — 2001

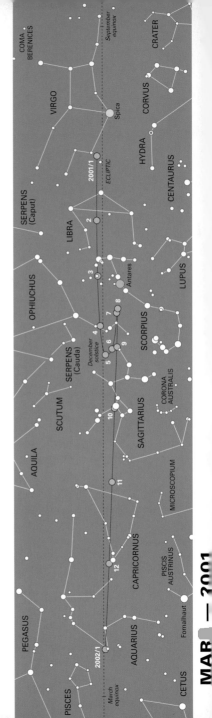

MAR — 2001

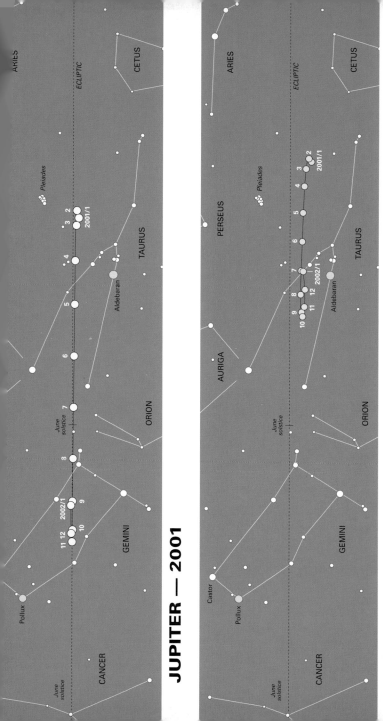

JUPITER — 2001

SATURN — 2001

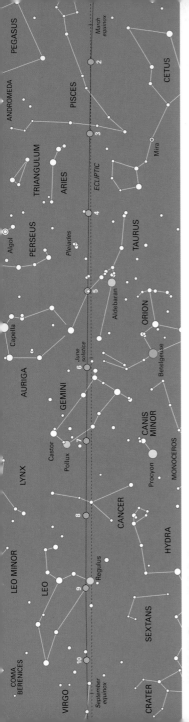

MARS — 2002

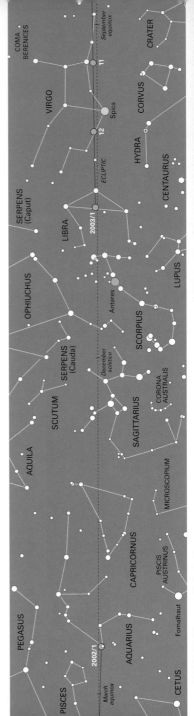

MARS — 2002

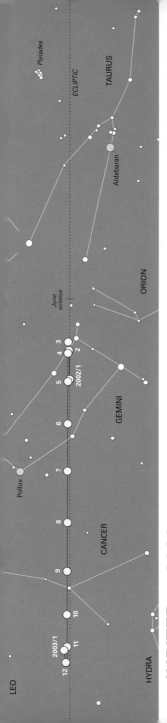

JUPITER — 2002

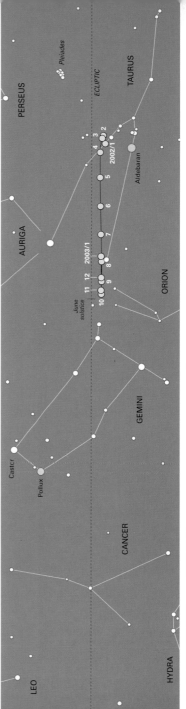

SATURN — 2002

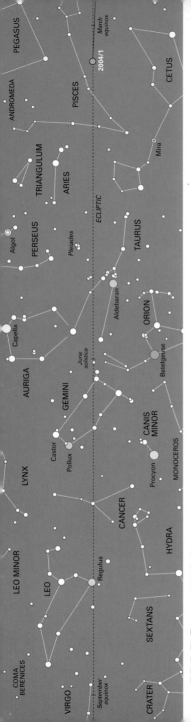

MARS — 2003

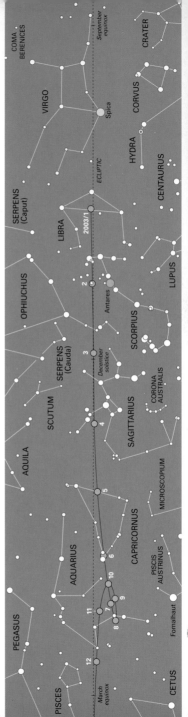

MARS — 2003

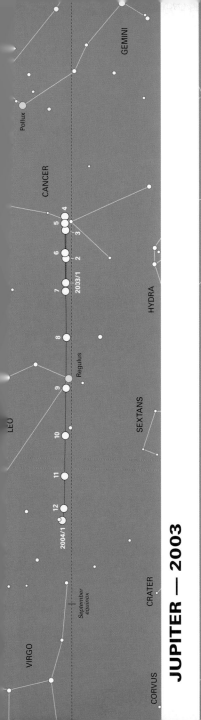

JUPITER — 2003

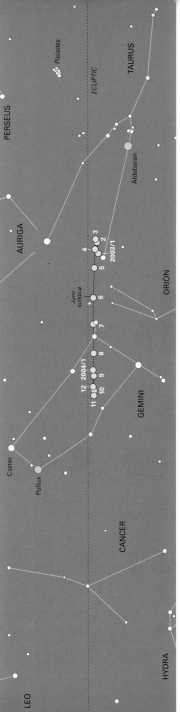

SATURN — 2003

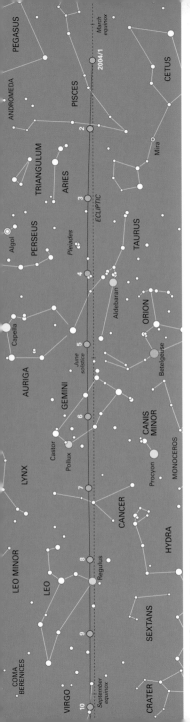

MARS — 2004

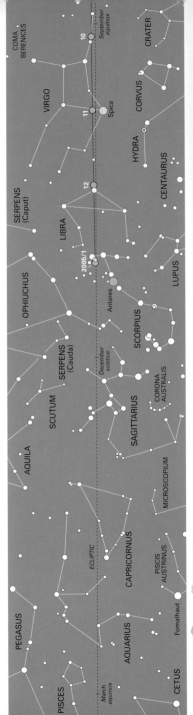

MARS — 2004

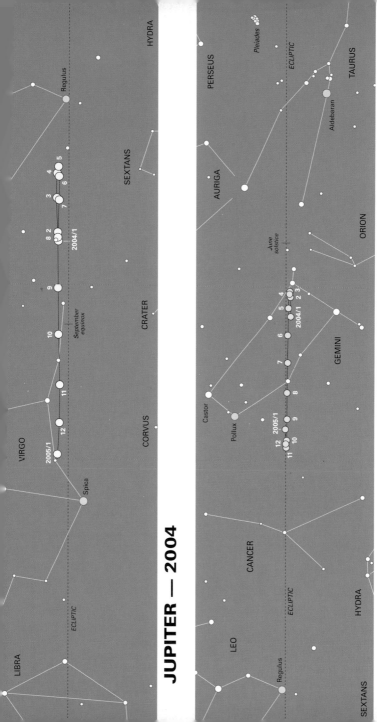

JUPITER — 2004

SATURN — 2004

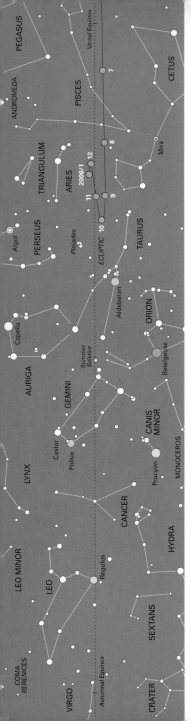

MARS — 2005

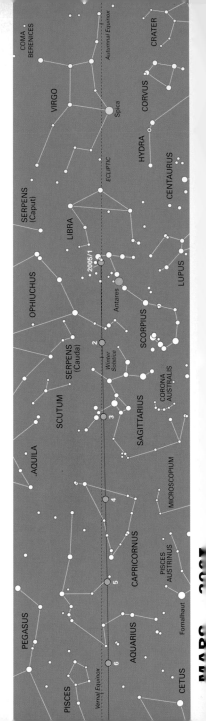

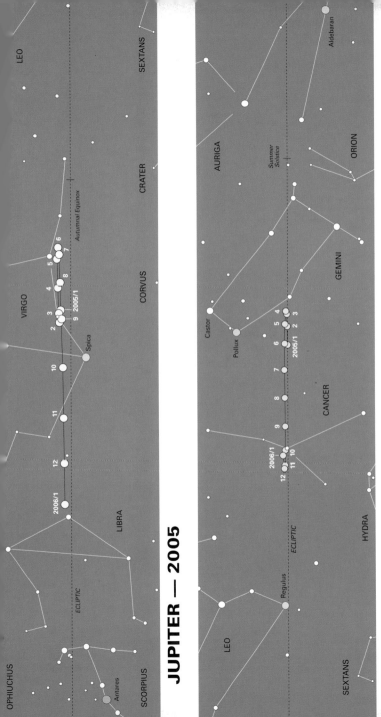

JUPITER — 2005

SATURN — 2005

Jupiter

Jupiter is the king of the planets, and the most fascinating of all to study with small instruments. Humble binoculars reveal the planet's cream-coloured disk and four main moons, called the Galilean satellites after the Italian scientist Galileo Galilei who discovered them in 1609. Some people with exceptionally acute vision can see the Galilean satellites with the naked eye as faint starlike points either side of the planet.

A small telescope brings into view some of the details of Jupiter's disk: dark belts of cloud parallel to the equator and an eye-shaped marking in the southern hemisphere known as the Great Red Spot, first observed in 1831 (a similar feature was seen as early as 1664). Careful study of these features reveals that Jupiter's rotation period varies with latitude, from 9 hours 50 minutes at the equator (the fastest rotation of any planet in the Solar System) to 9 hours 55 minutes at higher latitudes. In addition, the Great Red Spot drifts slightly in relation to its surroundings. These effects

Jupiter photographed by the Voyager 2 space probe in 1979. Light-coloured zones of cloud alternate with darker belts. In the southern hemisphere is the swirling Great Red Spot, with several white ovals farther south. (USGS)

demonstrate that the visible surface of Jupiter is not solid. We are looking at clouds, constantly seething and swirling, changing in colour and shape. Jupiter never appears the same twice. That is its attraction.

Jupiter is easy to see with the naked eye, reaching a maximum magnitude of −2.9 when at its closest to Earth, 590 million km away. Even when at its most distant from us it still manages to outshine every star except Sirius. Its brightness results from its highly reflective clouds and its imposing size – it is the largest planet in the Solar System. Examination of its outline gives added confirmation that it is not a solid planet: Jupiter has a bulging midriff. The diameter at its equator is nearly 143,000 km, but from pole to pole it measures 133,700 km. A line of 11 Earths would be needed to equal Jupiter's equatorial width. And even though Jupiter is composed largely of the lightest elements in the Universe, hydrogen and helium, its bulk is such that its mass is two and a half times as much as all the other planets put together.

Jupiter orbits the Sun every 11.9 years at an average distance of 778 million km, over five times the distance of the Earth, and is well placed for observation every 13 months. Positions of Jupiter for a five-year period are shown in the charts on pages 356–365.

Observers draw Jupiter on pre-printed blank outlines with an equatorial diameter of 64 mm and a polar diameter of 60 mm. After sketching in the main features, details can be added starting at the leading limb, where the planet's rotation of 6° every 10 minutes rapidly carries the features out of sight. Intensities can be estimated on a scale from 0 (the brightest) to 10 (black sky), as used in Europe, or in the reverse order (10 the brightest, 0 black sky), as used in the US. Most valuable of all are timings of the passage of various spots across the central meridian, the imaginary line running north–south down the centre of the planet's visible disk. Repeated timings will establish the rotation rate of spots, and reveal any drift in their longitude caused by winds in Jupiter's atmosphere. Even in this age of space probes and space telescopes, long-term series of such observations are useful.

Because Jupiter's cloud features are so impermanent and mobile, it is impossible to give more than a generalized description of the planet's appearance. Its disk is crossed by alternating bright and dark bands, termed zones and belts respectively. Frozen ammonia crystals form high, cold clouds in the bright zones, where gas ascends; the darker belts, where the gases descend, are lower and warmer (although 'warm' is only a relative term, for the temperature of the cloud tops is around −150°C). The colours of the belts can vary from yellow and brown to orange, red or even purple as a result of complex chemicals in the atmosphere of Jupiter, sulphur being one of the prominent colorants.

High-speed winds of up to 500 km per hour whip the edges of the zones and belts into turbulent eddies, giving them a scalloped appearance. The weather on Jupiter is unpredictable. Dark and light spots can suddenly erupt in the clouds, lasting for weeks or even decades before fading away. One of the main roles of amateur observers is to track these storms as they erupt and move around the planet.

Of all the markings on Jupiter, the most famous, and by far the most permanent, is the Great Red Spot. It certainly is great: 14,000 km wide and as much as 40,000 km long, enough to swallow three Earths. But it is not always red: most often it is pinkish, and sometimes it can fade to a colourless grey. Its colour is believed to be due either to red phosphorus or sulphur. By good fortune, the spot was particularly prominent when the Voyager 1 and 2 spacecraft reached the planet in 1979. Even now its nature is not fully understood, but it appears to be an upwardly spiralling column of gas similar to a hurricane on Earth, its top spreading out about 8 km above the surrounding cloud deck. Jupiter's other, smaller spots, including a series of long-lived white ovals, are believed to be similar swirling storm systems. As if to emphasize the storminess of the planet, the Voyagers and the more recent Galileo probe photographed massive flashes of lightning on Jupiter's night side, far larger than in any thunderstorm on Earth.

The key to Jupiter's meteorology is the fact that it gives off twice as much heat as it receives from the Sun. The planet was hot when it formed, and still retains some of that heat today. This internal store of heat drives the complex cloud systems of Jupiter, keeping the Great Red Spot and its smaller relatives alive for far longer than any storms can persist on Earth.

Interestingly, Jupiter has virtually the same chemical composition as the Sun: mostly hydrogen and helium. There is thought to be a rocky core about twice the size of the Earth at Jupiter's centre, but no space probe could ever land on it. Beneath the wispy high-altitude clouds of frozen ammonia are complex chemicals that give the dark belts their colour. Deeper still, temperatures are similar to those on Earth and clouds of water vapour condense. About 1000 km below the visible cloud tops, temperatures and pressures have increased to the point where hydrogen

Dark stains in Jupiter's clouds south of the Great Red Spot mark the sites where fragments of Comet Shoemaker–Levy 9 impacted the planet in July 1994. Photograph by the Hubble Space Telescope. (STScI/NASA)

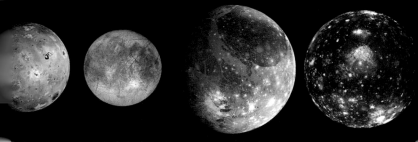

Family portrait of Jupiter's four largest satellites as seen by the Galileo space probe, to scale. From left: the sulphurous surface of volcanic Io; icy, cracked Europa; dark markings and bright impact craters on Ganymede; and heavily cratered Callisto, with the Valhalla impact basin at centre. (NASA)

is compressed into a liquid. The liquid hydrogen seas of Jupiter are about 20,000 km deep. Below, under the crushing pressure of 3 million Earth atmospheres, hydrogen is compressed into a superdense state with the properties of a metal; hence it is known as metallic hydrogen. Convection within the hot metallic hydrogen interior of Jupiter is thought to be responsible for the planet's intense magnetic field, 10 times stronger than the Earth's, extending out for 100 times Jupiter's radius into space. If the magnetic field around Jupiter were visible to the naked eye, it would appear over twice the size of the full Moon.

One of the most remarkable events in the history of planetary observation took place in 1994 when Comet Shoemaker–Levy 9, which had previously been captured into orbit around Jupiter and broken into more than 20 fragments, crashed into the planet, leaving dark markings in its clouds that could be seen through small telescopes. Over the ensuing months, the dark patches spread out into a band around Jupiter that took over a year to fade away. Crater chains on Jupiter's moons Callisto and Ganymede suggest that they, too, have been struck by fragments of disintegrated comets in the past.

Jupiter possesses a fascinating collection of satellites, like a mini Solar System. With the simplest optical aid the four largest, known as the Galilean satellites, can be seen performing a merry dance around Jupiter, changing position from night to night – sometimes out of sight behind the planet, sometimes transiting across its face and sometimes being eclipsed in its shadow.

The closest to Jupiter of the Galilean satellites is Io, 3630 km in diameter (slightly larger than our own Moon), orbiting every 42½ hours. Io is the most volcanically active body in the Solar System. The Voyager 1 probe in 1979 photographed eight volcanoes erupting simultaneously on it. Hundreds of other volcanic vents were visible, although not actually

Blowing its top – a volcano called Loki erupts at the limb of Jupiter's moon Io, sending clouds of sulphur 150 km into space, as photographed by Voyager 1 in 1979. The volcanoes of Io are named after fire gods. (USGS)

erupting. Those volcanoes erupt not just molten rock (lava) as on Earth but also sulphur which solidifies to form the garish red, orange and yellow of Io's surface.

Io is molten because it is caught in a gravitational tug of war between Jupiter and the other Galilean satellites; their opposing pulls release tidal energy that melts Io's interior. Io recycles its interior onto its surface, endlessly turning itself inside out. Some of the sulphur escapes and showers onto the innermost moon of Jupiter, Amalthea, giving it an orange coating. Amalthea is an irregularly shaped lump of rock only about 200 km in diameter, too faint to be seen in amateur telescopes.

Within the orbit of Amalthea the Voyager probes discovered a faint ring of dust, extending inwards to a mere 30,000 km above Jupiter's cloud tops. This tenuous ring of Jupiter is believed to result from the break-up of one or more tiny moons, two of which, Adrastea and Metis, were discovered by the Voyagers, orbiting at the ring's outer edge.

Moving outwards past Io we come to Europa, smallest of the Galilean satellites with a diameter of 3140 km. Europa is encased in a white shell of ice, veined with fine cracks probably produced by tidal forces. Underneath its fractured icy crust, Europa may have an ocean of liquid water.

Next in line from Jupiter is Ganymede, the largest and brightest of the satellites; what's more, at 5260 km in diameter it is the largest moon in the Solar System, larger even than the planet Mercury. Ganymede and the fourth of the Galilean satellites, Callisto, 4800 km in diameter, are both balls of rock and ice, rather like giant muddy snowballs. Callisto is saturated with impact craters, the largest 300 km in diameter and called Valhalla; it is similar to the large basins on the Moon and Mercury, surrounded by wave-like ridges. Ganymede is also cratered by impacts, but less heavily than Callisto, and exhibits a strange grooving on much of its surface, apparently formed by faulting and stressing of the moon's icy surface layers. Bright spots on Ganymede and Callisto mark places where recent impacts have exposed fresh ice.

The rest of Jupiter's 17 known moons are small and insignificant. Several of these – particularly the outer four, which move in highly elliptical, retrograde orbits – were probably once passing bodies that were captured by Jupiter's gravity.

Saturn

Bright rings girdle Saturn's equator, making it the most beautiful of the planets. Those distinctive rings can be spotted clearly in a small telescope; good binoculars, mounted steadily, will show the small outline of the planet elongated into an ellipse by the rings. Mounted binoculars should also pick out Saturn's largest moon, Titan, which orbits the planet every 16 days.

Saturn's rings can reflect more light than the body of the planet itself, so that at its best Saturn appears of magnitude −0.3, outshone only by Sirius and Canopus. Without the rings, Saturn would be no more than magnitude 0.7, less than half as bright. Strange to say, from time to time Saturn really can appear to be without rings. The reason is the tilt of the planet's axis. As Saturn orbits the Sun, the rings are sometimes tipped towards us, while at other times the rings are presented edge-on. So thin are the rings that when edge-on (which happens about every 15 years, as in 2009 and 2025) they disappear from view in even the largest telescopes on Earth. The brightness of Saturn therefore depends not only on its distance from Earth, but also on the aspect of the rings.

The ringed planet orbits the Sun every 29½ years at an average distance of 1430 million km, 9½ times farther than the Earth. Being so slow-moving, Saturn returns to opposition about two weeks later each year. Positions of Saturn for a five-year period are shown on pages 356–365.

Saturn photographed by Voyager 2 in 1981; note the Cassini Division and subtle detail in the rings. Saturn's moons Tethys, Dione and Rhea are visible as tiny dots beneath the planet, with the shadow of Tethys on the clouds. (USGS)

In many ways, Saturn is a smaller brother of Jupiter. Its equatorial diameter is 120,500 km, second only to that of Jupiter; its rotation period of 10¼ hours is second-fastest to that of Jupiter; and, like Jupiter, Saturn is composed mostly of hydrogen and helium. In one way, though, it is unique among the planets: its average density is less than that of water.

This remarkable fact comes about because the planet's mass is less than a third that of Jupiter, so its gravity is less and hence its central regions are not compressed as densely. There probably is a rocky central core, but the region surrounding the core, in which hydrogen is compressed into a liquid metallic form, extends out to only half the planet's radius, against three-quarters of the radius in the case of Jupiter. That is not sufficient to compensate for the low density of Saturn's outer layers, so the planet's overall density is a mere 70 per cent of that of water. Saturn's low density is apparent from its outline, which is even more squashed than that of Jupiter. Saturn's pole-to-pole diameter is 109,000 km, fully 10 per cent less than its equatorial diameter; with Jupiter the difference is 6 per cent.

Through a telescope Saturn appears as a tranquil, ochre-coloured disk, darker at the poles and with some dusky horizontal bands. When drawing Saturn, observers use a range of pre-printed blank forms to match the changing tilt of the rings. As with Jupiter, the main features are sketched in quickly and detail added before it rotates out of view. Observers can estimate the intensities of various features; in Europe, the scale runs from 1 (the brightest part of the rings) to 10 (black sky), whereas the US scale runs the opposite way, from 0 (black sky) to 8 (brightest part of the rings). Markings in the clouds can be timed as they cross the planet's central meridian, in the same way as with Jupiter.

Pleasing views of the planet can be obtained with small telescopes, but serious study of Saturn requires at least 200 mm aperture, for there are none of the swirling, multicoloured storm clouds that make Jupiter's globe so interesting in a small telescope, and no equivalent of the Great Red Spot. However, every 30 years or so a large white spot erupts in Saturn's northern hemisphere, when the planet's north pole is tilted at its maximum towards the Sun. The white spots are evidently storm clouds produced by solar heating. Such an outbreak occurred in 1990 and lasted a few months, with smaller spots appearing over the next few years.

White spots apart, this is not to say that Saturn lacks activity in its atmosphere; it is just that the cloud patterns are usually concealed by high-altitude haze. Cameras aboard the probes Voyager 1 and 2 which reached Saturn in 1980 and 1981 recorded low-contrast cloud swirls resembling those on Jupiter. The meteorology of both planets should be similar, for they both have internal sources of heat. Like Jupiter, Saturn radiates twice as much heat as it receives from the Sun, a legacy of its birth. However, since Saturn is farther from the Sun, its clouds are about 30°C colder than on Jupiter and form lower in the atmosphere. Tracking of cloud systems revealed that gales of up to 1800 km per hour blow on Saturn, over three times faster than on Jupiter.

Mimas, one of the smaller moons of Saturn, sports a huge crater with a central peak, formed by a major impact. The crater is called Herschel, after William Herschel, the astronomer who discovered Mimas in 1789. The rest of the moon's surface is saturated with smaller craters. This photograph was taken by Voyager 1 in 1980. (NASA)

Inevitably, the attention of the space probes focused on the glorious rings. As seen from Earth, the rings look like a continuous disk encircling the planet, but appearances are deceptive. The Dutchman Christiaan Huygens in 1655 was the first to realize that the rings are not solid, but consist of a swarm of tiny particles orbiting Saturn. The central part of the rings, known as Ring B, is the widest and brightest. It is separated from the outer, fainter Ring A by a 5000-km-wide gap known as Cassini's Division, visible with a 75-mm telescope. Extending inwards from Ring B towards the planet is the faintest ring of all, the transparent Ring C, also known as the crepe ring. A narrow gap called Encke's Division can be seen in Ring A, while observers looking through large telescopes under good conditions have reported apparent ripples within the rings, a sign that the density of ring material varies from place to place.

But even the best telescopic views did not prepare astronomers for the astounding wealth of detail that was revealed by space probes. Under the close scrutiny of the Voyagers' cameras, the rings broke up into thousands of narrow ringlets and gaps, like the ridges and grooves of a gramophone record. Some ringlets were not perfectly circular, but were elliptical in shape. Even Cassini's Division was not empty, but contained thread-like ringlets. A new outer ring, called the F Ring, appeared to consist of twisted strands like a rope.

Saturn's rings are extraordinarily thin in relation to their 270,000 km diameter. Voyager observations showed that the rings are no more than 100 m thick. On the same ratio of thickness to diameter, a gramophone record would be 5 km in diameter. The particles that make up the rings come in a wide range of sizes, from dust specks to lumps the size of a house or larger. Their composition is mostly frozen water, possibly mixed with dust, resembling loosely compacted snowballs. Saturn's rings could have originated in several ways. One is from material that was prevented from forming into a moon by the overpowering force of Saturn's gravity. Alternatively, they may be the remains of a former moon that strayed too close to the planet and broke up, or a moon that was broken up by the impact of a comet. In fact, the rings may be replenished from time to time by icy material from impacting comets. In some places, fine dust overlays the rings, apparently supported by electromagnetic forces in Saturn's magnetosphere, producing darker features known as spokes. Spoke-like features have been reported by ground-based observers, but it took the Voyager pictures to establish their reality.

The Voyagers discovered several moons of Saturn too small to be seen from Earth, bringing the total of known Saturnian moons to 18, with others suspected. One of these tiny moons, Pan, actually orbits in Encke's Division in Ring A. Another, Atlas, patrols the outer edge of Ring A. Two of the moons, Prometheus and Pandora, orbit either side of the F Ring, shepherding its particles.

Slightly farther out from Saturn, Janus and Epimetheus move along the same orbit, which initially confused astronomers who first spotted them from Earth in 1966. Orbit-sharing is common around Saturn. Tethys, a satellite visible from Earth, has two small siblings, Telesto and Calypso, moving on the same path. Dione, visible from Earth, shares its orbit with tiny Helene, another Voyager discovery.

The gravitational effect of some of these satellites helps maintain the gaps in Saturn's rings. For instance, the gravitational tug of Mimas pulls particles out of the Cassini Division. Mimas itself, like most of the moons of Saturn, is a dirty snowball of frozen water and rock. It sports a remarkable crater 135 km wide, larger than Copernicus on the Moon, fully a third of its own 390-km diameter. The impact that caused this grotesque feature, called Herschel, must have nearly shattered Mimas.

Iapetus, Saturn's outermost moon but one, is another strange body: one side is five times darker than the other. This harlequin effect is probably caused by dust knocked off Saturn's outermost moon, Phoebe, which is the darkest of Saturn's moons. The dark dust from Phoebe falls inwards to be swept up by Iapetus, coating its leading face, while bright ice on its trailing side remains exposed.

Saturn's largest moon, Titan, 5150 km in diameter, deserves to be considered as a planet in its own right. Bigger than Mercury (but slightly smaller than Jupiter's main moon Ganymede), it is the only moon to have a substantial atmosphere – in fact, the atmospheric pressure on the surface of Titan is 50 per cent greater than at sea level on Earth. Titan's atmosphere is 90 per cent nitrogen, with methane making up most of the rest. Clouds of orange smog top the atmosphere, obscuring the surface from the prying eyes of space probes. Despite that thick atmosphere, Titan's surface temperature is very low, about −180°C. Perhaps a rain of liquid methane falls from the rust-coloured sky into seas of methane on Titan's surface.

A space mission called Cassini is due to go into orbit around Saturn in 2004; it will drop a smaller probe, called Huygens, to land on the surface of Titan.

Uranus, Neptune and Pluto

Uranus should in theory be visible to the naked eye: at its brightest it is of magnitude 5.5. But it is so insignificant that it was never noticed by ancient astronomers, for whom the Solar System stopped at Saturn. Uranus was not discovered until March 13, 1781, when William Herschel spotted it through his telescope during a systematic survey of the skies. Once its position is known, Uranus can easily be followed in binoculars as it moves against the background stars. There is a certain fascination in seeing the blue-green disk of this distant planet through a telescope, but even with large apertures Uranus displays no detail.

Uranus is one of the four 'gas giants' of the outer Solar System, the others being Jupiter, Saturn and Neptune. Uranus has an equatorial diameter of 51,100 km, less than half that of Saturn but four times larger than the Earth. It lies 2900 million km from the Sun, 19 times farther than the Earth. Seasons would pass slowly on Uranus, for it takes 84 years to complete one orbit, but the seasons would be extreme, for Uranus

The green, featureless disk of Uranus, with its encircling rings, photographed by Voyager 2 in 1986. The brightest ring is the outermost one, the Epsilon ring. Background stars and moons can also be seen. (Erich Karkoschka/NASA)

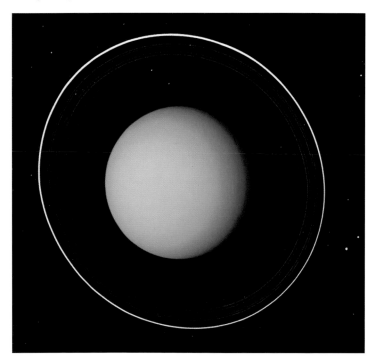

appears to have been knocked over onto its side: the planet's axial tilt is 98°, meaning that its axis of rotation lies almost in the plane of its orbit. Every 42 years, therefore, one of the poles of Uranus is pointing towards the Sun, while its opposite pole is in darkness for decades. In between times its equatorial region faces sunwards. During each 84-year orbit of Uranus the Sun can appear overhead at every latitude, something that never happens on any other planet. No one knows why Uranus should be lying on its side in this unique way; perhaps it suffered a collision with another large body long ago.

Uranus is celebrated for another reason: it was the second planet discovered to have rings. In 1977, astronomers watched as Uranus passed in front of a star. Unexpectedly, they noticed that the star winked on and off a number of times before and after it was obscured by the disk of Uranus, from which they deduced that Uranus was encircled by nine faint rings. The existence of these rings was confirmed, and their number increased to eleven, by the Voyager 2 space probe which flew past the planet in January 1986. Since then the rings have been photographed from Earth at infrared wavelengths.

The rings are narrow, mostly only a few kilometres wide, separated by gaps many times greater than their own width. They lie between 13,000 and 26,000 km above the cloud tops of Uranus, and are composed of dust and other debris from tiny moonlets that orbit among them. In addition to the rings, Uranus has five moons visible from Earth: Miranda, Ariel, Umbriel, Titania and Oberon, ranging in size from 500 to 1500 km in diameter. Eleven smaller moons were discovered by the Voyager 2 space probe and other, more distant moons have since been discovered from Earth, giving Uranus a larger family than any other planet. The moons, like the rings, move in nearly circular orbits around Uranus's crazily tilted equator, with the exception of the outer moons discovered from Earth, which have highly elliptical and inclined orbits, suggesting they have been captured since the planet's formation.

Disappointingly, even to the close-up eye of Voyager, the gaseous surface of the planet itself was bland, with scarcely any cloud features. In structure, Uranus is believed to have a rocky core surrounded by a mantle of ice, topped by an atmosphere of hydrogen and helium mixed with a fair proportion of methane, which gives the planet its greenish colour.

Neptune, the next-farthest planet from the Sun, is in some ways a twin of Uranus. It is slightly smaller – diameter 49,500 km – and shows the same featureless disk through amateur telescopes, with a stronger blue-green tinge because there is more methane in the atmosphere above its clouds. Magnitude 7.8 at brightest, it is completely invisible to the naked eye, but can be followed in binoculars when one knows where to look.

Unlike the accidental discovery of Uranus, Neptune's existence was predicted in advance. Astronomers found that Uranus was not keeping to its expected course, and some suspected that it was being pulled by the gravity of an as-yet-unseen planet. In France, mathematician Urbain Le

The icy blue of Neptune's clouds is disturbed by the Great Dark Spot, an anticyclone the size of the Earth, fringed with white methane cirrus. This photograph was taken by Voyager 2 in August 1989. (NASA)

Verrier calculated the new planet's position in 1846, beating the Englishman John Couch Adams who had been working on the same problem. On September 23 that year, astronomers at the Berlin Observatory found Neptune close to Le Verrier's predicted position.

Neptune crawls around the Sun in 165 years at an average distance of 4500 million km, 30 times farther away than the Earth. Being so distant from the Sun, it is a very cold and dark planet. Unlike bland Uranus, Neptune has been found to possess markings in its clouds similar to those of Jupiter. A cloud feature called the Great Dark Spot, similar in nature to Jupiter's Great Red Spot, was revealed by the cameras of Voyager 2 when it reached Neptune in August 1989. This spot gradually drifted towards the planet's equator and had disappeared by the time the Hubble Space Telescope looked at Neptune in 1994, although similar spots have appeared since.

But the planet's main distinction is its two curious outer satellites, Triton and Nereid. Triton, the largest of Neptune's eight known moons, is in a retrograde (east to west) orbit at a distance of 355,000 km. Tidal forces from Neptune mean that Triton's orbit is gradually shrinking, so that the moon will spiral closer to the planet until it is broken up in the distant future. A shattered Triton will form a far more substantial set of rings around Neptune than the thin, faint ones that currently encircle the planet, for Triton is 2700 km in diameter, over three-quarters the size of our own Moon. The outermost satellite of Neptune, Nereid, is much smaller, only 340 km across, but it has an exceptionally elliptical orbit that carries it between 1.4 and 9.7 million km from Neptune. Something seems to have disturbed the satellite system of Neptune.

Pluto is the odd man out among the planets – indeed, many astronomers doubt whether it deserves to be called a planet at all. It is by far the smallest planet. Pluto's diameter of under 2400 km is not only less than that of our own Moon but smaller even than Neptune's largest satellite, Triton; in fact, the icy surface of Triton, photographed by Voyager 2, probably looks much like that of Pluto. Triton and Pluto may both have wandered the outer reaches of the Solar System before Triton was captured by Neptune; the strange orbits of Triton and Nereid around Neptune could therefore be the consequences of that disruptive event. Certainly, Pluto has the oddest orbit of any planet: it crosses the orbit of Neptune, so that at times Neptune temporarily becomes the outermost planet of the Solar System, as was the case for 20 years between February 1979 and February 1999. Pluto's average distance from the Sun is 5900 million km.

Pluto is not alone in its 248-year-long orbit around the Sun. In 1978 astronomers discovered that it has a moon, since named Charon, half Pluto's diameter. Charon orbits its parent every 6.4 days, the same time that the planet takes to spin on its axis. Charon therefore hangs over one spot on the surface of Pluto, visible permanently from one hemisphere of the planet but invisible from the other.

Being so faint and insignificant, Pluto was not discovered until 1930. For decades prior to that, various astronomers had tried to predict where a planet beyond Neptune might lie, but without success. In the end, Pluto was found by Clyde Tombaugh at the Lowell Observatory in Arizona as the result of a deliberate round-the-sky photographic search for new planets. Tombaugh's search failed to detect any sign of a tenth planet lurking beyond Pluto, as have all subsequent searches. But what has been found is a swarm of icy bodies, like large comet nuclei, called the Kuiper Belt (sometimes, the Edgeworth–Kuiper Belt), which is an inner extension of the larger Oort Cloud of comets (see next section). Pluto can be thought of as the largest member of the Kuiper Belt.

Pluto's moon Charon is shown above and below the planet in these two images as it moves along its orbit every 6.4 days. The orbit is currently aligned almost edge-on to our line of sight. (Gemini Observatory)

Comets and Meteors

Comets are insubstantial bodies, loosely knit assemblages of frozen gas and dust that loop around the Sun on highly elongated orbits. They return to the inner Solar System at intervals ranging from a few years to many thousands of years, becoming visible to us on Earth as ghostly, glowing apparitions for a few weeks or months before receding back into obscurity.

When far from the Sun, a comet shines only by reflecting sunlight. At that stage it is small – usually no more than a few kilometres across – and faint. Approaching the Sun, the comet warms up, turning its icy surface into gas. Under the influence of the Sun's radiance the gases of the comet begin to fluoresce, in similar fashion to the gas in a neon tube, thereby considerably increasing the comet's brightness. Gas and dust released from the warming comet produce a halo or *coma* 100,000 km or so in diameter. At the centre of the coma is the nucleus, the only solid part of the comet, consisting of a 'dirty snowball' of ice, dust and perhaps some rock. In a large comet the nucleus may be a few tens of kilometres across, but most are only a kilometre or so wide. Well over a thousand million comet nuclei would be needed to equal the mass of the Earth.

Not all comets develop tails, but many do. One part of the tail consists of gas blown away from the comet's head by the solar wind of atomic particles streaming from the Sun. The other part of the tail is made up of dust particles liberated from the head by the evaporating gases. Comet tails always point away from the Sun. A comet's tail can extend for

Halley's Comet pursues an elliptical path around the Sun, steeply inclined to the orbits of the planets. It travels from between the orbits of Mercury and Venus out to beyond Neptune and back again every 76 years. (Wil Tirion)

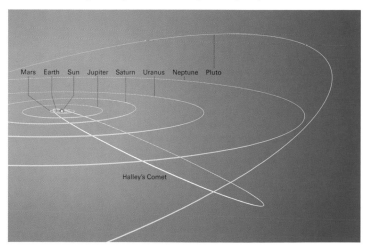

Mars Earth Sun Jupiter Saturn Uranus Neptune Pluto

Halley's Comet

Comet Hale–Bopp seen over Stonehenge in April 1997. Cassiopeia is at upper right and the Pleiades at upper left. The comet previously appeared some 4200 years ago, about the time that Stonehenge was built. (Paul Sutherland)

100 million km or so, further than the distance from the Earth to the Sun, as did the gas and dust tails of of Comet Hale–Bopp in 1997. Yet, for all its glorious appearance, the tail is less dense than a laboratory vacuum – stars shine through it undimmed. The tail of a comet gives it the appearance of speeding across the sky, but actually its movement against the stars is barely noticeable during the course of a night.

Two dozen or more comets may be visible through a telescope each year, and several can come within the range of binoculars, but only occasionally does one become bright enough to be prominent to the naked eye. Each year's batch of comets is a mixture of known specimens returning to the Sun and completely new discoveries. About a thousand comets have well-known orbits, and more are being discovered all the time. Dedicated amateur astronomers sweep the skies to discover new comets; each new comet is given its discoverer's name.

Comets with the longest orbital periods – more than a few centuries – are thought to come from an unseen swarm of thousands of millions of comets, the Oort Cloud, that envelops the Solar System at its dim outer edges, about a light year from the Sun. The gravitational influence of passing stars nudges comets from this cloud into new orbits that bring them towards the Sun. An inner comet cloud, termed the Kuiper Belt, lies just beyond the orbit of Pluto. Most so-called *periodic comets*, which orbit the Sun in under 200 years, are thought to come from the Kuiper Belt rather than the Oort Cloud.

The comet of shortest known period is Encke's Comet, which orbits the Sun every 3.3 years. It is so old that it has lost most of its gas and dust, and is too faint to see with the naked eye. Most famous of all is Halley's Comet, named after the English astronomer Edmond Halley, who calculated its orbit in 1705. Halley's Comet returns every 76 years or so, and last appeared in 1985–6. Its orbit takes it from 88 million km from the Sun (between the orbits of Mercury and Venus) out to 5300 million km (beyond Neptune).

Dust lost from a comet disperses into space. The Earth and other planets are continually sweeping up cometary dust. When a particle of cometary dust comes whizzing into the atmosphere, it burns up by friction at a height of about 100 km, producing a streak of light known as a shooting star or meteor. The whole event is over quite literally in a flash, usually lasting less than a second. On any clear night, a few meteors are visible each hour as particles of dust dash at random to their deaths in the atmosphere.

Such random meteors are termed *sporadic*. Occasionally, though, the Earth crosses the orbit of a comet and encounters a dense swarm of dust. This gives rise to a so-called meteor shower, in which meteors may be visible coming from one direction in the sky at the rate of dozens per hour. The area of sky from which the meteors seem to come is known as the *radiant*. However, this is not the best point to look at, since the meteors are approaching head-on from this direction and so their trails are shortest; the longest trails will be seen up to 90° from the radiant.

Members of a meteor shower appear to diverge from a small area of sky termed the radiant. This diagram shows the radiant of the Lyrid meteors, which lies in the constellation Lyra, near the bright star Vega. (Wil Tirion)

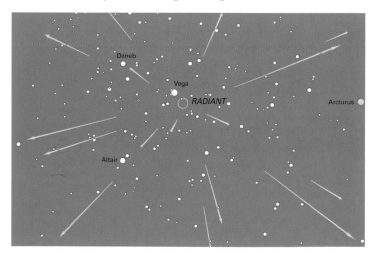

A meteor shower is named for the constellation in which the radiant lies. For instance, the Perseids, an abundant shower of bright meteors which the Earth encounters each August, seem to radiate from Perseus; the Geminids from Gemini; and so on. One historical oddity concerns the Quadrantids, which come from an area in Boötes that was once part of the now-defunct constellation of Quadrans Muralis, the wall quadrant.

The strength of a meteor shower is measured by its *zenithal hourly rate* (ZHR), which is the number of meteors that an individual observer would see if the radiant were directly overhead in a dark sky. Since the radiant is seldom, if ever, at the zenith, the number of meteors actually seen per hour will be less than the theoretical ZHR. In addition, bright moonlight and sky glow from artificial lights will wash out the fainter meteors, again reducing the observed ZHR.

Amateur astronomers make valuable observations of meteor showers, counting the number of meteors visible with the naked eye and estimating their brightness. Comfort is essential for a meteor watch, which may last several hours: wrap up warm and recline in a deckchair or on a sunbed.

Typical meteors are of magnitude 2 or 3, but some are brighter than the brightest stars and the occasional spectacular example, termed a *fireball*, can cast shadows. Some meteors seem to split up as they fall, and some leave trains of glowing gas which take several seconds to fade. The table shows the main meteor showers visible each year. The ZHR is only a guide, and can vary considerably from year to year.

One extreme case is the Leonids, normally a modest shower, which bursts into life at 33-year intervals when its parent comet, Tempel–Tuttle, returns to perihelion. An intense storm of Leonids was seen over the United States in 1966 with as many as 100,000 in an hour, like celestial snowflakes, and astronomers in Europe saw rates of about 2000 an hour in 1999.

MAJOR METEOR SHOWERS

Shower	Limits of activity	Date of maximum	Maximum rate (ZHR)
Quadrantids	January 1–6	January 3–4	100
Lyrids	April 19–25	April 21–22	10
Eta Aquarids	May 1–10	May 5	35
Delta Aquarids	July 15–August 15	July 28–29	20
Perseids	July 23–August 20	August 12–13	80
Orionids	October 16–27	October 20–22	25
Taurids	October 20–November 30	November 4	10
Leonids	November 15–20	November 17–18	10
Geminids	December 7–15	December 13–14	100

Asteroids and Meteorites

Between Mars and Jupiter orbits a belt of rubble known as the asteroids or minor planets. The asteroids were unknown until 1801, when the Italian astronomer Giuseppe Piazzi discovered Ceres, the largest of them, although astronomers had previously speculated that an unknown planet might exist in the suspiciously wide gap that separates Mars and Jupiter.

Two centuries after Piazzi's initial discovery, well over 10,000 asteroids had been found and that number was expected to double in five years, so fast are the discoveries now coming. However, even if all the asteroids were rolled together, they would make a body less than half the size of the Moon. The asteroids are not, as has been suggested, the remnants of a former planet that was disrupted; they are merely the leftovers from the formation of the other planets.

Ceres itself is 940 km in diameter and takes 4.6 years to orbit the Sun. Although Ceres is the largest asteroid, it is made of dark rock and so is not the brightest. That honour goes to Vesta, 580 km in diameter, which is composed of paler rock; at times it is just visible to the naked eye. Vesta is the second-largest asteroid; the third-largest is Pallas, diameter 540 km. These asteroids, and several others, are bright enough to be followed in binoculars.

Our first close-up view of an asteroid came in 1991 when the space probe Galileo, on its way to Jupiter, flew past the asteroid Gaspra. Galileo's photographs showed that Gaspra is an irregularly shaped, rocky body about 17 km long, and pitted with impact craters. Gaspra resembles the moons of Mars, Phobos and Deimos, thus strengthening the suspicion that those moons are captured asteroids. (For a photograph, see page 354.) Two years later, Galileo flew past asteroid Ida, 55 km long, discovering that it had a tiny moon, now named Dactyl. In February 2000 the Near Earth Asteroid Rendezvous (NEAR) spacecraft went into orbit around the highly elongated asteroid Eros, 33 km long and 13 km wide, to study its cratered surface in detail.

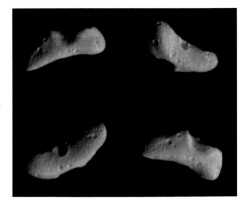

Four images of the asteroid Eros rotating, photographed from above its north pole in February 2000 by the Near Earth Asteroid Rendezvous space probe. As with most objects in the Solar System, the direction of rotation is from west to east, i.e. anticlockwise; the sequence of frames is top left, top right, bottom left, bottom right. (NASA)

Ninety-five per cent of asteroids orbit in the main belt between Mars and Jupiter, but there are some notable exceptions. One group of asteroids, known as the Trojans, moves along the same orbit as Jupiter. Most significant from our point of view are the near-Earth asteroids, which have orbits that bring them close to or even across the path of the Earth. Some of these objects may be the nuclei of extinct comets. Asteroids of this type must have hit the Earth in the past, and others may do so in future, with devastating effect. Searching for, and keeping track of, such potentially hazardous asteroids (PHAs), as they are called, is now a major international project.

Objects that do reach the surface of the Earth are termed meteorites. Most meteorites are believed to be chips off asteroids, but a few rare specimens have compositions that show they come from the surfaces of the Moon and Mars.

Over 10,000 meteorites are estimated to land on Earth each year, but only about a dozen are actually picked up; the rest fall in uninhabited areas or into the sea. Most meteorites observed to fall are of the stony kind, and if not picked up immediately they soon weather away – at least, they do under normal conditions. But in Antarctica, scientists have found large numbers of ancient meteorites, including many rare types, preserved in pristine condition by the natural deep-freeze of the ice cap. Another rich source is the deserts of Earth, where the arid conditions preserve the meteorites.

Stony meteorites are usually termed *chondrites*, because they contain mineral-rich blobs known as chondrules. The small percentage of stones without chondrules are called *achondrites*. The most interesting class of stony meteorite are the carbonaceous chondrites; these have a high carbon content, and are believed to be among the most primitive rocks in existence, virtually unchanged since the formation of the Solar System. Some carbonaceous chondrites may be fragments of comets.

A small intermediate group of meteorites is the stony-irons, which consist of about 50 per cent iron and 50 per cent rock. But the other main group of meteorites apart from the chondrites is the irons. These contain 90 per cent iron and about 10 per cent nickel. The largest known meteorite is iron, and weighs about 60 tonnes. It must have hit the Earth at a relatively slow speed, for it did not dig a crater and did not break up. It lies where it fell in ancient times, near Grootfontein in Namibia. By contrast, the largest stony meteorite weighs 1.7 tonnes, and is one of a group of meteorites that fell near the city of Jilin, China, in 1976.

One particularly large meteorite, estimated to have weighed a quarter of a million tonnes, crashed into the Arizona desert about 25,000 years ago, digging out the now-famous Meteor Crater, 1.2 km in diameter. (Strictly, it should be called Meteorite Crater, but the old name is too well entrenched to be changed now.) Most of the meteorite was destroyed on impact, but enough fragments remain scattered around the crater to show that the meteorite was made of iron.

Astronomical Instruments and Observing

Binoculars and telescopes serve two main purposes: they collect more light than the human eye, and they magnify objects. In astronomy, the first of these two aspects is paramount. Astronomers seek large telescopes not because they magnify more, but because their increased light-gathering power brings fainter objects into view and allows finer details to be distinguished.

Telescopes come in two main types: refractors, which use a main lens (known as the object glass or objective) to collect light, and reflectors, which collect light with a mirror. Astronomical telescopes have interchangeable eyepieces to give different magnifications. Binoculars are a modified form of refractor in which the light path is folded by prisms to make them more compact. Telescopes that combine lenses and mirrors are becoming increasingly popular; these are known as catadioptric systems, and are best thought of as a modified form of reflector.

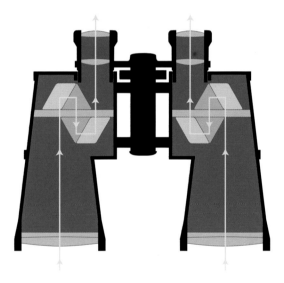

In binoculars, the path of the light is folded by prisms. (Wil Tirion)

The first optical equipment that most astronomers own is a pair of binoculars. Indeed, they are virtually indispensable for any observer, even those who own a substantial telescope, for a binocular sweep can pick up objects for the telescope to home in on. Binoculars bear markings such as the following: 8×30, 8×40, 7×50, 10×50. In each case, the first figure gives the magnification, and the second figure is the aperture (in millimetres) of the front lenses.

Any of the just-mentioned combinations of magnification and aperture are suitable for astronomy. Binoculars with powers greater than about ×10 are difficult to hold steady, for the image jumps around with each tremble of the hand; instead, they need to be mounted on a stand. Remember also that the higher the magnifying power for a given aperture, the fainter the image and the narrower the field of view. Binoculars of modest power give breathtaking wide-angle views of the heavens that telescopes cannot match.

Binoculars have the advantage that they are relatively inexpensive. A small telescope – one with an aperture of 50–60 mm (2–2.4 inches) – can cost several times more than a pair of binoculars, with little gain in light grasp. What telescopes do offer is higher magnification and a tripod mounting to support the instrument. Unfortunately, in many mass-produced telescopes the mounting is not as steady as might be wished, and the image frequently vibrates for several seconds each time the telescope is moved. As a selling point, some small telescopes offer eyepieces with impressively high powers of over ×200. But such high magnifications applied to a small telescope produce images so faint that little can be seen, and are therefore a complete waste. As a good general rule, the maximum usable magnification on a telescope is ×20 for each 10 mm of aperture (×50 per inch). These pitfalls aside, there are many serviceable small telescopes available which will bring into view a wide selection of the celestial sights mentioned in this book.

For more ambitious observers, a refractor of 75 mm (3 inches) aperture is the minimum required to begin serious work. Telescopes greater than 75 mm in aperture are usually reflectors, because size for size they are

Below: The path of light rays through a refracting telescope.
Bottom: A reflecting telescope of Newtonian design, the most common type used by amateurs. D is the telescope's aperture, F the focal length. (Wil Tirion)

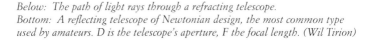

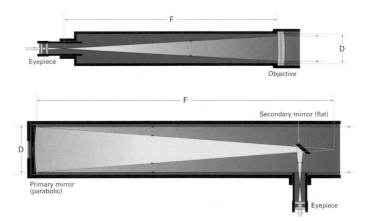

cheaper to make than refractors. Popular sizes of reflectors are 150 mm (6 inch) and 220 mm (8½ inch) aperture, which should bring into view virtually all the targets for observation listed in this book. As for cost, remember that a telescope is a precision optical instrument, and so you must expect to pay at least as much as you would for any similar instrument such as a good camera.

Reflectors used by amateurs are usually built to the design invented in 1668 by Isaac Newton. In the Newtonian reflector, light collected by the concave main mirror is bounced back up the tube to a smaller secondary mirror, which diverts the light into an eyepiece at the side of the tube. Inevitably, the secondary mirror blocks some of the incoming light from the main mirror, but this shadowing effect of the secondary is not significant and does not adversely affect the image. Another reflector design is the Cassegrain, in which the light is reflected back from the secondary through a hole bored in the main mirror. Large professional telescopes frequently use the Cassegrain design.

In *catadioptric* systems, which combine lenses and mirrors, the incoming light passes through a thin glass plate at the front of the telescope before falling onto the main mirror and being reflected as in a Cassegrain. The advantage of this design is that the overall length of the telescope can be made very much shorter than in a conventional reflector. The consequent saving in weight and space, allied with increased portability, compensates for the higher cost of a catadioptric telescope.

A telescope requires a mounting on which it can be supported and swivelled to point at various places in the sky. The quality of the mounting is just as important as the quality of the optics, for even the best telescope cannot be expected to show much if it is shaking all the time, or if you have difficulty steering it to track objects as the Earth rotates. Sturdiness and smoothness of movement are all-important in a telescope mounting.

The simplest form of mounting is the *altazimuth*. This has two axes, one horizontal and one vertical, which allow the telescope to be pivoted up and down (in altitude) and to pan from side to side (in azimuth). In an altazimuth mount, the telescope is usually held in a fork on top of a tripod. The very smallest refractors have tripods so short that they have to be placed on a table. Such instruments are little more than toys and they are quite unsuitable for astronomy, where it must be remembered that the observer is looking upwards most of the time. A number of refractors get over this problem by providing prisms known as star diagonals to fit at the eyepiece end. These bend the light so that the observer can look down into the eyepiece, which is certainly more convenient. For Newtonian reflectors, tall tripods are not necessary since the eyepiece is near the top of the tube, and so is normally quite accessible from a standing position.

Convenient additions to an altazimuth mount are small knobs, known as slow-motion controls, which can be turned to move the telescope

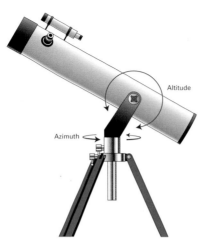

An altazimuth mounting, the simplest form of support for a telescope. The telescope is held in a cradle which allows it to rock up and down (in altitude) and swivel from side to side (in azimuth). (Wil Tirion)

slightly in each axis. This is useful both for centring an object in the field of view, and for tracking it as the Earth spins. When high powers are in use, the Earth's rotation can carry an object out of the telescope's field of view in a remarkably short time.

A popular version of the altazimuth for large Newtonian reflectors is the Dobsonian mount, named after the American amateur John Dobson who devised it. In a Dobsonian, the telescope tube is made of lightweight material so that its centre of gravity lies near the mirror end. The tube is supported at its lower (mirror) end by a wooden box with a Formica base that swings around in azimuth on pads made of Teflon (the slippery plastic material used in non-stick frying pans). Formica glides smoothly on Teflon, making a simple yet steady bearing. The tube pivots up and down in altitude on a shaft supported by more Teflon pads. Despite its deceptive simplicity, the Dobsonian has proved a very effective mounting. Cheapness and portability have ensured its widespread adoption.

Best of all for tracking objects is an *equatorial* mount, which has a main axis that is aligned parallel to the Earth's axis. This main axis is known as the polar axis because it points to the celestial pole (either north or south, depending which hemisphere the observer is in); it governs the telescope's movement in right ascension. The telescope pivots in declination on a second axis at right angles to the first, called (logically enough) the declination axis. Equatorial mounts are usually fitted with a small motor that slowly turns the equatorial axis at exactly the same speed as the

Earth rotates. Once the telescope is pointed at a celestial object and the drive motor is running, the object will remain fixed firmly in the field of view for as long as the observer wishes. Experience soon demonstrates the desirability of a stationary image if you are trying to split a faint double star or make a drawing of a planet.

To help with locating objects, a smaller sighting telescope called a *finder* is usually mounted on the side of the main tube. A finder is commonly a refractor of small aperture and low power, with a wide field of view; cross-wires in its eyepiece, or sometimes a concentric series of faintly illuminated rings, ensure the target is accurately centred.

A few words need to be said about what you can expect to see with telescopes of different sizes, and why. Anyone looking through an astronomical telescope for the first time is usually surprised to find that the image is upside down. There is a simple practical reason for this: to turn the image the right way up, an extra lens would have to be inserted in the eyepiece. Every time light passes through a lens (or is reflected off a mirror), some light is lost. Astronomical objects are usually so faint that any loss of light is undesirable, and in any case the additional lens is unnecessary. For astronomical purposes it does not really matter which way up the image appears, so the extra lens is left out and the image remains inverted. Some telescopes come equipped with so-called terrestrial eyepieces which turn the image the right way up for everyday viewing.

When you make an observation you should record the date, the time, the instrument, the viewing conditions and the magnification. Unlike

In an equatorial mounting, one axis (termed the polar axis) is set up so that it is parallel to the Earth's axis and hence points towards the celestial pole; the other axis (the declination axis) is at right angles to it. (Wil Tirion)

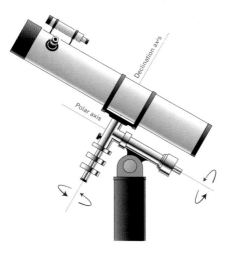

The 4.2-m William Herschel reflector on La Palma in the Canary Islands is one of the world's most powerful telescopes. In this photograph, the telescope can be seen inside its dome. During the exposure the dome was turned, with its slit open, to give a transparent effect. (Royal Greenwich Observatory)

binoculars and spyglass telescopes, which have fixed eyepieces, the eyepieces in astronomical telescopes are interchangeable, offering a range of magnifications to suit the object under study. For instance, a star cluster or galaxy will require low magnification; planets are best viewed with medium magnification; and to split a close double star you will need your highest powers.

The magnification of an eyepiece depends on both its own focal length and the focal length of the telescope. To work out the magnification of an eyepiece in a particular telescope requires a little straightforward arithmetic. You simply divide the focal length of the telescope by the focal length of the eyepiece. The result tells you the power of that eyepiece. It is easy to see that, the shorter the focal length, the higher the magnification. Eyepieces have their focal length marked on them. An additional lens, called a Barlow lens, can be used to increase the magnification of any eyepiece (usually doubling it), making available a wider range of powers from a given set of eyepieces.

Manufacturers often describe their telescopes in terms of focal ratio, such as $f/6$ or $f/8$. The focal ratio is the focal length of the lens or mirror divided by its aperture. If you don't know the focal length of your tele-scope you can easily find out by multiplying its aperture by the focal ratio stated by the manufacturers. For instance, a 100-mm telescope with a focal ratio of $f/6$ has a focal length of 600 mm; if its focal ratio is $f/8$, its focal length is longer, at 800 mm. For a 150-mm telescope of $f/6$ or $f/8$, the focal length would be 900 mm or 1200 mm respectively.

Now, assume that you have an eyepiece of focal length 20 mm. In a telescope of 600 mm focal length it will give a magnification of 600 divided by 20, which is ×30. That is quite a low power for astronomical purposes. In a telescope of 1200 mm focal length, the same eyepiece will give a magnification of ×60. An eyepiece of half the focal length, 10 mm, will give twice the magnification. Note that the aperture of the telescope is irrelevant in these calculations; the focal length is the sole figure governing magnification.

Where the telescope's aperture becomes crucial is in the matters of light grasp and resolution of detail. All other things being equal, a larger aperture shows fainter stars and finer detail than a smaller aperture, but exactly how faint the stars are and how fine the detail is depends on atmospheric conditions, optical quality and the observer's eyesight. Prac-tical experience shows that the faintest stars likely to be visible through amateur telescopes of various apertures are as follows:

TELESCOPIC LIMITING MAGNITUDES

Aperture		Limiting magnitude
50 mm	(2 inches)	11.2
60 mm	(2.4 inches)	11.6
75 mm	(3 inches)	12.1
100 mm	(4 inches)	12.7
150 mm	(6 inches)	13.6
220 mm	(8.5 inches)	14.4

Incidentally, your eyes need time to become accustomed to the dark before you can hope to see the faintest objects; when you go out at night from a bright room allow at least 10 minutes for your eyes to become dark adapted. One useful trick when trying to glimpse faint objects is to use *averted vision* – that is, to look to the side of the object under study so that its light falls on the outer, more sensitive part of the retina.

The resolution, or resolving power, of a telescope is expressed in seconds of arc (″). One second of arc is a very small quantity, equivalent to the size a coin would appear when several kilometres away. A tele-scope's resolution governs the detail that one can see on the Moon or planets, and the closeness of double stars that can be separated. The theoretical limits of resolution for telescopes of various apertures are

TELESCOPIC LIMITS OF RESOLUTION

Aperture		Limit of resolution (seconds of arc)
50 mm	(2 inches)	2.3
60 mm	(2.4 inches)	1.9
75 mm	(3 inches)	1.5
100 mm	(4 inches)	1.1
150 mm	(6 inches)	0.8
220 mm	(8.5 inches)	0.5

shown in the table above. Under exceptional conditions, the limits of magnitude and resolution tabulated here may be surpassed; but in many cases, particularly when observing from cities, they will not be achieved.

Finally, we come to the atmosphere itself. Professional observatories are sited on high mountain tops to get above as much of the atmosphere as possible, but most amateurs are stuck with conditions in their own back yard, all too often including urban haze and glare from streetlights. There are two aspects of the atmosphere to take into account: its clarity and its steadiness. A good index of atmospheric clarity is the magnitude of the faintest stars that can be seen overhead by the naked eye. The zenithal limiting magnitude should always be noted when observing meteors, because it affects the hourly rates. Atmospheric clarity needs to be at its best when you are looking for faint objects, particularly nebulae, galaxies and comets.

On the other hand, the steadiness of the atmosphere, known as the *seeing*, is paramount for planetary and double star observation. Turbulence in the atmosphere produces a boiling effect of the telescopic image, drastically reducing the resolution. Hot air rising from neighbouring buildings can be a particularly annoying localized form of turbulence. Ironically, on a crystal-clear night after a rainstorm the seeing can be particularly atrocious, whereas slightly misty nights, when the clarity is low, can be the steadiest.

Seeing is estimated on a five-point scale: 1, perfect; 2, good; 3, moderate; 4, poor; 5, very bad. If a close double star cannot be clearly split on a night of indifferent seeing, examine it again on a better night.

A useful addition to the modern astronomer's armoury is one of the special filters designed to enhance the visibility of faint objects such as nebulae, galaxies and comets. Such filters screw into the barrel of the eyepiece and come in two main types: one type blocks out light pollution, thereby increasing the contrast of the object against the night sky; more extreme types block out all wavelengths except those which nebulae and comets emit most strongly. For a description of filters designed for safe viewing of the Sun, see page 293.

Basic astrophotography

A camera can be used as an astronomical instrument. Photographs of the sky can be taken with any camera capable of time exposures. That usually means a single lens reflex (SLR) camera, since most automatic compact cameras do not have a time-exposure facility. Depending on the length of the exposure, it is possible to photograph constellations, graceful curving star trails, groupings of planets and perhaps the occasional meteor.

Firstly, the camera must be loaded with fast film. Whether it is colour or black-and-white doesn't matter. Black-and-white has the advantages that it is more economical and can be easily processed at home, but colour transparencies are the most convenient for beginners. Colour negative film is not so good, as it requires special printing. Very fast colour slide films, rated ISO 1000 or faster, are available and these can produce beautiful results.

It is important to use the widest aperture on the camera lens. Camera apertures are quoted in terms of f/stops. Most camera lenses will open up to f/2.8 and many even wider, up to f/1.4. Cameras often have interchangeable lenses. A wide-angle lens will photograph more sky than a standard lens, but the images will be smaller and the constellation shapes will appear distorted towards the edges. Finally, the camera focus must be set to infinity.

A useful accessory is a cable release, which enables the shutter to be opened and closed without shaking the camera. For time exposures the camera needs to be on the setting marked 'B' (this, incidentally, stands for 'bulb', a legacy of the days when cameras were operated by air bulbs). On the B setting, the button must remain depressed to keep the shutter open, so the cable release must be locked for the length of the time exposure. Some cameras have a 'T' setting (for 'time'); on this, one press of the button opens the shutter, and a second press closes it.

The camera must be firmly mounted for the duration of the exposure. If there is no suitable tripod, the camera can be wedged on the ground by pieces of wood, bricks or stones. One good way of avoiding camera shake is to hold a piece of card over the lens while the shutter is opened. Once all vibrations have died down, the card can be moved away. The card must be put back in front of the lens before the shutter is closed at the end of the exposure. When the camera is pointing in the rough direction of the area to be photographed (stars are so faint they will probably not be visible in the viewfinder), you are ready to shoot.

The exposure used depends on what you are trying to achieve, and on the darkness of the sky. A good way to start is by photographing star trails, achieved by leaving the shutter open for a length of time. As the rotation of the Earth carries the stars across the sky, they leave trails of light on the film. If the camera is pointed in the direction of the pole star trails will come out noticeably curved, as seen on page 18, whereas on an exposure of the equatorial region of the sky the star trails will appear straight. Colour film will register an attractive range of star colours.

In an area with lots of streetlights, you should restrict the exposure to five minutes, or else the sky brightness will build up on the film and fog the image. Under darker skies, exposure of half an hour or longer will be possible without fogging. The only way to find out for certain is by experimenting.

If the exposure is kept short – 15 to 20 seconds – the Earth will not have rotated far enough to produce noticeable trails, and the constellations will appear as they do to the naked eye. Stars down to 6th magnitude will be recorded. Photographs taken with short exposures in twilight, or with moonlight illuminating the surrounding landscape, can be particularly attractive. From time to time, several planets may form interesting groupings which are worth capturing with a short exposure. Always take a succession of photographs with different exposures to be sure of getting the best result.

Use of a telephoto lens can sometimes add drama to a scene – for instance, a partially eclipsed Moon rising over a distant skyline. But you must remember that the effect of the Earth's rotation shows up more quickly through a telephoto, so exposures should be restricted to no more than a few seconds to avoid trailing of the images.

When a bright meteor shower such as the Perseids or Geminids is due, try a series of long exposures with a wide-angle lens in the hope of catching some. Meteors dart across the sky so quickly that only the brightest will register on your film, and many of them will in any case fall outside the camera's field of view. But the excitement of photographing the brilliant streak of a meteor more than compensates for the disappointment of the other frames that show nothing but star trails.

If a camera is placed on an equatorial mount it can be guided to prevent star trails from forming. Guided exposures of only a few minutes' duration will capture stars far fainter than those visible to the naked eye. Most exciting of all is to use a camera to photograph what can be seen through a telescope. An SLR camera is essential for this. The camera lens must be removed completely and the camera body attached to the eyepiece mount by an adaptor. In effect, the entire telescope is acting like a super telephoto lens. With such a system, craters on the Moon, the belts of Jupiter and the rings of Saturn can be photographed. Exposures range from a fraction of a second for the Moon to a second or more for the planets. Even longer exposures, of several minutes' duration, can reveal delicate details of faint nebulae and galaxies.

Increasingly, amateurs are now turning from photographic film to electronic chips known as CCDs (short for charge-coupled devices). A CCD is a light-sensitive silicon chip, as used in digital cameras, on which an image builds up and can be read-off electronically into a computer; the chip is then ready for another exposure. The advantages are that CCDs are far more sensitive to light than is photographic film, thereby significantly reducing exposure times, and the results can be computer-processed to bring out details.

INDEX

All named stars of mag. 2.0 and brighter are included in the Index. Star designations such as α (alpha) Centauri and η (eta) Carinae will be found under the relevant constellations. Page numbers in **bold type** refer to photographs or illustrations.